Theory and Applications of Information Geometry

Theory and Applications of Information Geometry

Edited by
Brayan Cox

www.willfordpress.com

Published by Willford Press,
118-35 Queens Blvd., Suite 400,
Forest Hills, NY 11375, USA

ISBN: 978-1-64728-530-2

Cataloging-in-Publication Data

Theory and applications of information geometry / edited by Brayan Cox.
p. cm.
Includes bibliographical references and index.
ISBN 978-1-64728-530-2
1. Geometry, Differential. 2. Mathematical statistics. 3. Information theory in mathematics. I. Cox, Brayan.
QA641 .T44 2023
516.36--dc23

For information on all Willford Press publications
visit our website at www.willfordpress.com

Contents

Preface

This book has been an outcome of determined endeavour from a group of educationists in the field. The primary objective was to involve a broad spectrum of professionals from diverse cultural background involved in the field for developing new researches. The book not only targets students but also scholars pursuing higher research for further enhancement of the theoretical and practical applications of the subject.

Information geometry refers to an interdisciplinary field that studies probability theory and statistics by using methods of differential geometry. It is a general framework of dual affine connected with Riemannian manifolds. Information geometry involves the study of decision-making by using geometry which also includes modeling-fitting and pattern matching. It offers a novel approach that can be applied in many fields including physical and information sciences. Its primary goal is to research the geometrical structures that can be presented in the manifold connected with a set of probability distributions of a statistical model. Information geometry is also used to measure the degree of information integration among various terminals of a causal dynamical system. This book aims to shed light on the theory and applications of information geometry. It will also provide interesting topics for research, which interested readers can take up. This book is a resource guide for experts as well as students.

It was an honour to edit such a profound book and also a challenging task to compile and examine all the relevant data for accuracy and originality. I wish to acknowledge the efforts of the contributors for submitting such brilliant and diverse chapters in the field and for endlessly working for the completion of the book. Last, but not the least; I thank my family for being a constant source of support in all my research endeavours.

Editor

1

Information Geometry of Spatially Periodic Stochastic Systems

Rainer Hollerbach [1,*] and Eun-jin Kim [2]

[1] Department of Applied Mathematics, University of Leeds, Leeds LS2 9JT, UK
[2] School of Mathematics and Statistics, University of Sheffield, Sheffield S3 7RH, UK
[*] Correspondence: rh@maths.leeds.ac.uk

Abstract: We explore the effect of different spatially periodic, deterministic forces on the information geometry of stochastic processes. The three forces considered are $\mathbf{f}_0 = \sin(\pi x)/\pi$ and $\mathbf{f}_\pm = \sin(\pi x)/\pi \pm \sin(2\pi x)/2\pi$, with $\mathbf{f}_-$ chosen to be particularly flat (locally cubic) at the equilibrium point $x = 0$, and $\mathbf{f}_+$ particularly flat at the unstable fixed point $x = 1$. We numerically solve the Fokker–Planck equation with an initial condition consisting of a periodically repeated Gaussian peak centred at $x = \mu$, with μ in the range $[0, 1]$. The strength D of the stochastic noise is in the range 10^{-4}–10^{-6}. We study the details of how these initial conditions evolve toward the final equilibrium solutions and elucidate the important consequences of the interplay between an initial PDF and a force. For initial positions close to the equilibrium point $x = 0$, the peaks largely maintain their shape while moving. In contrast, for initial positions sufficiently close to the unstable point $x = 1$, there is a tendency for the peak to slump in place and broaden considerably before reconstituting itself at the equilibrium point. A consequence of this is that the information length $\mathcal{L}_\infty$, the total number of statistically distinguishable states that the system evolves through, is smaller for initial positions closer to the unstable point than for more intermediate values. We find that $\mathcal{L}_\infty$ as a function of initial position μ is qualitatively similar to the force, including the differences between $\mathbf{f}_0 = \sin(\pi x)/\pi$ and $\mathbf{f}_\pm = \sin(\pi x)/\pi \pm \sin(2\pi x)/2\pi$, illustrating the value of information length as a useful diagnostic of the underlying force in the system.

Keywords: stochastic processes; Fokker–Planck equation; information length

---

1. Introduction

It is of interest to apply the idea of a metric to problems involving stochastic processes, e.g., [1–6]. Given a metric, the differences between different Probability Density Functions (PDFs) can be quantified, with different metrics focusing on a range of aspects, and hence most suitable for various applications. Fisher information [7] yields a metric where distance is measured in units of the PDF's width. The distance in the Fisher metric is thus dimensionless, and represents the number of statistically different states [8].

By extending the statistical distance in [8] to time-dependent situations, we recently introduced a way of quantifying information changes associated with time-varying PDFs [9–16]. We first compare two PDFs separated by an infinitesimal increment in time, and consider the corresponding infinitesimal distance. Integrating in time gives the total number of statistically distinguishable states that the system passes through, called the *information length* $\mathcal{L}$, e.g., [6–8,14]. Another interpretation of $\mathcal{L}$ that can be useful is as a measure of the total elapsed time in units of an 'information-change' dynamical timescale.

We start by defining the dynamical time $\tau(t)$ as

$$\mathcal{E} \equiv \frac{1}{[\tau(t)]^2} = \int \frac{1}{p(x,t)} \left[\frac{\partial p(x,t)}{\partial t}\right]^2 dx. \tag{1}$$

That is, $\tau(t)$ is the characteristic timescale over which the information changes, and quantifies the PDF's correlation time. Alternatively, $1/\tau$ quantifies the (average) rate of change of information in time. A PDF that evolves such that $\mathcal{E}$ is constant in time is referred to as a geodesic, along which the information propagates at a uniform rate [6]. The information length $\mathcal{L}(t)$ is then defined by

$$\mathcal{L}(t) = \int_0^t \frac{dt_1}{\tau(t_1)} = \int_0^t \sqrt{\int dx \frac{1}{p(x,t_1)} \left[\frac{\partial p(x,t_1)}{\partial t_1}\right]^2} dt_1. \tag{2}$$

which can be interpreted as measuring time in units of τ. It is important to note that $\mathcal{L}$ has no dimension (unlike entropy) and represents the total number of statistically different states that a system passes through in time between 0 and t. If we know the parameters that determine the PDF $p(x,t)$, $\mathcal{E}$ and $\mathcal{L}$ in Equations (1) and (2) can be written in terms of the Fisher metric tensor defined in the statistical space spanned by those parameters. However, it is not always possible to have access to the parameters that govern PDFs, for instance, in the case of PDFs calculated from data. The merit of Equations (1) and (2) is thus that $\mathcal{E}$ and $\mathcal{L}$ can be directly calculated from PDFs even without knowing the parameters governing the PDFs, nor the Fisher metric. For instance, $\mathcal{L}$ was calculated from PDFs of music data in [10]. In the work here, we first compute time-dependent PDFs by solving the Fokker–Planck equation numerically, and then calculate $\mathcal{E}$ and $\mathcal{L}$ from these PDFs as additional diagnostics.

Unlike quantities such as entropy, relative entropy, Kullback–Leibler divergence, or Jensen divergence, information length is a Lagrangian measure, that is, it includes the full details of the PDF's evolution, and not just the initial and final states. $\mathcal{L}_\infty$, the total information length over the entire evolution, is then particularly useful to quantify the proximity of any initial PDF to a final attractor of a dynamical system. In previous work [12,15] we explored these aspects of $\mathcal{L}$ for restoring forces that were power-laws in the distance to the attractor. For instance, for the Ornstein–Uhlenbeck process, which is a linear relaxation process, we showed that $\mathcal{L}$ consists of two parts: the first is due to the movement of the mean position measured in units of the width of the PDF, and the second is due to the entropy change. Thus, the total entropy change that is often discussed in previous works (e.g., [17]) contributes only partially to $\mathcal{L}$. Importantly, for the Ornstein–Uhlenbeck process, $\mathcal{L}_\infty$ increases linearly from the stable equilibrium point (with its minimum value at the stable equilibrium point) with the mean position of the initial PDFs regardless of the strength of the stochastic noise and the width of the initial PDFs. The linear relation indicates that a linear process preserves a linearity of the underlying process. Heseltine & Kim [18] shows that this linear relation is lost for other metrics (e.g., Kullback–Leibler divergence, Jensen divergence). Note that $\mathcal{L}$ is related to the integral of the square root of the infinitesimal relative entropy (see Appendix A). In comparison, for a chaotic attractor, $\mathcal{L}_\infty$ varies sensitively with the mean position of a narrow initial PDF, taking its minimum value at the most unstable point [9]. This sensitive dependence of $\mathcal{L}_\infty$ on the initial PDF is similar to a Lyapunov exponent.

These results highlight $\mathcal{L}_\infty$ as an alternative diagnostic to understand attractor structures of dynamical systems. It is this attractor structure that we are interested in in this paper. We thus focus on the relaxation problem as in [9,12,15,18] by considering periodic deterministic forces and elucidate the importance of the initial condition and its interplay with the deterministic forces in the relaxation and thus attractor structure.

2. Model

We consider the following nonlinear Langevin equation:

$$\frac{dx}{dt} = -f(x) + \xi. \tag{3}$$

Here x is a random variable; $f(x)$ is a deterministic force; ξ is a stochastic forcing, which for simplicity can be taken as a short-correlated Gaussian random forcing as follows:

$$\langle \xi(t)\xi(t') \rangle = 2D\delta(t - t'), \tag{4}$$

where the angular brackets represent the average over ξ, $\langle \xi \rangle = 0$, and D is the strength of the forcing.

In [15] we considered the choice $f(x) = x^n$ and investigated how varying the degree of nonlinearity $n = 3, 5, 7$ affects the system. In this work we take $f(x)$ to be periodic in x, and explore some of the new effects this can create. The three choices of $f(x)$ we consider are

$$\mathbf{f_0} = \sin(\pi x)/\pi, \qquad \mathbf{f_\pm} = \sin(\pi x)/\pi \pm \sin(2\pi x)/2\pi. \tag{5}$$

Figure 1 shows these profiles, which are all anti-symmetric in x, and periodic on the interval $x \in [-1, 1]$. All three choices have $x = 0$ as an attractor, and $x = 1$ as an unstable fixed point. The particular combinations of harmonics for $\mathbf{f_\pm}$ were chosen so that they are locally cubic rather than linear at either $x = 0$ (for $\mathbf{f_-}$) or $x = 1$ (for $\mathbf{f_+}$). In applications such a Brownian motors many specific choices of $f(x)$ are considered to model particular physics. However, as noted in the introduction, we are here more interested in attractor structures in the relaxation problem, in particular, how initial conditions and stochastic noise interact with deterministic forces and the role of the asymmetry of the deterministic force and the stable and unstable fixed points on the local dynamics.

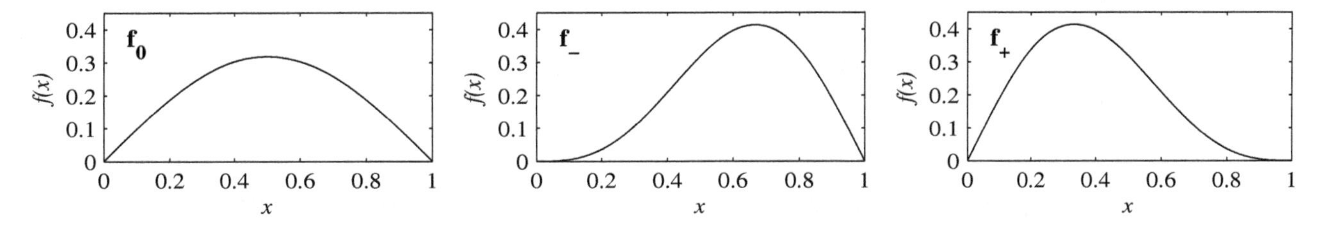

Figure 1. The three profiles $\mathbf{f_0}$, $\mathbf{f_-}$ and $\mathbf{f_+}$. Note how $\mathbf{f_-}$ is chosen to be flat at the attractor $x = 0$, and $\mathbf{f_+}$ at the unstable fixed point $x = 1$. All three choices are anti-symmetric in x, and periodic with period 2.

Comparing these three periodic functions with the previous choices, two significant differences stand out. First, for $f(x) = x^n$ with $n = 3, 5, 7$, all initial conditions are pushed directly toward the origin, and there are no unstable fixed points. It is therefore of particular interest to see how the choices here behave for initial conditions near $x = 1$. Second, $f(x) = x^{3,5,7}$ all curve upward (that is, have $f'' > 0$ for all $x > 0$), whereas the choices here have different combinations of curvatures, which will turn out to have clearly identifiable effects.

The Fokker–Planck equation [19,20] corresponding to Equation (3) is

$$\partial_t p(x,t) = \partial_x (f(x)p) + D\partial_{xx} p. \tag{6}$$

In [15] we solved the corresponding equation by finite-differencing in x. For the periodic systems considered here, it is more convenient to start with the Fourier expansion

$$p(x,t) = a_0(t) + \sum_{k=1}^{K} \Big(a_k(t) \cos(k\pi x) + b_k(t) \sin(k\pi x) \Big). \tag{7}$$

The coefficients a_k and b_k are then time-stepped using second-order Runge-Kutta. The term $\partial_x(f(x)p)$ is separated out into the relevant Fourier components using a fast Fourier transform. (For the very simple choices of $f(x)$ considered here, consisting of at most two Fourier modes, it would be straightforward to do this separation analytically, and thereby do the entire calculation purely in Fourier space, but the code was developed with more general choices for $f(x)$ in mind, where this approach becomes increasingly cumbersome as the number of harmonics in $f(x)$ increases. For such more general choices of $f(x)$ the FFT approach is most convenient).

Resolutions in the range $K = 2^{11} - 2^{14}$ are used, and carefully checked to ensure fully resolved solutions. Time-steps were in the range 10^{-4}–10^{-5}, and were again varied to ensure proper accuracy. Another useful test of the numerical implementation is to monitor the coefficient a_0: this is time-stepped along with the others, but must in fact remain constant if the total probability $\int p\,dx = 2a_0$ is to remain constant. It was found that if the initial condition is correctly set to have $a_0 = 0.5$, then this was maintained throughout the entire subsequent evolution.

The initial conditions are of the form

$$p(x,0) = \frac{1}{\sqrt{2\pi\,D_0}} \exp\left[-\frac{(x-\mu)^2}{2D_0}\right], \tag{8}$$

that is, Gaussians centred at $x = \mu$ and having half-width scaling as $\sqrt{D_0}$. We are interested in the range $\mu \in [0,1]$; by symmetry the range $\mu \in [-1,0]$ would behave the same, simply approaching $x = 0$ from the other direction.

This initial condition is also periodic, on the same $x \in [-1,1]$ interval as the entire problem. For the purposes of actually implementing Equation (8), it was most convenient to consider the range as being $x \in [-0.5, 1.5]$. In particular, for $\mu \in [0,1]$ and the values of D_0 considered here, Equation (8) yields results at $x = -0.5$ and $x = 1.5$ that are different, but both are so vanishingly small that the discrepancy does not need to be smoothed out in defining the initial condition. If instead Equation (8) were implemented on either $x \in [0,2]$ or $x \in [-1,1]$, then μ near either 0 or 1 would be more awkward to handle correctly.

In [15] we also used a Gaussian initial condition, with $D_0 = 10^{-6}$, and then explored the regime $D = 10^{-6}$ to 10^{-9}. Here we are again interested in the regime $D_0 \geq D$, which allows at least the initial parts of the evolution to be nondiffusive. Having the initial peak be so narrow that $D_0 < D$ can also be interesting in other contexts (e.g., [21]), but diffusive effects are then necessarily important from the outset, which would obscure some of the dynamics of interest here. We therefore focus on the range $D_0 = 10^{-2}$ to 10^{-4}, and $D = 10^{-4}$ to 10^{-6}.

3. Results

Figure 2 shows how the peak amplitudes evolve in time for the three choices $\mathbf{f_0}$, $\mathbf{f_-}$ and $\mathbf{f_+}$. Starting with the initial position $\mu = 0.5$ in the top row, we see that the solutions for $\mathbf{f_0}$ and $\mathbf{f_+}$ equilibrate to their final values on very rapid timescales, involving relatively little variation with D. In contrast, the timescales for $\mathbf{f_-}$ are much longer, and vary substantially with D. Comparing the $\mathbf{f_-}$ results here with Figure 1 in [15], we see that $\mathbf{f_-}$ is exactly analogous to the previous $f(x) = x^3$. This is because for $x \leq 0.5$ the shape of $\mathbf{f_-}$ is very close to a cubic. Similarly, for $x \leq 0.5$ the shape of $\mathbf{f_0}$ is still reasonably close to linear, and the evolution is therefore essentially like the linear Ornstein–Uhlenbeck process $f(x) = x$, for which an exact analytic solution exists [21].

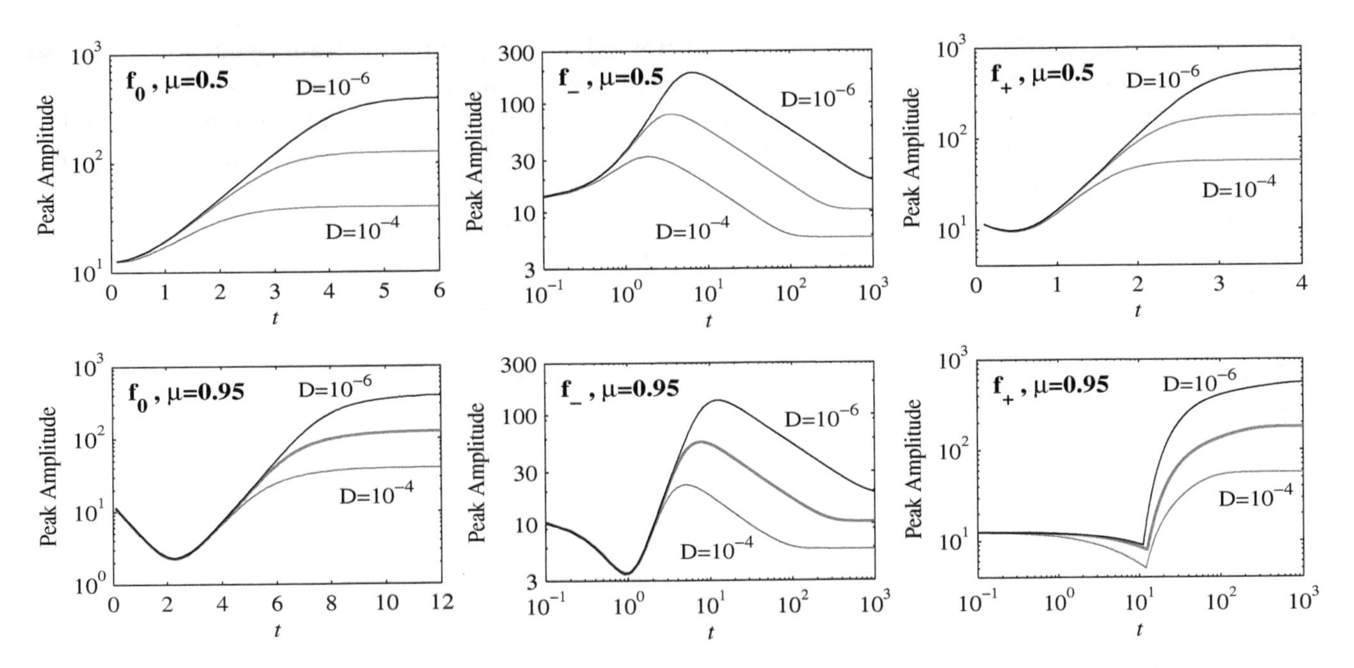

Figure 2. Peak amplitudes as functions of time, for the three choices $\mathbf{f}_*$ as labelled. The initial positions are at $\mu = 0.5$ in the top row, and $\mu = 0.95$ in the bottom row, with initial widths $D_0 = 10^{-3}$ in all cases. $D = 10^{-4}$ to 10^{-6} as labelled, also colour-coded as blue, red, black. Note also that some panels have t on a linear scale, indicating a very rapid adjustment process, whereas others have t on a logarithmic scale, corresponding to much slower dynamics.

It is only $\mathbf{f}_+$ whose shape is already substantially different from either linear or cubic even on the interval $x \leq 0.5$, being close to linear for $x \leq 0.2$ but strongly curved for $0.2 < x \leq 0.5$. Correspondingly $\mathbf{f}_+$ also shows a new effect, namely an initial reduction in the peak amplitudes. This effect becomes even more pronounced for $\mathbf{f}_0$ and $\mathbf{f}_-$ and the initial position $\mu = 0.95$, in the bottom row of Figure 2. This reduction in the peak amplitudes is *not* caused by diffusive spreading but is a consequence of the non-diffusive ($\xi = D = 0$) evolution resulting from the interplay between an initial PDF and the deterministic force. We note in particular how $D = 10^{-4}$ to 10^{-6} yield identical reductions in amplitudes here. It is worth comparing this with the non-diffusive evolution in [15] where the opposite behavior—an initial increase in peak amplitudes (the same effect as seen here for $\mathbf{f}_-$, $\mu = 0.5$)—was observed. The interplay between the initial PDF and the deterministic force is elaborated below.

If $f(x)$ is such that it increases more rapidly than linearly, i.e., curves upward, then those parts of any initial condition furthest from the origin are pushed toward it fastest, whereas those parts closest move more slowly. The result is that an initial Gaussian peak bunches up on itself, causing the amplitude to increase. In contrast, if $f(x)$ curves downward the opposite effect occurs, and an initial Gaussian peak is spread out, even before diffusion starts to play a role. Eventually of course the peak moves sufficiently close to the origin that the behaviour is as before, explaining why the behaviour at later times is similar to the previous $\mu = 0.5$ results.

Finally, the behaviour for $\mathbf{f}_+$ with $\mu = 0.95$ is yet again different, namely an initial reduction in amplitude up to $t \approx 10$, followed by an abrupt increase. This is caused by a fundamentally new peak forming at the origin, rather than the initial peak moving toward it. Note also that time here is on a logarithmic scale, corresponding to a very slow equilibration process, unlike the previous case $\mathbf{f}_+$ with $\mu = 0.5$.

Figures 3–5 illustrate these various behaviours in more detail, showing the actual PDFs at different times for $\mathbf{f}_0$, $\mathbf{f}_-$ and $\mathbf{f}_+$, respectively. Starting with $\mathbf{f}_0$, we see how the peak initially located at $\mu = 0.95$ becomes broader as it moves toward the origin, an effect again not caused by diffusion, but rather by the curvature of $\mathbf{f}_0$ at these values of x. Note for example how the solutions at $t = 3$ or 4 have much steeper leading edges (nearer to the origin) than trailing edges, caused by the trailing edges moving so much slower. Another feature to note is how parts of the solution reach the origin coming from the 'other' direction. That is, if the initial condition is a peak centred at $\mu = 0.95$, and having half-width 0.07 (corresponding to $D_0 = 10^{-3}$), then a small but non-negligible portion of the initial condition is in the range $x \geq 1$, as seen also in Figure 3. For this part of the initial condition the nearest attractor is $x = 2$ rather than $x = 0$. Viewed on the interval $x \in [-0.2, 1.2]$, this part therefore approaches from negative x values, as seen at times $t = 5$ and 6. (The interval $x \in [1.2, 1.8]$ or equivalently $x \in [-0.8, -0.2]$ is not shown in these figures because the amplitudes are rather small there, due to the PDFs being very spread out as they traverse this range). Finally, between $t = 6$ and 9 we see how the two peaks coming from negative and positive x values combine to form the single final equilibrium consisting of a Gaussian centred at the origin.

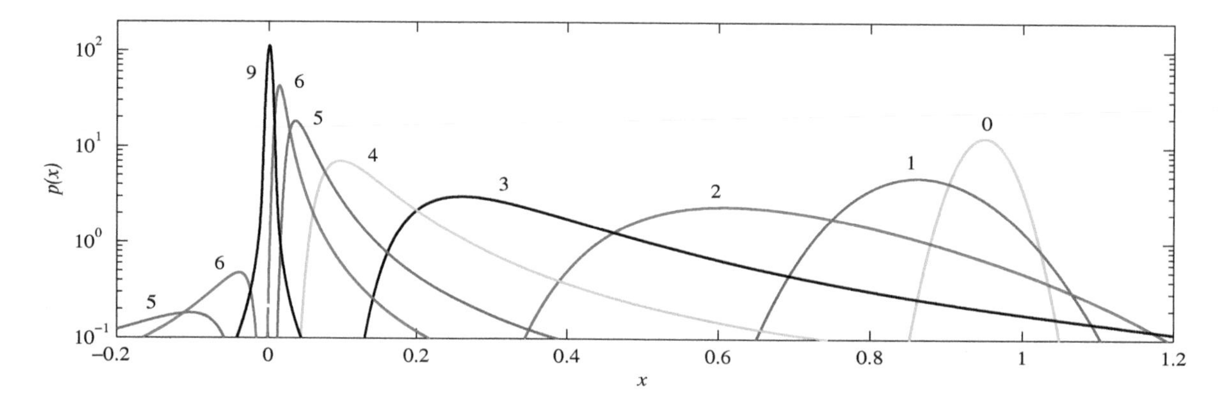

Figure 3. The solutions $p(x,t)$ for $\mathbf{f_0}$, $D = 10^{-5}$, and initial condition $\mu = 0.95$ and $D_0 = 10^{-3}$. The numbers beside individual curves indicate the times, from $t = 0$ to 9. The different colours are for clarity only, but do not indicate a specific colour $\leftrightarrow$ time relationship.

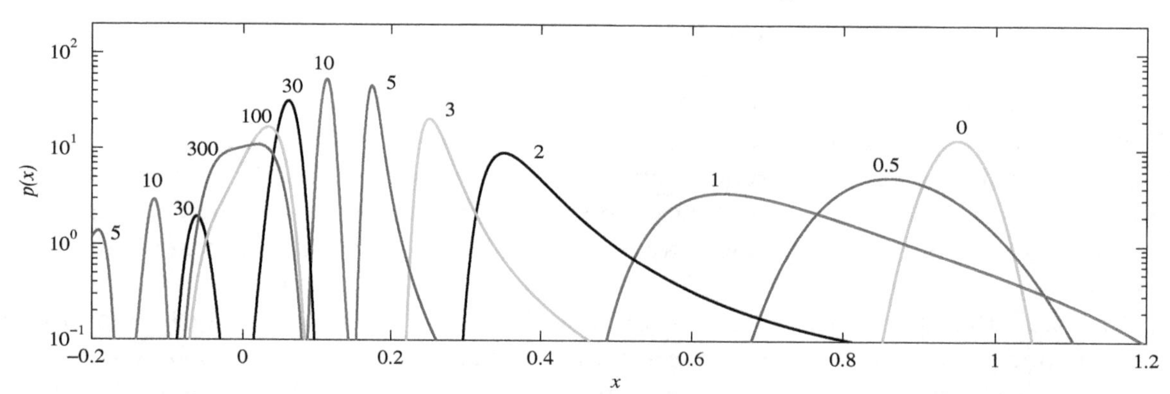

Figure 4. The solutions $p(x,t)$ for $\mathbf{f_-}$, $D = 10^{-5}$, and initial condition $\mu = 0.95$ and $D_0 = 10^{-3}$. The numbers beside individual curves indicate the times, from $t = 0$ to 300. Different colours are again only to help distinguish between the different lines.

Figure 4 shows the corresponding solutions for $\mathbf{f_-}$. For small durations the behaviour is very similar to that seen in Figure 3, except that it happens roughly twice faster (e.g., compare $t = 0.5$ in Figure 4 with $t = 1$ in Figure 3). This is readily understandable by noting that the slope of $\mathbf{f_-}$ near $x = 0.95$ is roughly twice that of $\mathbf{f_0}$, yielding faster evolution. The later evolution is much slower though, with the merging of the two peaks only occurring between $t = 30$ and 100, and even $t = 300$ still displaying some asymmetry, and hence not yet the final quartic profile. This is the same very slow final adjustment process previously analysed in detail in [15], and is caused by $\mathbf{f_-}$ being cubic rather than linear near the origin.

Figure 5 shows the solutions for $\mathbf{f_+}$. We see the behaviour alluded to above, of an abrupt transition from one peak to another. Because $\mathbf{f_+}$ is so flat near $x = 1$, there is hardly any tendency to push the initial peak away. Instead, it simply broadens out, slumping as it spreads. A new peak then forms at the origin, overtaking the original one in amplitude around $t \approx 10$, as previously noted in Figure 2. Note though that long after this time a significant portion of the original peak still remains near $x = 1$, and this portion only fades away on very long timescales; $x = 1$ is an unstable fixed point, but $\mathbf{f}_+$ is so small everywhere near $x = 1$ that there is very little tendency to push the solutions away from there.

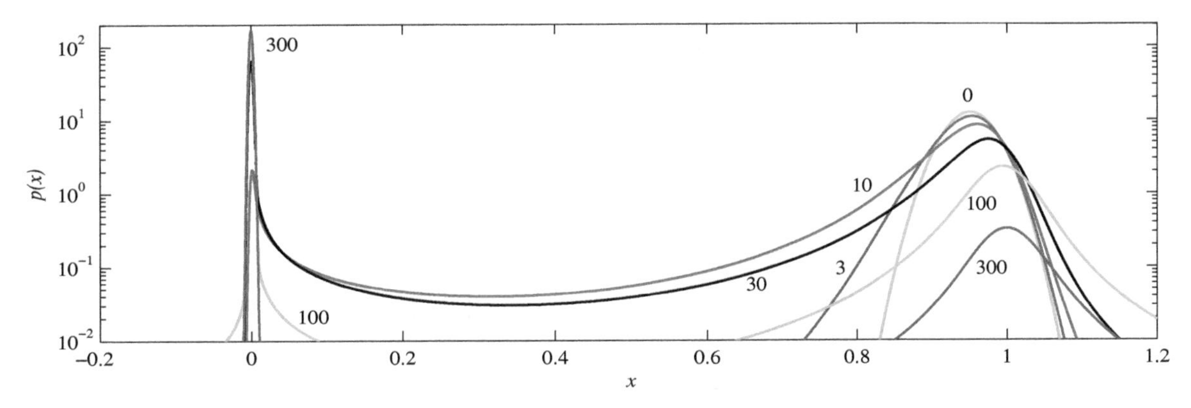

Figure 5. The solutions $p(x,t)$ for $\mathbf{f}_+$, $D = 10^{-5}$, and initial condition $\mu = 0.95$ and $D_0 = 10^{-3}$. The numbers beside individual curves indicate the times, from $t = 0$ to 300.

As noted in the introduction, we are particularly interested in the effects that these various different types of behaviour have on the information length quantities $\mathcal{E}(t)$ and $\mathcal{L}_\infty$. Figure 6 shows $\mathcal{E}(t)$ for the same solutions as before in Figure 2. We see that $\mathcal{E}$ is initially uniform, and independent of D (provided D is sufficiently small in comparison with D_0), corresponding to the 'geodesic' behaviour first identified by [6]. For some configurations, $\mathcal{E}$ then immediately transitions to an exponential decay, whereas for others it first has a power-law decay before ultimately decaying exponentially. Correspondingly, the timescales to achieve $\mathcal{E} \leq 10^{-8}$ also vary dramatically, as seen by the various linear and logarithmic scales for t. Different scaling regimes signify fundamentally different dynamics.

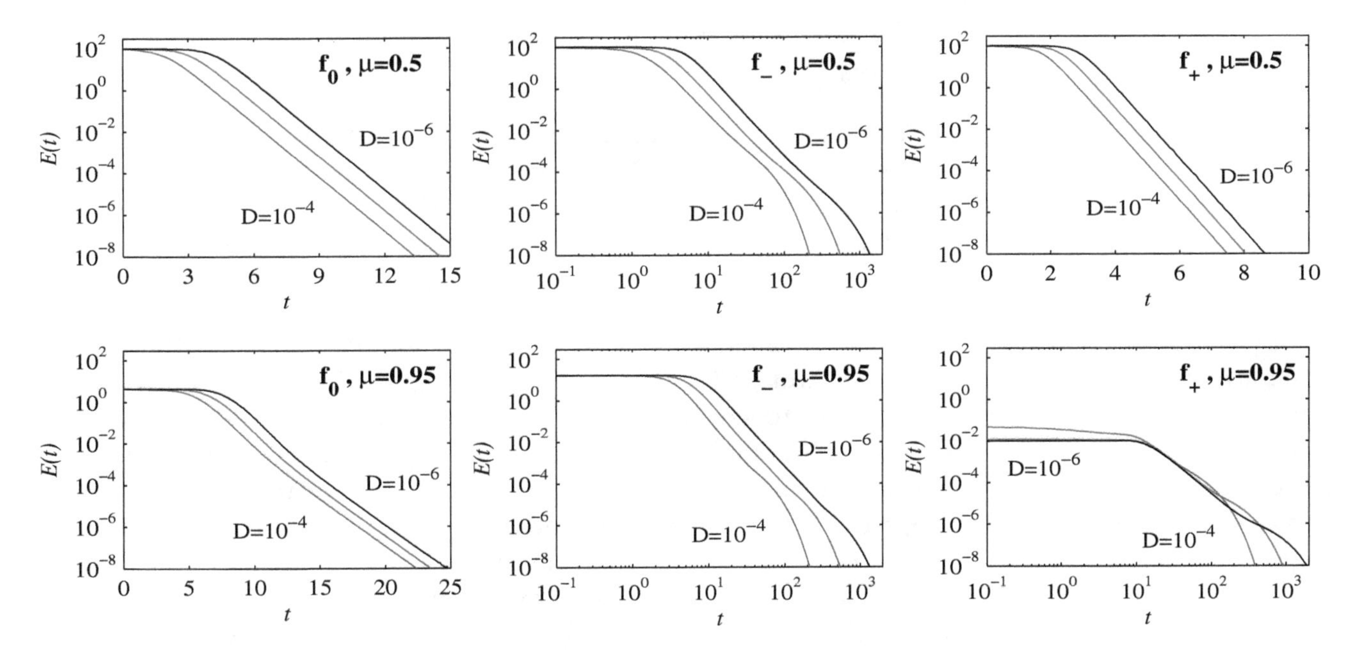

Figure 6. $\mathcal{E}$ as a function of time, for the six configurations as in Figure 2. Note again how the scale for t is sometimes linear and sometimes logarithmic.

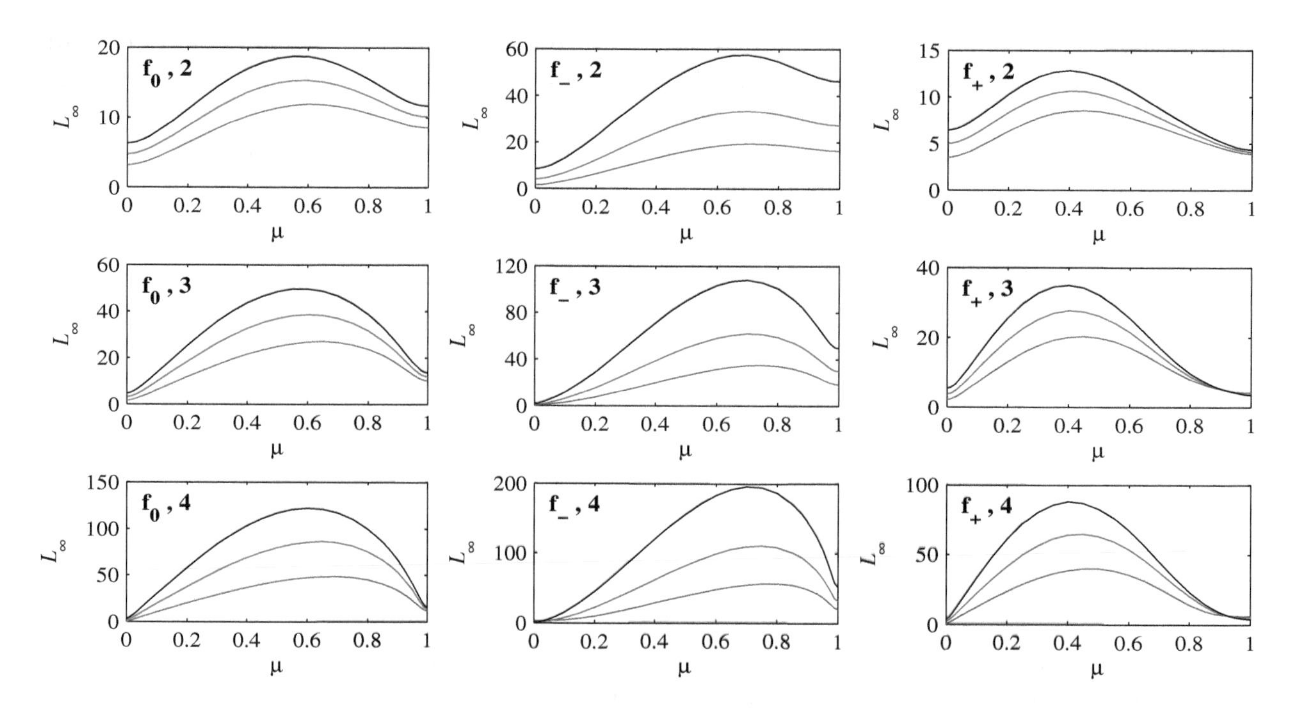

Figure 7. The total information length $\mathcal{L}_\infty$ as a function of initial position μ. The labels $\mathbf{f}_*, \mathbf{n}$ indicate the particular profile $\mathbf{f}_*$, and the initial width $D_0 = 10^{-n}$. Within each panel the three curves are $D = 10^{-4}$ (lowest, blue), $D = 10^{-5}$ (middle, red), and $D = 10^{-6}$ (top, black).

Figure 7 shows how $\mathcal{L}_\infty$ varies with μ, for $D_0 = 10^{-2}$ to 10^{-4}, and $D = 10^{-4}$ to 10^{-6} within each panel. It is interesting to note how the shapes generally mimic the corresponding functions $\mathbf{f_0}$, $\mathbf{f}_-$ and $\mathbf{f}_+$. The largest values always occur for intermediate values of μ, even though larger values correspond to initial conditions that have farther to travel to reach the origin. Such initial conditions also spread out much more though, as seen above, and according to the interpretation of information length, this should indeed reduce $\mathcal{L}$. Very close to $x = 1$ the $\mathcal{L}_\infty$ values are particularly small, because having peaks collapse in place and reform at the new location is an informationally very efficient way to move, as seen also in other contexts [13,22,23].

Finally, Figure 8 shows the time, call it T_∞, needed for $\mathcal{E}$ to drop to 10^{-8}. The precise cutoff $\mathcal{E} = 10^{-8}$ is of course somewhat arbitrary, but as seen in Figure 6 is sufficiently small to be in the exponential decay regime in all cases. This is therefore a convenient measure of the time taken to reach $\mathcal{L}_\infty$, and any even smaller cutoff would only add small increments to T_∞ (and essentially nothing to $\mathcal{L}_\infty$).

Starting with $\mathbf{f_0}$, we note first that T_∞ is on a linear scale, meaning that each reduction of D by a factor of 10 only adds a constant amount to T_∞. This is the same effect already seen in Figure 2, where smaller D requires slightly longer to settle in to the final states. Equivalently, smaller D in Figure 6 remains in the flat, geodesic regime for slightly longer times. The other feature to note for $\mathbf{f_0}$ is the behaviour near $\mu = 1$, where T_∞ increases strongly, and increasingly abruptly for smaller D_0. This can be understood by noting that if $\sqrt{D_0} \ll 1 - \mu$, the initial condition Equation (8) is essentially zero at $x = 1$, whereas if μ is within $\sqrt{D_0}$ of 1, Equation (8) does have a non-negligible component at $x = 1$. Therefore, if $\sqrt{D_0} \ll 1 - \mu$ the initial peak will simply move monotonically toward the origin, which occurs on a rapid timescale, whereas if $1 - \mu \leq \sqrt{D_0}$ the evolution will include a significant component of the slumping-in-place behaviour, which we saw only happens on slower timescales.

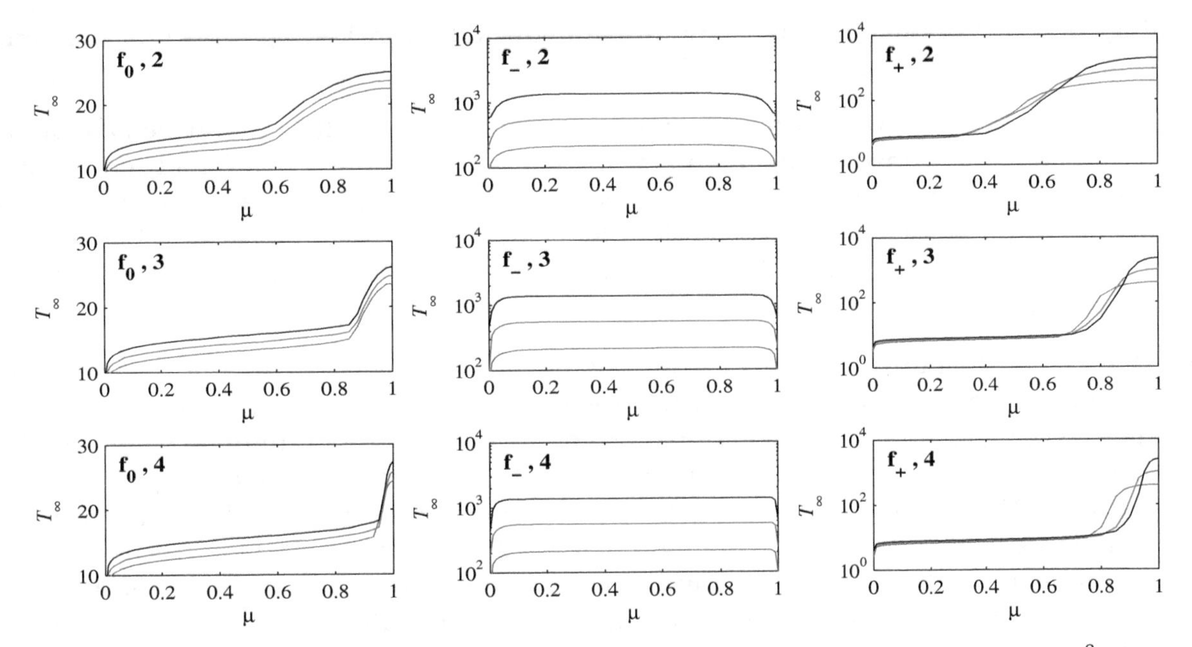

Figure 8. As in Figure 7, but now showing the time T_∞ that it takes to achieve $\mathcal{E} = 10^{-8}$, that is, a measure of the time it takes to reach $\mathcal{L}_\infty$. Note how T_∞ has a linear scale for $\mathbf{f_0}$, but logarithmic scales for $\mathbf{f_-}$ and $\mathbf{f_+}$. The colour-coding is again $D = 10^{-4}$ blue, $D = 10^{-5}$ red, $D = 10^{-6}$ black.

For $\mathbf{f_-}$, the scale for T_∞ is logarithmic, so that each reduction of D by a factor of 10 increases T_∞ by a factor of $\sqrt{10}$. For intermediate values of μ, T_∞ is also essentially independent of μ. The equilibration time is completely dominated by the final settling-in time, just as in the cubic case in [15], and the initial motion of the peak toward the origin is negligible in comparison. For very small values of μ the behaviour is different, with much smaller values of T_∞. If $\mu \leq \sqrt{D_0}$, the peak is essentially at the origin already, making the adjustment quicker. Finally, there is a similar end-effect for μ sufficiently close to 1; if $1 - \mu \leq \sqrt{D_0}$, the initial peak is essentially at the unstable fixed point, and the evolution is the slumping-in-place behaviour, which has a faster final adjustment than if the peak moves toward the origin and then adjusts its shape there (but still scaling as $D^{-1/2}$).

Finally, $\mathbf{f_+}$ is qualitatively similar to $\mathbf{f_0}$, in the sense that T_∞ is a monotonically increasing function of μ. Indeed, for intermediate values of μ the behaviour is virtually identical to $\mathbf{f_0}$, with T_∞ increasing by a constant amount every time D is decreased by a constant factor. (This is simply not visible because T_∞ is on a logarithmic rather than linear scale here). Because $\mathbf{f_0}$ and $\mathbf{f_+}$ are both linear near the origin, the extremely slow final adjustment that happens for $\mathbf{f_-}$ does not apply to either of them, leaving only this much weaker dependence on D. The behaviour near $\mu = 1$, with the very strong increase in T_∞, and again more abruptly for smaller D_0, is again because this is the regime where the slumping-in-place behaviour occurs. Also, because $\mathbf{f_+}$ is so much flatter near $x = 1$ than either of $\mathbf{f_0}$ of $\mathbf{f_-}$, this slumping-in-place behaviour is much slower for $\mathbf{f_+}$ than for the other choices (recall again how long the peak at $x = 1$ lasts in Figure 5). This explains why T_∞ is on a logarithmic scale for $\mathbf{f_+}$ but on a linear scale for $\mathbf{f_0}$, even though for intermediate values of μ they exhibit the same (weak) scaling with D.

4. Conclusions

The results presented here extend our previous work [12,15] to the deterministic forces that are periodic in space. This naturally allows for forces $f(x)$ that curve in opposite directions in different regions, as well as unstable fixed points. The deterministic force can also be adjusted to be particularly flat at either the stable equilibrium or the unstable fixed points, which both turn out to have important consequences, with either choice yielding particularly long timescales, scaling as $D^{-1/2}$. The interesting consequences of the interplay between an initial PDF and a deterministic force on the PDF evolution and scalings of different quantities were discussed in detail by considering three types of periodic deterministic forces and comparing the results.

In particular, we computed how the rate of information change and the resulting total information length $\mathcal{L}_\infty$ depend on the position of an initial Gaussian peak. We found that for all choices of $f(x)$, the unstable fixed points yield comparatively small $\mathcal{L}_\infty$, even though they are farthest away from the final equilibrium points. It is particularly interesting that $\mathcal{L}_\infty$ as a function of initial position qualitatively follows $f(x)$, indicating the close connection between the information geometry and the underlying forcing.

Finally, we note that this work can be extended in many different directions, including: (1) If the initial condition is not one Gaussian peak for every period of $f(x)$, but only one peak for some much larger (tending to infinity) number of periods of $f(x)$, then one can study how this initial peak gradually spreads out, how that compares with pure diffusion, and what effect the precise shape of $f(x)$ might have [24–26]. This is also related to so-called anomalous diffusion [27–29], which can be considerably more general though, involving fractional derivatives. (2) If the force also includes a constant component tending to push the initial condition in a particular direction, it is of interest to study situations where the constant force is comparable to the periodic component of the force [30–32]. (3) Allowing the force to vary in time as well as space is relevant to so-called Brownian motors [33–36]. Work in some of these areas is currently ongoing.

Author Contributions: The underlying ideas were developed by E.K. and R.H. Numerical calculations were done by R.H. The paper was written by R.H. and E.K.; both authors have read and approved the final manuscript.

Appendix A

We now recall how $\tau(t)$ and $\mathcal{L}(t)$ in Equations (1) and (2) are related to the relative entropy (Kullback–Leibler divergence) [6,11]. We consider two nearby PDFs $p_1 = p(x,t_1)$ and $p_2 = p(x,t_2)$ at time $t = t_1$ and t_2, and the limit of a very small $\delta t = t_2 - t_1$ to do Taylor expansion of $D[p_1,p_2] = \int dx\, p_2 \ln(p_2/p_1)$ by using

$$\frac{\partial}{\partial t_1} D[p_1,p_2] = -\int dx\, p_2 \frac{\partial_{t_1} p_1}{p_1}, \tag{A1}$$

$$\frac{\partial^2}{\partial t_1^2} D[p_1,p_2] = \int dx\, p_2 \left\{ \frac{(\partial_{t_1} p_1)^2}{p_1^2} - \frac{\partial_{t_1}^2 p_1}{p_1} \right\}, \tag{A2}$$

$$\frac{\partial}{\partial t_2} D[p_1,p_2] = \int dx \left\{ \partial_{t_2} p_2 + \partial_{t_2} p_2 \left[\ln p_2 - \ln p_1\right] \right\}, \tag{A3}$$

$$\frac{\partial^2}{\partial t_2^2} D[p_1,p_2] = \int dx \left\{ \partial_{t_2}^2 p_2 + \frac{(\partial_{t_2} p_2)^2}{p_2} + \partial_{t_2}^2 p_2 \left[\ln p_2 - \ln p_1\right] \right\}. \tag{A4}$$

In the limit $t_2 \to t_1 = t$ ($p_2 \to p_1 = p$), Equations (A1)–(A4) give us

$$\begin{aligned} \lim_{t_2 \to t_1} \frac{\partial}{\partial t_1} D[p_1,p_2] &= \lim_{t_2 \to t_1} \frac{\partial}{\partial t_2} D[p_1,p_2] = \int dx \partial_t p = 0, \\ \lim_{t_2 \to t_1} \frac{\partial^2}{\partial t_1^2} D[p_1,p_2] &= \lim_{t_2 \to t_1} \frac{\partial^2}{\partial t_2^2} D[p_1,p_2] = \int dx \frac{(\partial_t p)^2}{p} = \frac{1}{\tau^2}. \end{aligned} \tag{A5}$$

Up to $O((dt)^2)$ ($dt = t_2 - t_1$), Equation (A5) and $D(p_1,p_1) = 0$ lead to

$$D[p_1,p_2] = \frac{1}{2} \left[\int dx \frac{(\partial_t p(x,t))^2}{p(x,t)} \right] (dt)^2, \tag{A6}$$

and thus the infinitesimal distance $dl(t_1)$ between t_1 and $t_1 + dt$ as

$$dl(t_1) = \sqrt{D[p_1, p_2]} = \frac{1}{\sqrt{2}} \sqrt{\int dx \frac{(\partial_{t_1} p(x, t_1))^2}{p(x, t_1)}} dt. \quad \text{(A7)}$$

By summing $dt(t_i)$ for $i = 0, 1, 2, ..., n-1$ (where $n = t/dt$) in the limit $dt \to 0$, we have

$$\lim_{dt \to 0} \sum_{i=0}^{n-1} dl(idt) = \lim_{dt \to 0} \sum_{i=0}^{n-1} \sqrt{D[p(x, idt), p(x, (i+1)]} \, dt \propto \int_0^t dt_1 \sqrt{\int dx \frac{(\partial_{t_1} p(x, t_1))^2}{p(x, t_1)}} = \mathcal{L}(t), \quad \text{(A8)}$$

where $\mathcal{L}(t)$ is the information length. Thus, $\mathcal{L}$ is related to the sum of infinitesimal relative entropy. Note that $\mathcal{L}$ is a Lagrangian distance between PDFs at time 0 and t and sensitively depends on the particular path that a system passed through reaching the final state. In contrast, the relative entropy $D[p(x, 0), p(x, t)]$ depends only on PDFs at time 0 and t and thus does not tell us about intermediate states between initial and final states.

References

1. Jordan, R.; Kinderlehrer, D.; Otto, F. The variational formulation of the Fokker–Planck equation. *SIAM J. Math. Anal.* **1998**, *29*, 1–17. [CrossRef]
2. Gibbs, A.L.; Su, F.E. On choosing and bounding probability metrics. *Int. Stat. Rev.* **2002**, *70*, 419–435. [CrossRef]
3. Lott, J. Some geometric calculations on Wasserstein space. *Commun. Math. Phys.* **2008**, *277*, 423–437. [CrossRef]
4. Takatsu, A. Wasserstein geometry of Gaussian measures. *Osaka J. Math.* **2011**, *48*, 1005–1026.
5. Costa, S.; Santos, S.; Strapasson, J. Fisher information distance. *Discrete Appl. Math.* **2015**, *197*, 59–69. [CrossRef]
6. Kim, E.; Lee, U.; Heseltine, J.; Hollerbach, R. Geometric structure and geodesic in a solvable model of nonequilibrium process. *Phys. Rev. E* **2016**, *93*, 062127. [CrossRef] [PubMed]
7. Frieden, B.R. *Science from Fisher Information*; Cambridge University Press: Cambridge, UK, 2004.
8. Wootters, W.K. Statistical distance and Hilbert space. *Phys. Rev. D* **1981**, *23*, 357–362. [CrossRef]
9. Nicholson, S.B.; Kim, E. Investigation of the statistical distance to reach stationary distributions. *Phys. Lett. A* **2015**, *379*, 83–88. [CrossRef]
10. Nicholson, S.B.; Kim, E. Structures in sound: Analysis of classical music using the information length. *Entropy* **2016**, *18*, 258. [CrossRef]
11. Heseltine, J.; Kim, E. Novel mapping in non-equilibrium stochastic processes. *J. Phys. A* **2016**, *49*, 175002. [CrossRef]
12. Kim, E.; Hollerbach, R. Signature of nonlinear damping in geometric structure of a nonequilibrium process. *Phys. Rev. E* **2017**, *95*, 022137. [CrossRef] [PubMed]
13. Hollerbach, R.; Kim, E. Information geometry of non-equilibrium processes in a bistable system with a cubic damping. *Entropy* **2017**, *19*, 268. [CrossRef]
14. Kim, E.; Lewis, P. Information length in quantum systems. *J. Stat. Mech.* **2018**, 043106. [CrossRef]
15. Hollerbach, R.; Dimanche, D.; Kim, E. Information geometry of nonlinear stochastic systems. *Entropy* **2018**, *20*, 550. [CrossRef]
16. Hollerbach, R.; Kim, E.; Mahi, Y. Information length as a new diagnostic in the periodically modulated double-well model of stochastic resonance. *Phys. A* **2019**, *525*, 1313–1322. [CrossRef]
17. Van den Broeck, C.; Esposito, M. Three faces of the second law. II. Fokker–Planck formulation. *Phys. Rev. E* **2010**, *82*, 011144. [CrossRef] [PubMed]
18. Heseltine, J; Kim, E. Comparing information metrics for a coupled Ornstein–Uhlenbeck process. **2019**, in preparation.
19. Risken, H. *The Fokker–Planck Equation: Methods of Solution and Applications*; Springer: Berlin, Germany, 1996.
20. Klebaner, F. *Introduction to Stochastic Calculus with Applications*; Imperial College Press: London, UK, 2012.

21. Kim, E.; Hollerbach, R. Time-dependent probability density function in cubic stochastic processes. *Phys. Rev. E* **2016**, *94*, 052118. [CrossRef]
22. Kim, E.; Tenkès, L.-M.; Hollerbach, R.; Radulescu, O. Far-from-equilibrium time evolution between two gamma distributions. *Entropy* **2017**, *19*, 511. [CrossRef]
23. Kim, E.; Hollerbach, R. Geometric structure and information change in phase transitions. *Phys. Rev. E* **2017**, *95*, 062107. [CrossRef]
24. Dean, D.S.; Gupta, S.; Oshanin, G.; Rosso, A.; Schehr, G. Diffusion in periodic, correlated random forcing landscapes. *J. Phys. A* **2014**, *47*, 372001. [CrossRef]
25. Dean, D.S.; Oshanin, G. Approach to asymptotically diffusive behavior for Brownian particles in periodic potentials: Extracting information from transients. *Phys. Rev. E* **2014**, *90*, 022112. [CrossRef] [PubMed]
26. Sivan, M.; Farago, O. Probability distribution of Brownian motion in periodic potentials. *Phys. Rev. E* **2018**, *98*, 052117. [CrossRef]
27. Metzler, R.; Jeon, J.-H.; Cherstvy, A.G.; Barkai, E. Anomalous diffusion models and their properties: Non-stationarity, non-ergodicity, and ageing at the centenary of single particle tracking. *Phys. Chem. Chem. Phys.* **2014**, *16*, 24128–24164. [CrossRef] [PubMed]
28. Zaburdaev, V.; Denisov, S.; Klafter, J. Lévy walks. *Rev. Mod. Phys.* **2015**, *87*, 483–530. [CrossRef]
29. Evangelista, L.R.; Lenzi, E.K. *Fractional Diffusion Equations and Anomalous Diffusion*; Cambridge University Press: Cambridge, UK, 2018.
30. Risken, H.; Vollmer, H.D. Brownian motion in periodic potentials; nonlinear response to an external force. *Z. Phys. B* **1979**, *33*, 297–305. [CrossRef]
31. Lindner, B.; Schimansky-Geier, L.; Reimann, P.; Hänggi, P.; Nagaoka, M. Inertia ratchets: A numerical study versus theory. *Phys. Rev. E* **1999**, *59*, 1417–1424. [CrossRef]
32. Guérin, T.; Dean, D.S. Universal time-dependent dispersion properties for diffusion in a one-dimensional critically tilted potential. *Phys. Rev. E* **2017**, *95*, 012109. [CrossRef]
33. Reimann, P. Brownian motors: Noisy transport far from equilibrium. *Phys. Rep.* **2002**, *361*, 57–265. [CrossRef]
34. Vorotnikov, D. Analytical aspects of the Brownian motor effect in randomly flashing ratchets. *J. Math. Biol.* **2014**, *68*, 1677–1705. [CrossRef]
35. Frezzato, D. Dissipation, lag, and drift in driven fluctuating systems. *Phys. Rev. E* **2017**, *96*, 062113. [CrossRef] [PubMed]
36. Ethier, S.N.; Lee, J. The tilted flashing Brownian ratchet. *Fluct. Noise Lett.* **2019**, *18*, 1950005. [CrossRef]

Information Geometry of Non-Linear Stochastic Systems

Rainer Hollerbach [1,*], **Donovan Dimanche** [2,3] **and Eun-jin Kim** [2]

[1] Department of Applied Mathematics, University of Leeds, Leeds LS2 9JT, UK
[2] School of Mathematics and Statistics, University of Sheffield, Sheffield S3 7RH, UK; donovan.dimanche@gmail.com (D.D.); e.kim@sheffield.ac.uk (E.K.)
[3] Institut National des Sciences Appliquées de Rouen, 76801 Saint-Étienne-du-Rouvray CEDEX, France
* Correspondence: rh@maths.leeds.ac.uk

Abstract: We elucidate the effect of different deterministic nonlinear forces on geometric structure of stochastic processes by investigating the transient relaxation of initial PDFs of a stochastic variable x under forces proportional to $-x^n$ ($n = 3, 5, 7$) and different strength D of δ-correlated stochastic noise. We identify the three main stages consisting of nondiffusive evolution, quasi-linear Gaussian evolution and settling into stationary PDFs. The strength of stochastic noise is shown to play a crucial role in determining these timescales as well as the peak amplitude and width of PDFs. From time-evolution of PDFs, we compute the rate of information change for a given initial PDF and uniquely determine the information length $\mathcal{L}(t)$ as a function of time that represents the number of different statistical states that a system evolves through in time. We identify a robust geodesic (where the information changes at a constant rate) in the initial stage, and map out geometric structure of an attractor as $\mathcal{L}(t \to \infty) \propto \mu^m$, where μ is the position of an initial Gaussian PDF. The scaling exponent m increases with n, and also varies with D (although to a lesser extent). Our results highlight ubiquitous power-laws and multi-scalings of information geometry due to nonlinear interaction.

Keywords: stochastic processes; Fokker-Planck equation; information length

1. Introduction

There is increasing interest in a metric on probability from theoretical and practical considerations, with different metrics proposed depending on the question of interest (e.g., [1–10] and further references therein). Theoretically, the assignment of an appropriate metric to probability enables us to mathematically quantify the difference among different Probability Density Functions (PDFs), providing a beautiful conceptual link between a stochastic process and geometry. At a practical level, it can be utilized for optimising various desired outcomes. For instance, the Wasserstein metric has been extensively studied for the optimal transport problem to minimize transport cost, typically taken to increase quadratically with distance between two locations [1]. The Fisher (also called Fisher-Rao) metric has recently been used for optimization, including the minimization of entropy production [11], parameter estimation [12], controlling population [13], understanding the arrow of time [14], and analysing the convexity of the relative entropy [15].

However, compared with the Wasserstein metric, whose application has established itself as a branch of applied mathematics, the geometric structure associated with the information change in the Fisher metric and its utility have been explored much less. Unlike the Wasserstein distance, the Fisher metric yields a hyperbolic geometry in the upper half-plane (e.g., [9,10]) where the distance is measured in units of the width of the PDF. That is, the distance in the Fisher metric is dimensionless and represents the number of different statistical states. Consequently, for a Gaussian PDF, statistically distinguishable states are determined by the standard deviation; two PDFs that have the same standard deviation and differ in peak positions by less than one standard deviation are statistically indistinguishable (e.g., [11,16–22]).

By extending the Fisher metric to time-dependent problems, we recently introduced a system-independent way of quantifying information change associated with time-evolution of PDFs [13,23–30]. (Note that we use information for statistically different states, refraining from the debate on the exact definition of information, e.g., [5,31].) The key idea is to define an infinitesimal distance at any time by comparing two PDFs at adjacent times and sum these distances. The total distance gives us the number of statistically different states that a system passes through in time, and is called information length $\mathcal{L}$. While the detailed derivation of $\mathcal{L}$ is given in [5,13,16,23–30], it is useful to highlight that $\mathcal{L}$ is a measure of the total elapsed time in units of a dynamical timescale for information change. To show this, we define the dynamical time $\tau(t)$ as follows:

$$\mathcal{E} \equiv \frac{1}{[\tau(t)]^2} = \int \frac{1}{p(x,t)} \left[\frac{\partial p(x,t)}{\partial t}\right]^2 dx. \tag{1}$$

Here, $\tau(t)$ is the characteristic timescale over which the information changes. Having units of time, $\tau(t)$ quantifies the correlation time of a PDF. Alternatively, $1/\tau$ quantifies the (average) rate of change of information in time. A particular path which gives a constant valued $\mathcal{E}$ is a geodesic along which the information propagates at the same speed [13]. $\mathcal{L}(t)$ is then defined by measuring the total elapsed time t in units of τ as:

$$\mathcal{L}(t) = \int_0^t \frac{dt_1}{\tau(t_1)} = \int_0^t \sqrt{\int dx \frac{1}{p(x,t_1)} \left[\frac{\partial p(x,t_1)}{\partial t_1}\right]^2} dt_1. \tag{2}$$

$\mathcal{L}(t)$ is a Lagrangian quantity (unlike entropy or relative entropy), uniquely defined as a function of time t for a given initial PDF, and represents the total number of statistically distinguishable states that a system evolves through. It thus provides a very convenient methodology for measuring the distance between $p(x,t)$ and $p(x,0)$ continuously in time for a given $p(x,0)$. One of the utilities of $\mathcal{L}(t)$ is to quantify the proximity of any initial PDF to a final attractor of a dynamical system. We note that for a linear process, we can express the information length in terms of a metric tensor g_{ij} using the two parameters $\beta = 1/2\langle(x-\langle x\rangle)^2\rangle$ and $\langle x\rangle$ (e.g., [13]). However, even in the case where the control parameters are not known, we can still define $\mathcal{L}$ as long as we can compute time-dependent PDFs (e.g., from data [24]).

Traditionally, concepts such as bifurcations, Lyapunov exponent or a distance to an equilibrium point are commonly used to understand dynamical systems (e.g., see [32]). An intriguing question arises as to how to define a distance between a point, say x, and an equilibrium which is not a point but a limit cycle, or chaotic attractor. One interesting possibility is to consider a narrow initial PDF around x and to measure the total information length $\mathcal{L}(t \to \infty) = \mathcal{L}_\infty$ as the initial PDF evolves toward the equilibrium PDF. $\mathcal{L}_\infty$ offers a Lagrangian distance that depends on the trajectory/history of the system (e.g., time-dependent PDF), being uniquely defined as a function of time for a given initial PDF. This enables us to map out the attractor structure by measuring $\mathcal{L}_\infty$ for different locations of a narrow initial PDF. In a chaotic system, $\mathcal{L}_\infty$ changes abruptly when a different initial PDF is used, as shown in Figure 4 in [24], where the very spiky curve represents a sensitive dependence of $\mathcal{L}_\infty$ on $x(t=0)$, the location of a very narrow initial PDF. This sensitive dependence of $\mathcal{L}_\infty$ on $x(t=0)$ means that a small change in the initial condition causes a large difference in a path that a system evolves through and thus $\mathcal{L}_\infty$. This is quite similar to the sensitive dependence of the Lyapunov exponent on the initial condition. That is, our $\mathcal{L}$ provides a new methodology to test chaos. Furthermore, Figure 4 in [24] shows small $\mathcal{L}_\infty$ for unstable points, demonstrating that unstable points are more similar to chaotic attractors.

The purpose of this paper is to consider the effect of different orders of nonlinear interaction on the geometric structure by considering the case when the equilibrium is a stable point [26,33]. The remainder of this paper is organised as follows: Section 2 introduces the basic model. Section 3 derives exact analytic solutions in the absence of stochastic noise, as well as asymptotic scalings for the timescales, peak amplitudes and widths once noise plays a role. Section 4 presents numerical results, and shows how they compare with the analytic scalings. Section 5 summarizes the results.

2. Model

We consider the following nonlinear Langevin equation:

$$\frac{dx}{dt} = -\gamma x^n + \xi. \tag{3}$$

Here, x is a random variable and ξ is a stochastic forcing, which for simplicity can be taken as a short-correlated (or δ-correlated, white-noise) random forcing as follows:

$$\langle \xi(t)\xi(t') \rangle = 2D\delta(t-t'), \tag{4}$$

where the angular brackets represent the average over ξ, $\langle \xi \rangle = 0$, and D is the strength of the forcing. The parameter γ is a positive constant. n is the order of nonlinearity, which we take to be an odd integer to make $x = 0$ an attractor, and also preserve a reflectional symmetry (under $x \to -x$) of the system.

We note that there are two possible interpretations of Equation (3). Specifically, consider the linear case where $\frac{\partial V}{\partial x} \propto x$. The first interpretation is to view x as a velocity, in which case the force term would give a frictional force (e.g., Equation (1.2) in [34]) as $\frac{dv}{dt} = -\gamma v + \xi$, with γ as a frictional (damping) constant. The second interpretation is to take the overdamped limit of the coupled equations for x and v by dropping d^2x/dt^2 (e.g., Equation (3.130) in [34]) to obtain one equation for x from the first line in Equation (3.130). In this case, x represents the position and γ^{-1} would be the frictional constant, and a harmonic potential would correspond to $n = 1$.

For a deterministic system with $\xi = 0$, the solution to Equation (3) is readily obtained as

$$x(t) = \frac{x_0}{[1 + \gamma(n-1)t x_0^{n-1}]^{\frac{1}{n-1}}}, \tag{5}$$

where x_0 is the initial condition $x(t = 0)$.

The corresponding Fokker–Planck (FP) equation is [34]

$$\partial_t p(x,t) = \partial_x(\gamma x^n p) + D\partial_{xx} p. \tag{6}$$

We first discuss some analytic limiting cases in Section 3, and then present numerical solutions in Section 4.

3. Analytic Solutions

3.1. Exact Solutions for $D = 0$

In the absence of the stochastic noise $D = 0$, the diffusion term in the Fokker–Planck Equation (6) is zero. In this case, the PDF does not have a stationary solution, but continues to change in time due to the linear/nonlinear force. For instance, for $n = 1$, the PDF's width becomes exponentially narrow in time as the fluctuation (as well as the mean value) decreases exponentially due to $\gamma > 0$ (see below). To obtain an analytic solution to Equation (6), we use the method of characteristics. Specifically, we rewrite Equation (6) with $D = 0$ in terms of the total derivative along the characteristic as

$$\frac{dp}{dt} \equiv \frac{\partial p}{\partial t} + \frac{dx}{dt}\frac{\partial p}{\partial x} = n\gamma x^{n-1} p. \tag{7}$$

Here, the characteristic is given by

$$\frac{dx}{dt} = -\gamma x^n, \tag{8}$$

which is Equation (3) without the stochastic noise ξ. Thus, the characteristic with the initial condition $x(t=0) = x_0$ satisfies Equation (5), which can also be written as

$$x_0 = \frac{x(t)}{[1-\gamma(n-1)t[x(t)]^{n-1}]^{\frac{1}{n-1}}}. \tag{9}$$

Along these characteristics, we rewrite Equation (7) as

$$\frac{dp}{p} = n\gamma x^{n-1}dt = n\gamma \frac{x_0^{n-1}dt}{[1+\gamma(n-1)tx_0^{n-1}]} = \frac{n}{n-1}d\left[\ln\left[1+\gamma(n-1)tx_0^{n-1}\right]\right]. \tag{10}$$

The integration of Equation (10) then gives us

$$p(x,t) = p(x_0,0)[1+\gamma(n-1)tx_0^{n-1}]^{\frac{n}{n-1}} = p(x_0,0)[1-\gamma(n-1)tx^{n-1}]^{-\frac{n}{n-1}}. \tag{11}$$

For simplicity, we consider an initial Gaussian PDF localised around μ as

$$p(x_0,0) = \sqrt{\frac{\beta_0}{\pi}}e^{-\beta_0(x_0-\mu)^2}. \tag{12}$$

Then, we can find $p(x,t)$ at a later time from Equations (9), (11) and (12) as

$$p(x,t) = \sqrt{\frac{\beta_0}{\pi}}\,\phi^n\, e^{-\beta_0(x\phi-\mu)^2}, \tag{13}$$

where

$$\phi = [1-\gamma(n-1)tx^{n-1}]^{-\frac{1}{n-1}}. \tag{14}$$

The maximum amplitude of p occurs at that x where $x\phi - \mu = 0$ in Equation (13), with the value

$$p_{\max}(t) = \sqrt{\frac{\beta_0}{\pi}}\left[1+\gamma(n-1)t\mu^{n-1}\right]^{\frac{n}{n-1}}. \tag{15}$$

The linear case $n=1$ can be obtained by taking the limit $n \to 1$ in Equations (13) and (14). One finds that $\phi = e^{\gamma t}$ so that

$$\lim_{n\to 1} p(x,t) = \sqrt{\frac{\beta}{\pi}}e^{-\beta(x-\mu e^{-\gamma t})^2}, \tag{16}$$

where $\beta = \beta_0 e^{2\gamma t}$. That is, the width of the PDF ($\propto \beta^{-1/2}$) as well as the peak position decrease exponentially. However, the linear case can be solved analytically even when $D \neq 0$ [33], so we are here more interested in the nonlinear cases $n = 3,5,7$.

Given $p(x,t)$ in Equation (13), we can further calculate $\mathcal{E}(t)$ as follows

$$\begin{aligned}\mathcal{E}(t) &= \gamma^2 \int dx\,(x\phi)^{2(n-1)}\,[n-2\beta_0 x\phi(x\phi-\mu)]^2\,p(x,t) \\ &= \gamma^2 \int dx_0\, x_0^{2(n-1)}\,[n-2\beta_0 x_0(x_0-\mu)]^2\,p(x_0,0),\end{aligned} \tag{17}$$

where we use $x_0 = x\phi$, $\frac{dx_0}{dx} = \phi^n$, $p(x_0,0) = p(x,t)\frac{dx}{dx_0}$, and $p(x_0,0)$ is the initial Gaussian PDF in Equation (12). Interestingly, Equation (17) is independent of time, with constant τ. That is, without a stochastic noise, the evolution of PDFs follows a geodesic due to the scaling relation satisfied along the characteristic. For example, we can show that, for $n = 1$, $\mathcal{E} = 2\gamma^2(1+\beta_0\mu^2)$ and for $n = 3$, $\mathcal{E} = \gamma^2\mu^4\left[2\beta_0\mu^2 + 24 + \frac{99}{2\beta_0\mu^2} + \frac{21}{2(\beta_0\mu^2)^2}\right]$. From these analyses, we can infer that for sufficiently large β_0 such that $\beta_0\mu^{n-1} \gg 1$, $\mathcal{E} \propto \beta_0\mu^{2n}$ to leading order.

3.2. *Approximate Solutions for $D \neq 0$*

According to the previous results, the peak amplitudes increase indefinitely in time. The corresponding widths also decrease, and eventually become so narrow that any non-zero D will start to play a significant role, and will start to broaden the PDFs again. However, during this phase of the evolution, the PDF width is still smaller than its mean position, so that we can use a quasi-linear analysis to approximate Equation (3) to leading order in $O(x'/\langle x\rangle)$ as

$$\frac{dx'}{dt} \approx -n\gamma\langle x(t)\rangle^{n-1}x' + \xi, \tag{18}$$

where $x'(t) = x(t) - \langle x(t)\rangle$ is the fluctuation, with $\langle x'(t)\rangle = 0$. x' in Equation (18) satisfies the Ornstein–Uhlenbeck process with the effective damping constant

$$\gamma_e(t) = n\gamma\langle x(t)\rangle^{n-1}. \tag{19}$$

Thus, in this stage, the PDFs remain essentially Gaussian and evolve as

$$p(x,t) = \sqrt{\frac{\beta}{\pi}}e^{-\beta(x-\langle x\rangle)^2}, \tag{20}$$

where

$$\langle x\rangle = \frac{\mu}{[1+\gamma(n-1)\mu^{n-1}t]^{\frac{1}{n-1}}}, \tag{21}$$

$$\frac{1}{2\beta} = \frac{e^{-2G(t)}}{2\beta_0} + \frac{D(1-e^{-2G(t)})}{\gamma_e}, \tag{22}$$

$$G(t) = \int_0^t dt'\gamma_e(t') = \frac{n}{n-1}\ln[1+\gamma(n-1)\mu^{n-1}t]. \tag{23}$$

The term $e^{-2G}/2\beta_0$ in Equation (22) represents the narrowing of the initial PDF width as discussed in the previous section. However, note that, when $1/2\beta_0 \ll D/\gamma_e$, a PDF can maintain the same width set by the constant value D/γ_e at the initial stage; that is, there is no nondiffusive evolution phase. See [33] for such a case with $n = 3$. For the $1/2\beta_0 > D/\gamma_e$ case, we consider here, the transition from the nondiffusive evolution phase to the quasi-linear Gaussian evolution phase occurs when $e^{-2G}/2\beta_0$ becomes comparable to D/γ_e, leading to the following criterion for the first transition timescale t_1:

$$2G \sim \ln\frac{\gamma_e}{2\beta_0 D} \sim \ln\frac{n\gamma\langle x\rangle^{n-1}}{2\beta_0 D}, \tag{24}$$

using $\gamma_e = n\gamma\langle x\rangle^{n-1}$. By using Equation (23) and $\langle x\rangle = \mu/[1+\gamma(n-1)\mu^{n-1}t_1]^{\frac{1}{n-1}}$, we obtain

$$\left[1+\gamma(n-1)\mu^{n-1}t_1\right]^{\frac{3n-1}{n-1}} \sim \frac{\gamma n}{2\beta_0 D}\mu^{n-1}. \tag{25}$$

For large time $\gamma(n-1)\mu^{n-1}t_1 \gg 1$, Equation (25) thus yields

$$t_1 \propto D^{-\frac{n-1}{3n-1}}\mu^{-\frac{2n(n-1)}{3n-1}}. \tag{26}$$

Note in particular how both exponents are negative for $n = 3, 5, 7$, so that smaller D and/or μ yield a larger t_1. Next, for $D \neq 0$ and $\gamma(n-1)\mu^{n-1}t \gg 1$, we approximate Equation (22) as

$$\frac{1}{2\beta} \sim \frac{D}{\gamma_e} \sim \frac{n-1}{n} Dt, \tag{27}$$

using also $\gamma_e \propto t^{-1}$. Thus, the PDF width increases as $(Dt)^{1/2}$, while the maximum amplitude decreases as

$$p_{\max}(t) \propto \beta^{1/2} \propto (Dt)^{-1/2}, \tag{28}$$

typical of Brownian motion [34]. Using Equation (26) in either Equation (15) or (28)—t_1 after all is precisely the time when one scaling transitions to the other, so they should agree at that time—we obtain the scalings of the overall maximum amplitude p_{Max} reached throughout the entire evolution as

$$p_{\text{Max}} \propto D^{-\frac{n}{3n-1}} \mu^{\frac{n(n-1)}{3n-1}}. \tag{29}$$

3.3. Final Stationary Distribution

For even greater times, fluctuations become stronger while the mean values decrease. The quasi-linear analysis in the previous section is therefore eventually no longer applicable, and numerical solutions are crucial. The final stationary solution of Equation (6) can however be computed analytically from $D\partial_x p = -\gamma x^n p$ and becomes

$$p(x) = \left(\frac{\gamma}{D(n+1)}\right)^{1/(n+1)} \frac{n+1}{\Gamma(1/(n+1))} \exp\left(-\frac{\gamma}{D(n+1)} x^{n+1}\right). \tag{30}$$

Thus, the maximum amplitude

$$p_{\max} \propto D^{-\frac{1}{n+1}}, \tag{31}$$

while the width of the PDF is proportional to $D^{\frac{1}{n+1}}$. Finally, the transition from the intermediate stage in the previous section to this final stage occurs when the two formulas for the widths $(Dt)^{1/2}$ and $D^{\frac{1}{n+1}}$ become comparable, yielding the second transition timescale

$$t_2 \propto D^{-\frac{n-1}{n+1}}. \tag{32}$$

In the following section, we see how numerical solutions compare with some of these predictions such as the two timescales t_1 and t_2, as well as explore other aspects of the solutions for which no analytic predictions are possible.

4. Numerical Solutions

For $D \neq 0$, an exact analytic solution to the Fokker–Planck Equation (6) only exists for the linear case $n = 1$ [33]. For the nonlinear cases $n = 3, 5, 7$ that are the focus of this paper, we must resort to numerical solutions. Without loss of generality, we set $\gamma = 1$. The interval in x is fixed to be $[-1, 1]$, rather than the original $[-\infty, \infty]$. This may seem drastic, but actually involves no real loss of generality either, since any finite interval can always be mapped to $[-1, 1]$ by suitably rescaling x, t and D. As long as the initial condition (and D) are chosen such that p would be negligible outside $[-1, 1]$ anyway, solving the FP equation only on $[-1, 1]$, and with $p = 0$ boundary conditions is then equivalent in all essentials to the original infinite interval.

The numerical solution is done by second-order finite-differencing in x, with up to $M = 2 \times 10^6$ grid points. The time-stepping is also second-order accurate, with increments as small as $\Delta t = 10^{-6}$. Both M and Δt were varied to check the accuracy of the solutions. In the later stages, when the PDFs are evolving to increasingly broad profiles, M can also be decreased, and Δt increased, while still preserving accuracy. Regridding the solutions in this way is indeed crucial, since the final adjustment timescale t_2 is extremely long for small D, and could not be reached if M and Δt were kept fixed at their initial values.

In [33], we considered the initial condition

$$p = \frac{1}{\sqrt{\pi\, 10^{-8}}} \exp\left[-\frac{(x-0.7)^2}{10^{-8}}\right], \tag{33}$$

and then compared numerical solutions for $n = 3$ and $D = 10^{-3}$ to 10^{-7} with the corresponding analytical solutions for $n = 1$. That is, the initial peak was extremely narrow, and diffusion was greater than what we consider here. As a result of these choices, the peaks only became broader but never narrower; the nondiffusive regime in Section 3.1 simply does not exist in this case.

In contrast, in this work, we take the initial condition

$$p = \frac{1}{\sqrt{2\pi\, 10^{-6}}} \exp\left[-\frac{(x-\mu)^2}{2\times 10^{-6}}\right], \tag{34}$$

and $D = 10^{-6}$ to 10^{-9}. By starting with broader peaks and having smaller D, we do have an initial nondiffusive regime here, and are able to observe the narrowing of the peaks predicted by Equations (13) and (14). We begin by fixing the initial peak position $\mu = 0.65$; below, we also consider the range $\mu = [0.01, 0.75]$.

Figure 1 shows how the peak amplitudes evolve in time. We can clearly see the three regimes deduced in Section 3: the peaks initially increase, in excellent agreement with Equation (15), then they decrease in agreement with Equation (28), and finally they equilibrate to Equation (31). To more quantitatively compare the overall peak amplitudes p_{Max} and the corresponding times t_{Max} at which they occur with the analytic predictions given by Equations (26) and (29), we note first that $D = 10^{-6}$ is clearly not yet sufficiently small for there to be an initial nondiffusive regime at all (for this particular width of the initial condition). Even $D = 10^{-7}$ only follows the nondiffusive Equation (15) for a very brief time, not yet long enough to be in the $\gamma(n-1)\mu^{n-1}t \gg 1$ regime where Equations (26) and (29) are expected to apply. Table 1 therefore compares the $D = 10^{-8}$ and 10^{-9} cases, and uses them to extract scaling exponents of the form $D^{-\alpha}$. We see that the agreement of p_{Max} with Equation (29) is virtually perfect. The corresponding times t_{Max} are close to the expected scaling t_1, but are not fully in the asymptotic limit yet. This is hardly surprising, since even for $D = 10^{-9}$ the t_{Max} values in Table 1 only have $\gamma(n-1)\mu^{n-1}t_{\text{Max}} \sim 8$ (for all three n values).

To similarly test the scalings that are predicted for μ, further runs were done with fixed $D = 10^{-9}$, and $\mu = 0.55$ and 0.75. As Table 2 shows, the extracted exponents are again in very good (t_{Max}) and perfect (p_{Max}) agreement with the predicted asymptotic scalings. The interesting feature that larger initial positions μ have smaller times to reach the maximum peak amplitude is certainly very well reproduced. Note finally that the two μ values used to extract these exponents in Table 2 differ by less than a factor of two even, and would thus certainly not be enough to extract reliable scaling exponents if we did not already have robust analytic predictions. As we also show in more detail below, in principle, it would be possible to have the analytic predictions extend over an arbitrarily large range in μ, but that would require increasingly small D as well, which becomes numerically too time-consuming.

Returning to the $\mu = 0.65$ runs in Figures 1 and 2 shows the expectation values $\langle x\rangle = \int xp\,dx$, and how they compare with the nondiffusive result (Equation (5)). The agreement is close to perfect even after the first timescale t_1 is reached. It is only once the very long second timescale t_2 is reached that $\langle x\rangle$ approaches 0 exponentially, far more rapidly than the $t^{-\frac{1}{n-1}}$ power law scaling in (5).

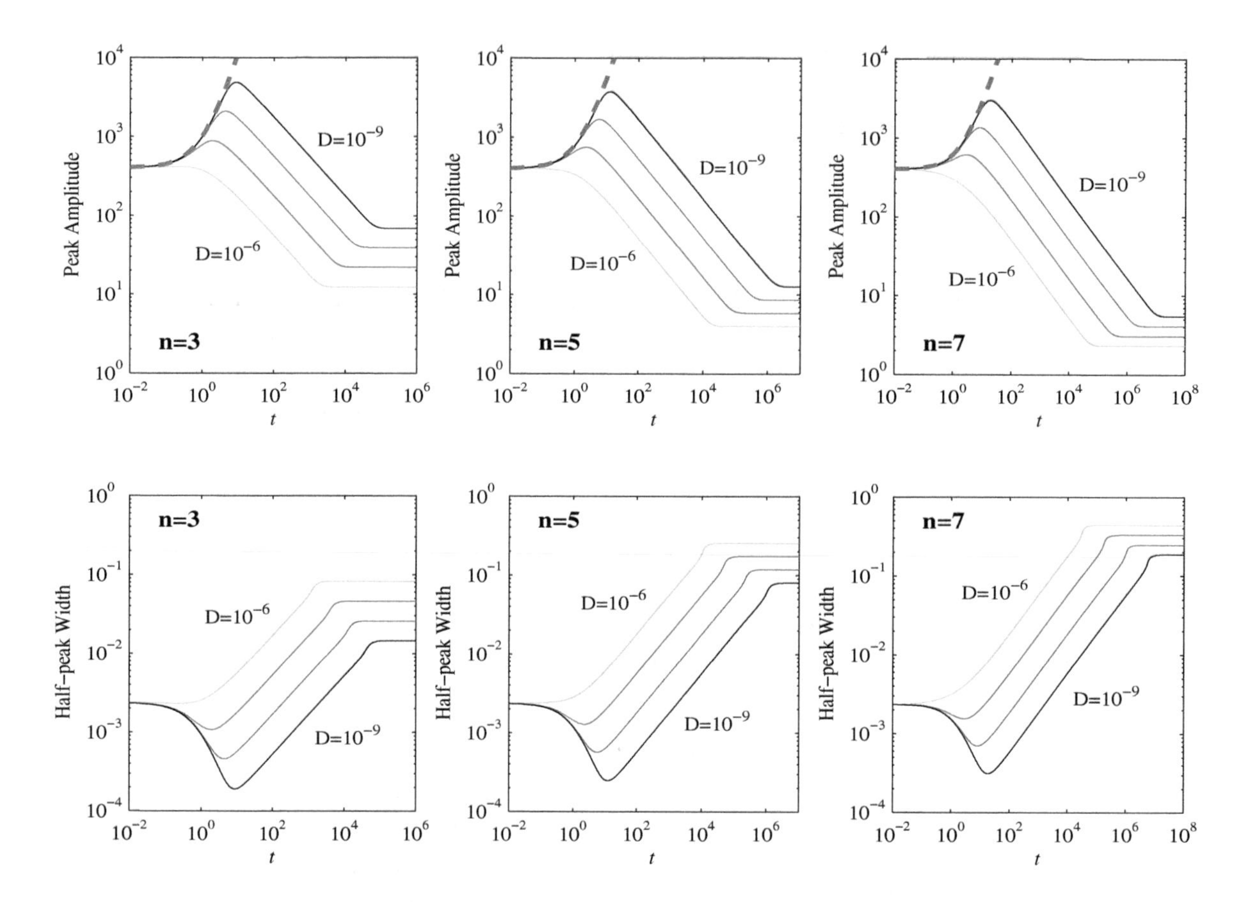

Figure 1. The top row shows the peak amplitudes as functions of time, for the initial condition in Equation (34) with $\mu = 0.65$, and $D = 10^{-6}$ to 10^{-9} as indicated. The three panels show $n = 3, 5, 7$, as labeled. The thick dashed (magenta) lines correspond to the analytic result (Equation (15)) that applies in the nondiffusive phase. The bottom row shows the equivalent widths at half-peak, which are inversely proportional to the peak amplitudes. The standard deviation $\langle (x - \langle x \rangle)^2 \rangle^{1/2}$ follows exactly the same pattern as the half-peak widths.

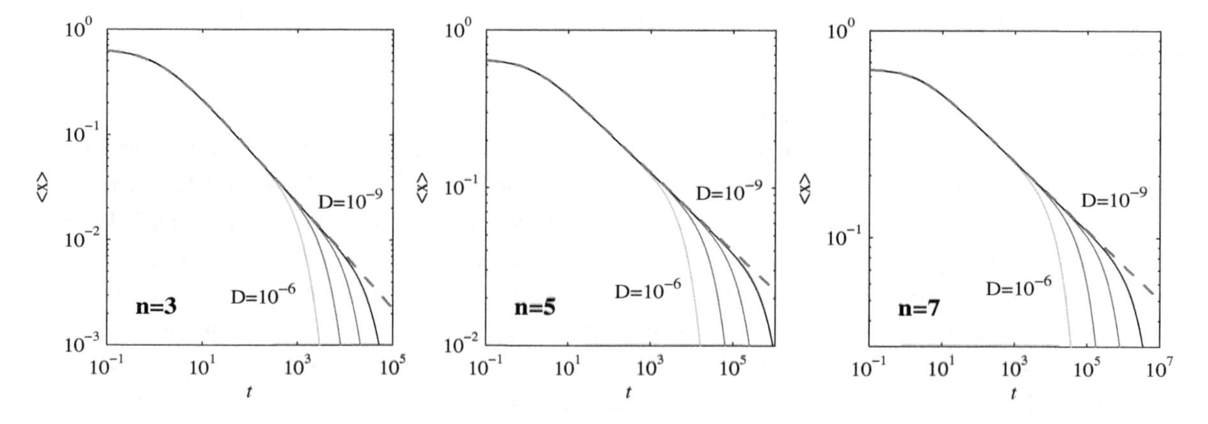

Figure 2. The mean values $\langle x \rangle$ as functions of time, for the solutions in Figure 1. The dashed (magenta) lines show Equation (5) that applies in the nondiffusive case.

Table 1. The first two rows show the overall peak amplitudes $p_{\rm Max}$ and the times $t_{\rm Max}$ at which they occur, for the results in Figure 1. $D = 10^{-8}$ and 10^{-9}, and $n = 3,5,7$ as indicated, and all results at $\mu = 0.65$. The row labeled "Exponent" uses the ratios of the two D values to extract scaling exponents of the form $D^{-\alpha}$. The final row compares these numerically deduced exponents with the analytic predictions from Equations (26) and (29). That is, the exponent α should equal $\frac{n-1}{3n-1}$ for $t_{\rm Max}$, and $\frac{n}{3n-1}$ for $p_{\rm Max}$.

	$n = 3$		$n = 5$		$n = 7$	
	$t_{\rm Max}$	$p_{\rm Max}$	$t_{\rm Max}$	$p_{\rm Max}$	$t_{\rm Max}$	$p_{\rm Max}$
$D = 10^{-8}$	4.42	2063	5.82	1661	8.17	1337
$D = 10^{-9}$	8.80	4888	12.55	3776	18.58	2998
Exponent	0.30	0.375	0.33	0.357	0.36	0.349
(26), (29)	0.25	0.375	0.29	0.357	0.30	0.350

Table 2. The first two rows are as in Table 1, but now for fixed $D = 10^{-9}$, $\mu = 0.55$ and 0.75, and $n = 3,5,7$ as indicated. The row labeled "Exponent" uses the ratios of the two μ values to extract scaling exponents of the form μ^{δ}. The final row compares these numerically deduced exponents with the analytic predictions from Equations (26) and (29). That is, the exponent δ should equal $-\frac{2n(n-1)}{3n-1}$ for $t_{\rm Max}$, and $\frac{n(n-1)}{3n-1}$ for $p_{\rm Max}$.

	$n = 3$		$n = 5$		$n = 7$	
	$t_{\rm Max}$	$p_{\rm Max}$	$t_{\rm Max}$	$p_{\rm Max}$	$t_{\rm Max}$	$p_{\rm Max}$
$\mu = 0.55$	11.17	4314	19.74	2975	35.87	2105
$\mu = 0.75$	7.18	5439	8.48	4631	10.46	4035
Exponent	−1.4	0.75	−2.7	1.43	−4.0	2.10
(26), (29)	−1.5	0.75	−2.9	1.43	−4.2	2.10

Figure 3 shows two commonly used diagnostic quantities, the skewness $\int[(x - \langle x \rangle)/\sigma]^3\, p\, dx$ and the kurtosis $\int[(x - \langle x \rangle)/\sigma]^4\, p\, dx$, where $\sigma = [\int(x - \langle x \rangle)^2\, p\, dx]^{1/2}$ is the standard deviation. Skewness is a measure of how asymmetric a PDF is about its peak, with a value of 0 indicating a symmetric peak. Kurtosis measures the flatness of a PDF, especially in comparison with a Gaussian, which has kurtosis $= 3$. We see that skewness ≈ 0 and kurtosis ≈ 3 is maintained all the way until the final equilibration timescale t_2 is reached, indicating that the PDFs remain largely Gaussian up to this time, as predicted in Section 3.2. For the final equilibrated profiles in Equation (30), the skewness is again 0, whereas the kurtosis has values that depend on n (the precise values can be evaluated analytically in terms of Γ functions, but are not particularly insightful). It is interesting though to note in Figure 3 that both skewness and kurtosis exhibit non-monotonic behavior during the final adjustment process, where the PDFs transition from being largely Gaussian to their final form. Note also how the maxima reached during this transition process are independent of D, and even broadly similar for the different values of n.

Figure 4 compares the PDFs at the times when the skewness reaches its maximum (negative) value with the final equilibrated profiles in Equation (30). The entire evolution is then clear: as long as the PDFs are well outside their final profiles, they remain essentially Gaussian, with widths as in Figure 1 and positions as in Figure 2. The skewness and kurtosis start to deviate significantly from their Gaussian values when the PDFs start to reach their final positions, with the maximum skewness as in Figure 4. The final rearrangement for $t > t_2$ then merely adjusts to the final profiles (30), but with relatively little further movement of the PDFs.

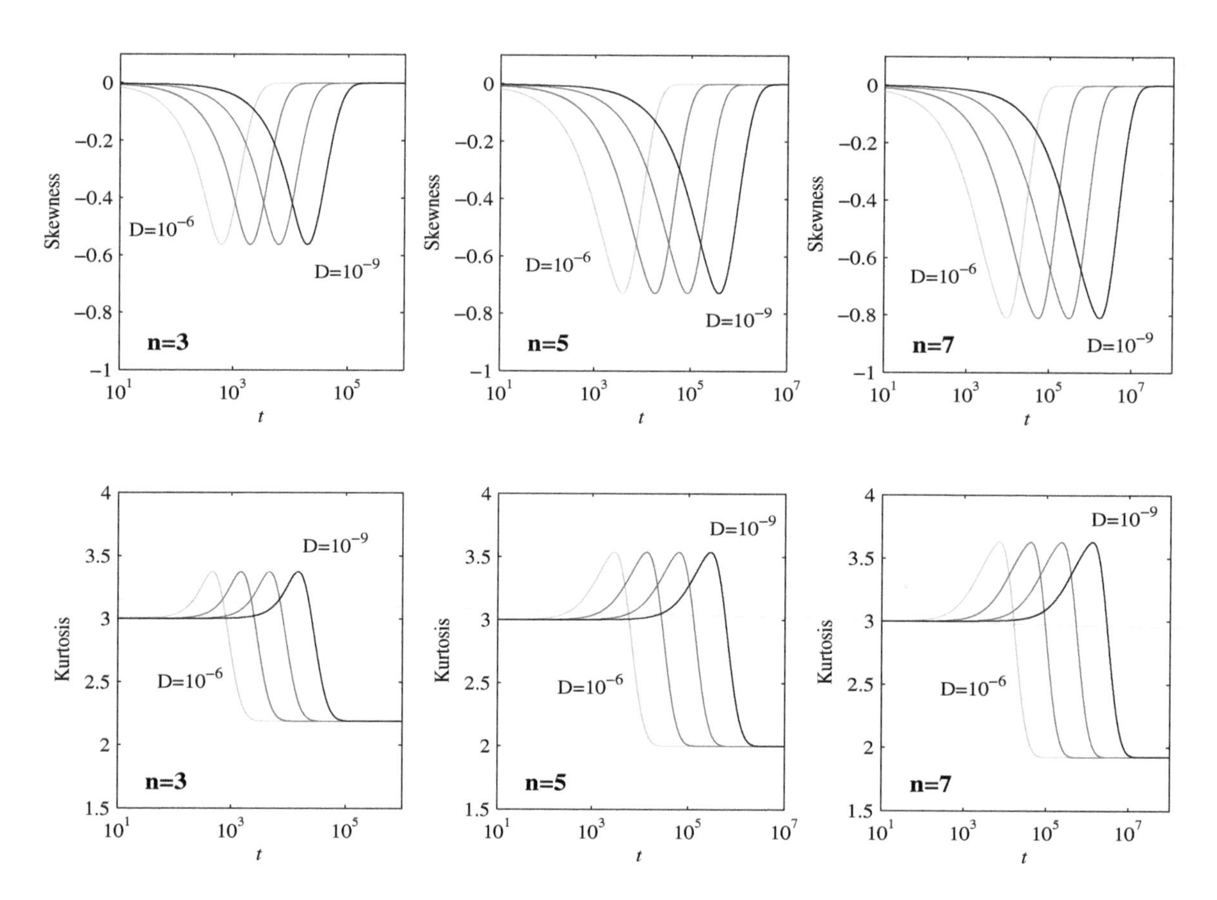

Figure 3. The top row shows the skewness $\int [(x - \langle x \rangle)/\sigma]^3\, p\, dx$, and the bottom row shows the kurtosis $\int [(x - \langle x \rangle)/\sigma]^4\, p\, dx$, as functions of time. The labeling of D and n is as in Figures 1 and 2. The peaks in both quantities occur at times in essentially perfect agreement with t_2 in (32).

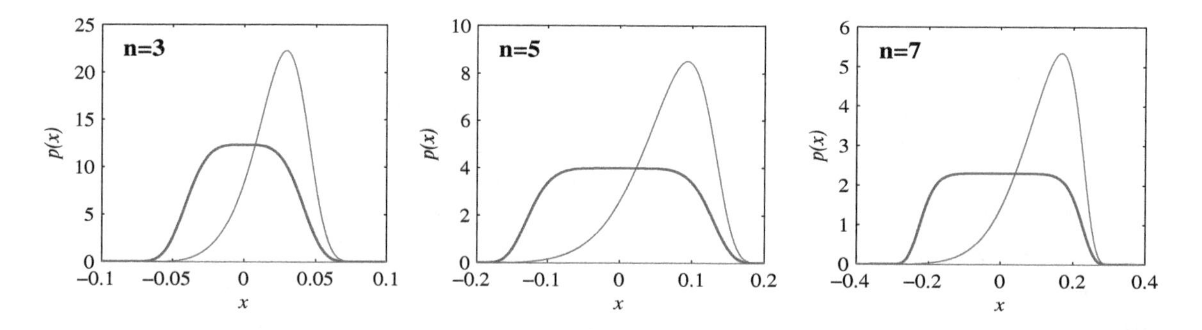

Figure 4. The heavy (blue) lines show the equilibrium profiles in Equation (30), for $D = 10^{-6}$, and $n = 3, 5, 7$ as indicated. The lighter (red) lines show $p(x)$ at the $t = 629$, 3930, 9794, for $n = 3, 5, 7$, respectively. As shown in Figure 3, these are the times when the skewness reaches its maximum negative values. Results for other values of D are identical, once x and $p(x)$ are rescaled as in Equation (31), and t is shifted as in Figure 3 to consistently have the correct skewness values.

Figure 5 shows the diagnostic quantities $\mathcal{E}(t)$ and $\mathcal{L}(t)$. As predicted by Equation (17), $\mathcal{E}$ remains constant in the initial nondiffusive phase. Once the first timescale t_1 is reached, $\mathcal{E}$ decreases as $t^{-\frac{3n-1}{n-1}} D^{-1}$. Once the second timescale t_2 is reached, $\mathcal{E}$ decreases exponentially. (This is not included in Figure 5, however, as $\mathcal{E} \propto D^{\frac{2(n-1)}{n+1}}$ at that point, and is thus already negligibly small.) The behavior of $\mathcal{L}$ is as expected: while $\mathcal{E}$ is constant, $\mathcal{L}$ increases linearly in time, and, once $\mathcal{E}$ starts decreasing, $\mathcal{L}$ levels off to its final value $\mathcal{L}_\infty$.

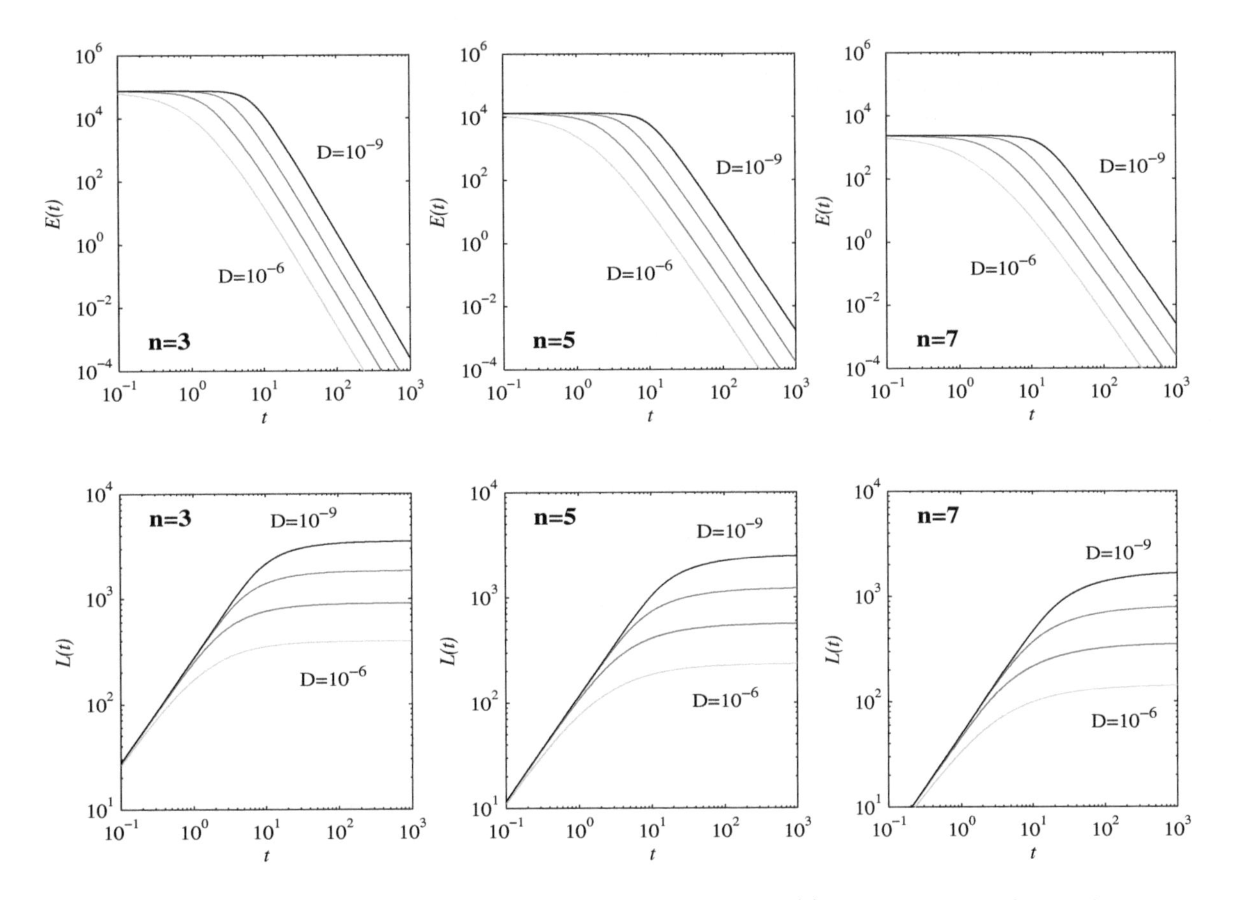

Figure 5. The top row shows $\mathcal{E}(t)$ and the bottom row shows $\mathcal{L}(t)$. The labeling of D and n is as in Figures 1–3. The peaks are initially located at $\mu = 0.65$, consistent with the initial plateau in $\mathcal{E}$ being reduced by a factor of $0.65^4 = 0.1785$ when comparing $n = 3$ and 5, and similarly $n = 5$ and 7.

Based on the scaling for $\mathcal{E}$, namely that $\mathcal{E} \propto \mu^{2n}$ initially, and the range t_1 for which this applies, we expect at least a lowest-order estimate for $\mathcal{L}_\infty$ to be given simply by

$$\mathcal{L}_\infty \propto \sqrt{\mathcal{E}}\, t_1 \propto D^{-\frac{n-1}{3n-1}}\, \mu^{\frac{n(n+1)}{3n-1}}. \tag{35}$$

Figure 6 shows numerically computed results, for the same $D = 10^{-6}$ to 10^{-9} range considered throughout, and $\mu = [0.01, 0.75]$. At least for sufficiently large μ, we do indeed see clear evidence of power-law behavior, with a negative exponent for D and a positive one for μ. Concentrating on the variation with μ, for $n = 3$, the slopes vary between 1.8 for $D = 10^{-6}$ and 1.6 for $D = 10^{-9}$; for $n = 5$, they vary between 2.7 and 2.3; and for $n = 7$, they vary between 3.5 and 2.9. By comparison, according to Equation (35), the exponent should be $\frac{n(n+1)}{3n-1}$, which yields 1.5, 2.1, 2.8 for $n = 3, 5, 7$, respectively. The agreement is thus quite good, especially when we recall (Tables 1 and 2) that even $D = 10^{-9}$ was not quite small enough yet to obtain precise agreement with the asymptotic formula for t_1, which in turn affects Equation (35) as well.

$\mathcal{L}_\infty$ as a function of μ provides a mapping from the physical distance μ (the distance between the peak position μ of an initial PDF and the final equilibrium point 0) to the information length (the total number of statistically different states of a system reaching a final stationary PDF from an initial PDF). From Equation (35), this mapping is linear for a linear force; that is, $\mathcal{L}_\infty$ is linearly proportional to the physical distance. This linear relation breaks down for nonlinear forces as $\mathcal{L}_\infty$ depends nonlinearly on the physical distance μ. Specifically, for $n = 3$, $\mathcal{L}_\infty \propto \mu^{1.5}$; for $n = 5$, $\mathcal{L}_\infty \propto \mu^{2.1}$; and, for $n = 7$, $\mathcal{L}_\infty \propto \mu^{2.8}$. Thus, we can envision that nonlinear forces affect the information geometry, changing a linear (flat) coordinate to a power-law (curved) coordinate. This is reminiscent of gravity changing a flat to a curved space-time. Furthermore, interestingly, Equation (35) shows that $\mathcal{L}_\infty$ is independent of D for the linear force, manifesting the information geometry as independent of the resolution (set by D). In contrast, $\mathcal{L}_\infty$ decreases with $n = 3, 5, 7$ as $D^{-1/4}$, $D^{-2/7}$, $D^{-3/10}$, suggesting that the information geometry is fractal, depending on the resolution. This would be equivalent to the I theorem of [14].

Finally, what about the small μ limit in Figure 6, which clearly does not follow the expected scaling (35)? The explanation is that the initial position μ is already within the $O(D^{\frac{1}{n+1}})$ width of the final distribution in Equation (30). In this limit, the behavior is different, and the peak merely spreads out, resulting in a small and relatively uniform $\mathcal{L}_\infty$. For any given μ, D must therefore satisfy $D \ll \mu^{n+1}$ to be in the regime where Equation (35) is expected to apply. That is, in principle, it would be possible to have Equation (35) apply over several orders of magnitude in μ, but D would have to be far smaller than is numerically feasible.

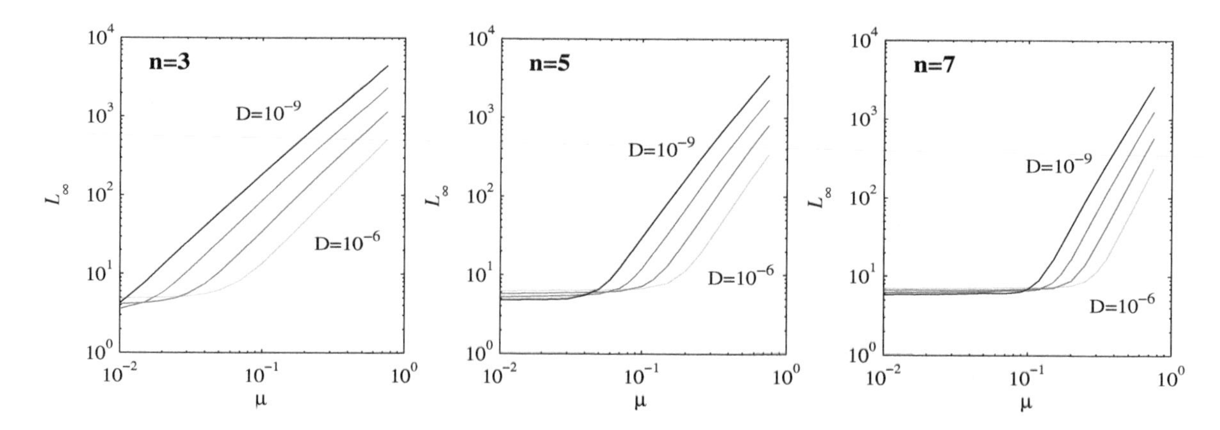

Figure 6. $\mathcal{L}_\infty$ as a function of the initial position μ, for $D = 10^{-6}$ to 10^{-9} and $n = 3, 5, 7$ as labeled.

5. Conclusions

The motivation of this research was to elucidate the link between force and geometry in a stochastic system. To this end, we investigated the transient relaxation of initial PDFs of a stochastic variable x under different nonlinear forces proportional to $-x^n$ ($n = 3, 5, 7$) and different strength D of δ-correlated stochastic noise. We identified the three main stages consisting of nondiffusive evolution, quasi-linear Gaussian evolution and settling into stationary PDFs. The strength of the stochastic noise is shown to play a crucial role in determining these timescales $t_1 \propto D^{-\frac{n-1}{3n-1}}$ and $t_2 \propto D^{-\frac{n-1}{n+1}}$, as well as the peak amplitude and width of PDFs. Note in particular how both timescales have (n-dependent) power-law scalings with D, in sharp contrast with the linear $n = 1$ case, where the entire evolution occurs on $t = O(1)$ timescales, with no dependence on D.

We further computed the rate of information change, and the information length $\mathcal{L}(t)$ representing the number of distinguishable statistical states that the system evolves through in time. We identified a robust geodesic (where the information changes at a constant rate) in the initial nondiffusive stage, and mapped out a geometric structure of an attractor as $\mathcal{L}_\infty \propto \mu^m$, where μ is the peak position of the initial Gaussian PDF. For sufficiently small D the exponent m was shown to be $\frac{n(n+1)}{3n-1}$, but still exhibits (moderate) variation with D even for values as small as 10^{-9}. This suggests that the geometry is curved by the nonlinear interaction, in contrast with the linear geometry of the Ornstein-Uhlenbeck process which has $m = 1$ and no variation with D. Our results thus highlight ubiquitous power-laws and multi-scalings of information geometry.

This work will form a basis for future investigation of a more general case of the superposition of nonlinear forces with different ns. Multiplicative noise would also be interesting, but the situation becomes somewhat more complicated as a deterministic force could appear as $-x^n$ as here, but the noise ξ could appear as ξx^m, with m not necessarily the same as n. Fractional noise would also be worth pursuing, although this would be challenging both analytically and numerically. Finally, it will also be interesting to extend our work to analyse a heterogeneous, linear system [35–37], and to explore applications of our work to estimators.

Author Contributions: Conceptualization, R.H. and E.-j.K.; Investigation, R.H., D.D. and E.-j.K.; Software, R.H.; Supervision, R.H. and E.-j.K.; Visualization, R.H. and D.D.; Writing—original draft, R.H. and E.-j.K.; and Writing—review and editing, R.H. and E.-j.K.

References

1. Gangbo, W.; McCann, R.J. Optimal maps in Monge's mass transport problem. *Comptes Rendus Acad. Sci. Paris* **1995**, *321*, 1653–1658.
2. Jordan, R.; Kinderlehrer, D.; Otto, F. The variational formulation of the Fokker-Planck equation. *SIAM J. Math. Anal.* **1998**, *29*, 1–17. [CrossRef]
3. Otto, F. The geometry of dissipative evolution equations: The porous medium equation. *Comm. Part. Diff. Equ.* **2001**, *26*, 101–174. [CrossRef]
4. Gibbs, A.L.; Su, F.E. On choosing and bounding probability metrics. *Int. Stat. Rev.* **2002**, *70*, 419–435. [CrossRef]
5. Frieden, B.R. *Science from Fisher Information*; Cambridge University Press: Cambridge, UK, 2004.
6. Lott, J. Some geometric calculations on Wasserstein space. *Commun. Math. Phys.* **2008**, *277*, 423–437. [CrossRef]
7. Takatsu, A. Wasserstein geometry of Gaussian measures. *Osaka J. Math.* **2011**, *48*, 1005–1026.
8. Ferradans, S.; Xia, G.-S.; Peyré, G.; Aujol, J.-F. Static and dynamic texture mixing using optimal transport. In *Lecture Notes in Computer Science*; Kuijper, A., Bredies, K., Pock, T., Bischof, H., Eds.; Springer: Berlin, Germany, 2013; Volume 7893.
9. Costa, S.; Santos, S.; Strapasson, J. Fisher information distance. *Discret. Appl. Math.* **2015**, *197*, 59–69. [CrossRef]
10. Chevallier, E.; Kalunga, E.; Angulo, J. Kernel density estimation on spaces of Gaussian distributions and symmetric positive definite matrices. *SIAM J. Imaging Sci.* **2017**, *10*, 191–215. [CrossRef]
11. Feng, E.H.; Crooks, G.E. Far-from-equilibrium measurements of thermodynamic length. *Phys. Rev. E* **2009**, *79*, 012104. [CrossRef] [PubMed]
12. Wilson, A.D.; Schultz, J.A.; Murphey, T.D. Trajectory synthesis for Fisher information maximization. *IEEE Trans. Robot.* **2014**, *30*, 1358–1370. [CrossRef] [PubMed]
13. Kim, E.; Lee, U.; Heseltine, J.; Hollerbach, R. Geometric structure and geodesic in a solvable model of nonequilibrium process. *Phys. Rev. E* **2016**, *93*, 062127. [CrossRef] [PubMed]
14. Plastino, A.R.; Casas, M.; Plastino, A. Fisher's information, Kullback's measure, and H-theorems. *Phys. Lett. A* **1998**, *246*, 498–504. [CrossRef]
15. Polettini, M.; Esposito, M. Nonconvexity of the relative entropy for Markov dynamics: A Fisher information approach. *Phys. Rev. E.* **2013**, *88*, 012112. [CrossRef] [PubMed]
16. Wootters, W.K. Statistical distance and Hilbert space. *Phys. Rev. D* **1981**, *23*, 357–362. [CrossRef]
17. Ruppeiner, G. Thermodynamics: A Riemannian geometric model. *Phys. Rev. A* **1979**, *20*, 1608–1613. [CrossRef]
18. Schlögl, F. Thermodynamic metric and stochastic measures. *Z. Phys. B Cond. Matt.* **1985**, *59*, 449–454. [CrossRef]
19. Braunstein, S.L.; Caves, C.M. Statistical distance and the geometry of quantum states. *Phys. Rev. Lett.* **1994**, *72*, 3439–3443. [CrossRef] [PubMed]
20. Nulton, J.; Salamon, P.; Andresen, B.; Anmin, Q. Quasistatic processes as step equilibrations. *J. Chem. Phys.* **1985**, *83*, 334–338. [CrossRef]
21. Sivak, D.A.; Crooks, G.E. Thermodynamic metrics and optimal paths. *Phys. Rev. Lett.* **2012**, *8*, 190602. [CrossRef] [PubMed]
22. Salamon, P.; Nulton, J.D.; Siragusa, G.; Limon, A.; Bedeaus, D.; Kjelstrup, S. A simple example of control to minimize entropy production. *J. Non-Equilib. Thermodyn.* **2002**, *27*, 45–55. [CrossRef]
23. Nicholson, S.B.; Kim, E. Investigation of the statistical distance to reach stationary distributions. *Phys. Lett. A* **2015**, *379*, 83–88. [CrossRef]

24. Nicholson, S.B.; Kim, E. Structures in sound: Analysis of classical music using the information length. *Entropy* **2016**, *18*, 258. [CrossRef]
25. Heseltine, J.; Kim, E. Novel mapping in non-equilibrium stochastic processes. *J. Phys. A* **2016**, *49*, 175002. [CrossRef]
26. Kim, E.; Hollerbach, R. Signature of nonlinear damping in geometric structure of a nonequilibrium process. *Phys. Rev. E* **2017**, *95*, 022137. [CrossRef] [PubMed]
27. Hollerbach, R.; Kim, E. Information geometry of non-equilibrium processes in a bistable system with a cubic damping. *Entropy* **2017**, *19*, 268. [CrossRef]
28. Kim, E.; Tenkès, L.-M.; Hollerbach, R.; Radulescu, O. Far-from-equilibrium time evolution between two gamma distributions. *Entropy* **2017**, *19*, 511. [CrossRef]
29. Tenkès, L.-M.; Hollerbach, R.; Kim, E. Time-dependent probability density functions and information geometry in stochastic logistic and Gompertz models. *J. Stat. Mech.* **2017**, *2017*, 123201. [CrossRef]
30. Kim, E.; Lewis, P. Information length in quantum system. *J. Stat. Mech.* **2018**, *2018*, 043106. [CrossRef]
31. Wilde, M.M. *Quantum Information Theory*; Cambridge University Press: Cambridge, UK, 2017; Volume 345, pp. 424–427.
32. Ott, E. *Chaos in Dynamical Systems*; Cambridge University Press: Cambridge, UK, 2002.
33. Kim, E.; Hollerbach, R. Time-dependent probability density function in cubic stochastic processes. *Phys. Rev. E* **2016**, *94*, 052118. [CrossRef] [PubMed]
34. Risken, H. *The Fokker-Planck Equation: Methods of Solution & Applications*; Springer: Berlin, Germany, 1996.
35. Cherstvy, A.G.; Chechkin, A.V.; Metzler, R., Anomalous diffusion and ergodicity breaking in heterogeneous diffusion processes. *New J. Phys.* **2010**, *15*, 083039. [CrossRef]
36. Andrey G.; Cherstvy, A.G.; Metzler, R. Population splitting, trapping, and non-ergodicity in heterogeneous diffusion processes. *Phys. Chem. Chem. Phys.* **2013**, *15*, 20220.
37. Sandev, T.; Schulz, A.; Kantz, H.; Iomin, A. Heterogeneous diffusion in comb and fractal grid structures. *Chaos Solitons Fractals* **2017**. [CrossRef]

3

Explicit Formula of Koszul–Vinberg Characteristic Functions for a Wide Class of Regular Convex Cones

Hideyuki Ishi

Graduate School of Mathematics, Nagoya University, Nagoya 464-8602, Japan; hideyuki@math.nagoya-u.ac.jp;

Academic Editors: Frédéric Barbaresco and Frank Nielsen

Abstract: The Koszul–Vinberg characteristic function plays a fundamental role in the theory of convex cones. We give an explicit description of the function and related integral formulas for a new class of convex cones, including homogeneous cones and cones associated with chordal (decomposable) graphs appearing in statistics. Furthermore, we discuss an application to maximum likelihood estimation for a certain exponential family over a cone of this class.

Keywords: convex cone; homogeneous cone; graphical model; Koszul–Vinberg characteristic function

---

1. Introduction

Let Ω be an open convex cone in a vector space $\mathcal{Z}$. The cone Ω is said to be regular if Ω contains no straight line, which is equivalent to the condition $\overline{\Omega} \cap (-\overline{\Omega}) = \{0\}$. In this paper, we always assume that a convex cone is open and regular. The dual cone Ω^* with respect to an inner product $(\cdot|\cdot)$ on $\mathcal{Z}$ is defined by:

$$\Omega^* := \{\, \xi \in \mathcal{Z}\,;\, (x|\xi) > 0 \ (\forall x \in \overline{\Omega} \setminus \{0\}) \,\}.$$

Then, Ω^* is again a regular open convex cone, and we have $(\Omega^*)^* = \Omega$. The Koszul–Vinberg characteristic function $\varphi_\Omega : \Omega \to \mathbb{R}_{>0}$ defined by:

$$\varphi_\Omega(x) := \int_{\Omega^*} e^{-(x|\xi)}\, d\xi \qquad (x \in \Omega)$$

plays a fundamental role in the theory of regular convex cones [1–4].

In particular, φ_Ω is an important function in the theory of convex programming [5], and it has also been studied recently in connection with thermodynamics [6,7]. There are several (not many) classes of cones for which an explicit formula of the Koszul–Vinberg characteristic function is known. Among them, the class of homogeneous cones [8–10] and the class of cones associated with chordal graphs [11] are particularly fruitful research objects. In this paper, we present a wide class of cones, including both of them, and give an explicit expression of the Koszul–Vinberg characteristic function (Section 3). Moreover, we get integral formulas involving the characteristic functions and the so-called generalized power functions, which are expressed as some product of powers of principal minors of real symmetric matrices (Section 4). After investigating the multiplicative Legendre transform of generalized power functions in Section 5, we study a maximum likelihood estimator for a Wishart-type natural exponential family constructed from the integral formula (Section 6).

A regular open convex cone $\Omega \subset \mathcal{Z}$ is said to be homogeneous if the linear automorphism group $GL(\Omega) := \{\, \alpha \in GL(\mathcal{Z})\,;\, \alpha\Omega = \Omega \,\}$ acts on Ω transitively. The cone $\mathcal{P}_n$ of positive definite $n \times n$ real symmetric matrices is a typical example of homogeneous cones. It is known [12–16] that every homogeneous cone is linearly isomorphic to a cone $\mathcal{P}_n \cap \mathcal{Z}$ with an appropriate subspace $\mathcal{Z}$ of the vector space $\mathrm{Sym}(n, \mathbb{R})$ of all $n \times n$ real symmetric matrices, where $\mathcal{Z}$ admits a specific block decomposition. Based on such results, our matrix realization method [15,17,18] has been developed for the purpose of the efficient study of homogeneous cones. In this paper, we present a generalization of matrix realization dealing with a wide class of convex cones, which turns out to include cones associated with chordal graphs. Actually, it was an enigma for the author that some formulas in [11,19] for the chordal graph resemble the formulas in [8,17] for homogeneous cones so much, and the mystery is now solved by the unified method in this paper to get the formulas. Furthermore, the techniques and ideas in the theory of homogeneous cones, such as Riesz distributions [8,20,21] and homogeneous Hessian metrics [4,18,22], will be applied to various cones to obtain new results in our future research.

Here, we fix some notation used in this paper. We denote by $\mathrm{Mat}(p,q,\mathbb{R})$ the vector space of $p\times q$ real matrices. For a matrix A, we write ${}^{t}A$ for the transpose of A. The identity matrix of size p is denoted by I_p.

2. New Cones $\mathcal{P}_{\mathcal{V}}$ and $\mathcal{P}^*_{\mathcal{V}}$

2.1. Setting

We fix a partition $n = n_1 + n_2 + \cdots + n_r$ of a positive integer n. Let $\mathcal{V} = \{\mathcal{V}_{lk}\}_{1\le k<l\le r}$ be a system of vector spaces $\mathcal{V}_{lk} \subset \mathrm{Mat}(n_l, n_k, \mathbb{R})$ satisfying

(V1) $A \in \mathcal{V}_{lk} \Rightarrow A\,{}^{t}A \in \mathbb{R}I_{n_l} \quad (1 \le k < l \le r)$,
(V2) $A \in \mathcal{V}_{lj},\ B \in \mathcal{V}_{kj} \Rightarrow A\,{}^{t}B \in \mathcal{V}_{lk} \quad (1 \le j < k < l \le r)$.

The integer r is called the rank of the system $\mathcal{V}$. We denote by n_{lk} the dimension of $\mathcal{V}_{lk}$. Note that some n_{lk} can be zero. Let $\mathcal{Z}_{\mathcal{V}}$ be the space of real symmetric matrices $x \in \mathrm{Sym}(n,\mathbb{R})$ of the form:

$$x = \begin{pmatrix} X_{11} & {}^{t}X_{21} & \cdots & {}^{t}X_{r1} \\ X_{21} & X_{22} & & {}^{t}X_{r2} \\ \vdots & & \ddots & \\ X_{r1} & X_{r2} & \cdots & X_{rr} \end{pmatrix} \quad \begin{pmatrix} X_{kk} = x_{kk} I_{n_k},\ x_{kk} \in \mathbb{R},\ k = 1,\dots,r \\ X_{lk} \in \mathcal{V}_{lk},\ 1 \le k < l \le r \end{pmatrix}, \tag{1}$$

and $\mathcal{P}_{\mathcal{V}}$ the subset of $\mathcal{Z}_{\mathcal{V}}$ consisting of positive definite matrices. Then, $\mathcal{P}_{\mathcal{V}}$ is a regular open convex cone in $\mathcal{Z}_{\mathcal{V}}$.

Example 1. *Let $r = 3$, and set $\mathcal{V}_{21} := \left\{ \begin{pmatrix} a & 0 \end{pmatrix} ; a \in \mathbb{R} \right\}$, $\mathcal{V}_{31} := \left\{ \begin{pmatrix} 0 & a \end{pmatrix} ; a \in \mathbb{R} \right\}$, and $\mathcal{V}_{32} := \mathbb{R}$. Then, $\mathcal{Z}_{\mathcal{V}}$ is the space of symmetric matrices x of the form:*

$$x = \begin{pmatrix} x_1 & 0 & x_4 & 0 \\ 0 & x_1 & 0 & x_5 \\ x_4 & 0 & x_2 & x_6 \\ 0 & x_5 & x_6 & x_3 \end{pmatrix}. \tag{2}$$

We shall see later that the cone $\mathcal{P}_{\mathcal{V}} = \mathcal{Z}_{\mathcal{V}} \cap \mathcal{P}_4$ is not homogeneous in this case, but admits various integral formulas, as well as explicit expression of the Koszul–Vinberg characteristic function.

2.2. Inductive Description of $\mathcal{P}_{\mathcal{V}}$

If the system $\mathcal{V} = \{\mathcal{V}_{lk}\}_{1\le k<l\le r}$ satisfies (V1) and (V2), any subsystem $\mathcal{V}_{\mathcal{I}} := \{\mathcal{V}_{lk}\}_{k,l\in\mathcal{I}}$ with $\mathcal{I} \subset \{1,\dots,r\}$ also satisfies the same conditions. In particular, the cone corresponding to the subsystem

$\{\mathcal{V}_{lk}\}_{2\le k<l\le r}$ will play an important role in this paper. Let us define $\mathcal{V}' := \{\mathcal{V}'_{lk}\}_{1\le k<l\le r-1}$ by $\mathcal{V}'_{lk} := \mathcal{V}_{l+1,k+1}$. Then, $\mathcal{V}'$ is a system of rank $r-1$. Any $x \in \mathcal{Z}_{\mathcal{V}}$ is written as:

$$x = \begin{pmatrix} x_{11}I_{n_1} & {}^{\mathrm{t}}U \\ U & x' \end{pmatrix} \qquad (x_{11} \in \mathbb{R},\, U \in \mathcal{W},\, x' \in \mathcal{Z}_{\mathcal{V}'}), \tag{3}$$

where:

$$\mathcal{W} := \left\{ U = \begin{pmatrix} X_{21} \\ \vdots \\ X_{r1} \end{pmatrix} ;\, X_{l1} \in \mathcal{V}_{l1}\ (1 < l \le r) \right\}. \tag{4}$$

If $x_{11} \neq 0$, then we have:

$$\begin{pmatrix} x_{11}I_{n_1} & {}^{\mathrm{t}}U \\ U & x' \end{pmatrix} = \begin{pmatrix} I_{n_1} & \\ x_{11}^{-1}U & I_{n-n_1} \end{pmatrix} \begin{pmatrix} x_{11}I_{n_1} & \\ & x' - x_{11}^{-1}U{}^{\mathrm{t}}U \end{pmatrix} \begin{pmatrix} I_{n_1} & x_{11}^{-1}{}^{\mathrm{t}}U \\ & I_{n-n_1} \end{pmatrix}. \tag{5}$$

Note that $U{}^{\mathrm{t}}U$ belongs to $\mathcal{Z}_{\mathcal{V}'}$ thanks to (V1) and (V2). Thus, we deduce the following lemma immediately from (5).

Lemma 1. (i) *Let $x \in \mathcal{Z}_{\mathcal{V}}$ as in (3). Then, $x \in \mathcal{P}_{\mathcal{V}}$ if and only if $x_{11} > 0$ and $x' - x_{11}^{-1}U{}^{\mathrm{t}}U \in \mathcal{P}_{\mathcal{V}'}$.*
(ii) *For $x \in \mathcal{P}_{\mathcal{V}}$, there exist unique $\tilde{U} \in \mathcal{W}$ and $\tilde{x} \in \mathcal{P}_{\mathcal{V}'}$ for which:*

$$\begin{aligned} x &= \begin{pmatrix} I_{n_1} & \\ \tilde{U} & I_{n-n_1} \end{pmatrix} \begin{pmatrix} x_{11}I_{n_1} & \\ & \tilde{x}' \end{pmatrix} \begin{pmatrix} I_{n_1} & {}^{\mathrm{t}}\tilde{U} \\ & I_{n-n_1} \end{pmatrix} \\ &= \begin{pmatrix} x_{11}I_{n_1} & x_{11}{}^{\mathrm{t}}\tilde{U} \\ x_{11}\tilde{U} & \tilde{x}' + x_{11}\tilde{U}{}^{\mathrm{t}}\tilde{U} \end{pmatrix}. \end{aligned} \tag{6}$$

(iii) *The closure $\overline{\mathcal{P}_{\mathcal{V}}}$ of the cone $\mathcal{P}_{\mathcal{V}}$ is described as:*

$$\overline{\mathcal{P}_{\mathcal{V}}} := \left\{ \begin{pmatrix} x_{11}I_{n_1} & x_{11}{}^{\mathrm{t}}\tilde{U} \\ x_{11}\tilde{U} & \tilde{x}' + x_{11}\tilde{U}{}^{\mathrm{t}}\tilde{U} \end{pmatrix} ;\, x_{11} \ge 0,\ \tilde{U} \in \mathcal{W},\ \tilde{x}' \in \overline{\mathcal{P}_{\mathcal{V}'}} \right\}.$$

2.3. The Dual Cone $\mathcal{P}_{\mathcal{V}}^*$

We define an inner product on the space $\mathcal{V}_{lk}$ by $(A|B)_{\mathcal{V}_{lk}} := n_l^{-1}\mathrm{tr}\, A{}^{\mathrm{t}}B$ for $A, B \in \mathcal{V}_{lk}$. Then, we see from (V1) that:

$$A{}^{\mathrm{t}}B + B{}^{\mathrm{t}}A = 2(A|B)_{\mathcal{V}_{lk}} I_{n_l}.$$

Gathering these inner products $(\cdot|\cdot)_{\mathcal{V}_{lk}}$, we introduce the standard inner product on the space $\mathcal{Z}_{\mathcal{V}}$ defined by:

$$(x|x') := \sum_{k=1}^{r} x_{kk}x'_{kk} + 2 \sum_{1\le k<l\le r} (X_{lk}|X'_{lk})_{\mathcal{V}_{lk}} \tag{7}$$

for $x, x' \in \mathcal{Z}_{\mathcal{V}}$ of the form (1). When $n_1 = n_2 = \cdots = n_r = 1$ (and only in this case), the standard inner product above equals the trace inner product $\mathrm{tr}\,(xx')$.

Let $\widetilde{\mathcal{W}}_k$ $(k = 1, \dots, r)$ be the vector space of $W \in \mathrm{Mat}(n, n_k, \mathbb{R})$ of the form:

$$W = \begin{pmatrix} 0_{n_1+\cdots+n_{k-1}, n_k} \\ X_{kk} \\ X_{k+1,k} \\ \vdots \\ X_{rk} \end{pmatrix} \qquad (X_{kk} = x_{kk} I_{n_k},\ x_{kk} \in \mathbb{R},\ X_{lk} \in \mathcal{V}_{lk},\ l > k).$$

Clearly, the space $\widetilde{\mathcal{W}}_k$ is isomorphic to $\mathbb{R} \oplus \sum_{l>k} \mathcal{V}_{lk}$, which implies $\dim \widetilde{\mathcal{W}}_k = 1 + q_k$ with $q_k := \sum_{l>k} n_{lk}$. Gathering orthogonal bases of $\mathcal{V}_{lk}$'s, we take a basis of $\widetilde{\mathcal{W}}_k$, so that we have an isomorphism $\widetilde{\mathcal{W}}_k \ni W \mapsto w = \mathrm{vect}(W) \in \mathbb{R}^{1+q_k}$, where the first component w_1 of w is assumed to be x_{kk}. Let us introduce a linear map $\phi_k : \mathcal{Z}_{\mathcal{V}} \to \mathrm{Sym}(1 + q_k, \mathbb{R})$ defined in such a way that:

$$(W^{\mathrm{t}} W | \xi) = {}^{\mathrm{t}} w \phi_k(\xi) w \qquad (\xi \in \mathcal{Z}_{\mathcal{V}},\ W \in \widetilde{\mathcal{W}}_k,\ w = \mathrm{vect}(W) \in \mathbb{R}^{1+q_k}). \tag{8}$$

It is easy to see that $\phi_r(\xi) = \xi_{rr}$ for $\xi \in \mathcal{Z}_{\mathcal{V}}$.

Theorem 1. *The dual cone $\mathcal{P}_{\mathcal{V}}^* \subset \mathcal{Z}_{\mathcal{V}}$ of $\mathcal{P}_{\mathcal{V}}$ with respect to the standard inner product is described as:*

$$\begin{aligned} \mathcal{P}_{\mathcal{V}}^* &= \{\, \xi \in \mathcal{Z}_{\mathcal{V}} \,;\ \phi_k(\xi) \textit{ is positive definite for all } k = 1, \dots, r \,\} \\ &= \{\, \xi \in \mathcal{Z}_{\mathcal{V}} \,;\ \det \phi_k(\xi) > 0 \textit{ for all } k = 1, \dots, r \,\}. \end{aligned} \tag{9}$$

Proof. We shall prove the statement by induction on the rank r. When $r = 1$, we have $\phi_1(\xi) = \xi_{11}$ and $\xi = \xi_{11} I_{n_1}$. Thus, (9) holds in this case.

Let us assume that (9) holds when the rank is smaller than r. In particular, the statement holds for $\mathcal{P}_{\mathcal{V}'}^* \subset \mathcal{Z}_{\mathcal{V}'}$, that is,

$$\begin{aligned} \mathcal{P}_{\mathcal{V}'}^* &= \{\, \xi' \in \mathcal{Z}_{\mathcal{V}'} \,;\ \phi'_k(\xi') \text{ is positive definite for all } k = 1, \dots, r-1 \,\} \\ &= \{\, \xi' \in \mathcal{Z}_{\mathcal{V}'} \,;\ \det \phi'_k(\xi') > 0 \text{ for all } k = 1, \dots, r-1 \,\}, \end{aligned}$$

where ϕ'_k is defined similarly to (8) for $\mathcal{Z}_{\mathcal{V}'}$. On the other hand, if:

$$\xi = \begin{pmatrix} \xi_{11} I_{n_1} & {}^{\mathrm{t}}V \\ V & \xi' \end{pmatrix} \qquad (\xi_{11} \in \mathbb{R},\ V \in \mathcal{W},\ \xi' \in \mathcal{Z}_{\mathcal{V}'}), \tag{10}$$

we observe that:

$$\phi_k(\xi) = \phi'_{k-1}(\xi') \qquad (k = 2, \dots, r).$$

Therefore, in order to prove (9) for $\mathcal{P}_{\mathcal{V}}^*$ of rank r, it suffices to show that:

$$\begin{aligned} \mathcal{P}_{\mathcal{V}}^* &= \{\, \xi \in \mathcal{Z}_{\mathcal{V}} \,;\ \xi' \in \mathcal{P}_{\mathcal{V}'}^* \text{ and } \phi_1(\xi) \text{ is positive definite} \,\} \\ &= \{\, \xi \in \mathcal{Z}_{\mathcal{V}} \,;\ \xi' \in \mathcal{P}_{\mathcal{V}'}^* \text{ and } \det \phi_1(\xi) > 0 \,\}. \end{aligned} \tag{11}$$

If $q_1 = 0$, then any element $\xi \in \mathcal{Z}_{\mathcal{V}}$ is of the form:

$$\xi = \begin{pmatrix} \xi_{11} I_{n_1} & \\ & \xi' \end{pmatrix},$$

which belongs to $\mathcal{P}_{\mathcal{V}}$ if and only if $\xi' \in \mathcal{P}_{\mathcal{V}'}$ and $\phi_1(\xi) = \xi_{11} > 0$, so that (11) holds.

Assume $q_1 > 0$. Keeping in mind that $\widetilde{\mathcal{W}}_1 \simeq \mathbb{R} \oplus \mathcal{W}$ and $\mathcal{W} \simeq \mathbb{R}^{q_1}$ by (4), we have for $\xi \in \mathcal{Z}_{\mathcal{V}}$ as in (10),

$$\phi_1(\xi) = \begin{pmatrix} \xi_{11} & {}^{\mathrm{t}}v \\ v & \psi(\xi') \end{pmatrix} \in \mathrm{Sym}(1+q_1, \mathbb{R}), \tag{12}$$

where $v = \mathrm{vect}(V) \in \mathbb{R}^{q_i}$ and $\psi : \mathcal{Z}_{\mathcal{V}'} \to \mathrm{Sym}(q_i, \mathbb{R})$ is defined in such a way that:

$$(U^{\mathrm{t}}U|\xi') = {}^{\mathrm{t}}u\psi(\xi')u \quad (\xi' \in \mathcal{Z}_{\mathcal{V}'}, U \in \mathcal{W}, u = \mathrm{vect}(U) \in \mathbb{R}^{q_1}). \tag{13}$$

On the other hand, for $x \in \mathcal{Z}_{\mathcal{V}}$ as in (6), we have:

$$(x|\xi) = x_{11}\xi_{11} + 2x_{11}{}^{\mathrm{t}}\tilde{u}v + x_{11}{}^{\mathrm{t}}\tilde{u}\psi(\xi')\tilde{u} + (\tilde{x}'|\xi). \tag{14}$$

Owing to Lemma 1 (iii), the element $\xi \in \mathcal{Z}_{\mathcal{V}}$ belongs to $\mathcal{P}_{\mathcal{V}}^*$ if and only if the right-hand side is strictly positive for all $x_{11} \geq 0$, $\tilde{U} \in \mathcal{W}$ and $\tilde{x}' \in \overline{\mathcal{P}_{\mathcal{V}'}}$ with $(x_{11}, \tilde{x}') \neq (0,0)$. Assume $\xi \in \mathcal{P}_{\mathcal{V}}^*$. Considering the case $x_{11} = 0$, we have $(\tilde{x}'|\xi') > 0$ for all $\tilde{x}' \in \overline{\mathcal{P}_{\mathcal{V}'}} \backslash \{0\}$, which means that $\xi' \in \mathcal{P}_{\mathcal{V}'}^*$. Then, the quantity in (13) is strictly positive for non-zero U because $U^{\mathrm{t}}U$ belongs to $\overline{\mathcal{P}_{\mathcal{V}}} \setminus \{0\}$. Thus, $\psi(\xi')$ is positive definite, and (14) is rewritten as:

$$(x|\xi) = x_{11}(\xi_{11} - {}^{\mathrm{t}}v\psi(\xi')^{-1}v) + x_{11}{}^{\mathrm{t}}(\tilde{u} + \psi(\xi')^{-1}v)\psi(\xi')(\tilde{u} + \psi(\xi')^{-1}v) + (\tilde{x}'|\xi'). \tag{15}$$

Therefore, we obtain:

$$\mathcal{P}_{\mathcal{V}}^* = \left\{ \xi \in \mathcal{Z}_{\mathcal{V}} ;\, \xi' \in \mathcal{P}_{\mathcal{V}'}^* \text{ and } \xi_{11} - {}^{\mathrm{t}}v\psi(\xi')^{-1}v > 0 \right\}. \tag{16}$$

On the other hand, we see from (12) that:

$$\phi_1(\xi) = \begin{pmatrix} 1 & {}^{\mathrm{t}}v\psi(\xi')^{-1} \\ & I_{q_1} \end{pmatrix} \begin{pmatrix} \xi_{11} - {}^{\mathrm{t}}v\psi(\xi')^{-1}v & \\ & \psi(\xi') \end{pmatrix} \begin{pmatrix} 1 & \\ \psi(\xi')^{-1}v & I_{q_1} \end{pmatrix}. \tag{17}$$

Hence, we deduce (11) from (16) and (17). □

We note that, if $q_1 > 0$, the $(1,1)$-component of the inverse matrix $\phi_1(\xi)^{-1}$ is given by:

$$(\phi_1(\xi)^{-1})_{11} = (\xi_{11} - {}^{\mathrm{t}}v\psi(\xi')^{-1}v)^{-1} \tag{18}$$

thanks to (17).

3. Koszul–Vinberg Characteristic Function of $\mathcal{P}_{\mathcal{V}}^*$

We denote by $\varphi_{\mathcal{V}}$ the Koszul–Vinberg characteristic function of $\mathcal{P}_{\mathcal{V}}^*$. In this section, we give an explicit formula of $\varphi_{\mathcal{V}}$.

Recall that the linear map $\psi : \mathcal{Z}_{\mathcal{V}'} \to \mathrm{Sym}(q_1, \mathbb{R})$ plays an important role in the proof of Theorem 1. We shall introduce similar linear maps $\psi_k : \mathcal{Z}_{\mathcal{V}} \to \mathrm{Sym}(q_k, \mathbb{R})$ for k such that $q_k > 0$. Let $\mathcal{W}_k$ be the subspace of $\widetilde{\mathcal{W}}_k$ consisting of $W \in \widetilde{\mathcal{W}}_k$ for which $w_1 = x_{kk} = 0$. Then, clearly, $\mathcal{W}_k \simeq \sum_{l>k}^{\oplus} \mathcal{V}_{lk}$ and $\dim \mathcal{W}_k = q_k$. If $q_k > 0$, using the same orthogonal basis of $\mathcal{V}_{lk}$ as in the previous section, we have the isomorphism $\mathcal{W}_k \ni W \mapsto w = \mathrm{vect}(W) \in \mathbb{R}^{q_k}$. Similarly to (8), we define ψ_k by:

$$(W^{\mathrm{t}}W|\xi) = {}^{\mathrm{t}}w\psi_k(\xi)w \qquad (\xi \in \mathcal{Z}_{\mathcal{V}},\ W \in \mathcal{W}_k,\ w = \mathrm{vect}(W) \in \mathbb{R}^{q_k}). \tag{19}$$

Then, we have:

$$\phi_k(\xi) = \begin{pmatrix} \xi_{kk} & {}^{\mathrm{t}}v_k \\ v_k & \psi_k(\xi) \end{pmatrix} \qquad (\xi \in \mathcal{Z}_{\mathcal{V}}), \tag{20}$$

where $v_k \in \mathbb{R}^{q_k}$ is a vector corresponding to the $\mathcal{W}_k$-component of ξ. If $\xi \in \mathcal{P}_{\mathcal{V}}^*$, we see from (19) that $\psi_k(\xi)$ is positive definite. In this case, we have:

$$\phi_k(\xi) = \begin{pmatrix} 1 & {}^t v_k \psi_k(\xi)^{-1} \\ & I_{q_k} \end{pmatrix} \begin{pmatrix} \xi_{kk} - {}^t v_k \psi_k(\xi)^{-1} v_k & \\ & \psi_k(\xi) \end{pmatrix} \begin{pmatrix} 1 & \\ \psi_k(\xi)^{-1} v_k & I_{q_k} \end{pmatrix}, \tag{21}$$

so that we get a generalization of (18), that is,

$$(\phi_k(\xi)^{-1})_{11} = (\xi_{kk} - {}^t v_k \psi_k(\xi)^{-1} v_k)^{-1}. \tag{22}$$

On the other hand, if $q_k = 0$, then $\phi_k(\xi)^{-1} = \xi_{kk}^{-1}$.

We remark that $\psi_1(\xi) = \psi(\xi')$, and that some part of the argument above is parallel to the proof of Theorem 1.

Theorem 2. *The Koszul–Vinberg characteristic function $\varphi_{\mathcal{V}}$ of $\mathcal{P}_{\mathcal{V}}^*$ is given by the following formula:*

$$\varphi_{\mathcal{V}}(\xi) = C_{\mathcal{V}} \prod_{k=1}^{r} (\phi_k(\xi)^{-1})_{11}^{1+q_k/2} \prod_{q_k>0} (\det \psi_k(\xi))^{-1/2} \qquad (\xi \in \mathcal{P}_{\mathcal{V}}^*), \tag{23}$$

where $C_{\mathcal{V}} := (2\pi)^{(N-r)/2} \prod_{k=1}^{r} \Gamma(1 + \frac{q_k}{2})$ and $N := \dim \mathcal{Z}_{\mathcal{V}}$.

Proof. We shall show the statement by induction on the rank as in the proof of Theorem 1. Then, it suffices to show that:

$$\varphi_{\mathcal{V}}(\xi) = (2\pi)^{q_1/2} \Gamma(1 + \frac{q_1}{2}) (\phi_1(\xi)^{-1})_{11}^{1+q_1/2} (\det \psi_1(\xi))^{-\mathrm{sgn}(q_1)/2} \varphi_{\mathcal{V}'}(\xi') \tag{24}$$

for $\xi \in \mathcal{P}_{\mathcal{V}}^*$ as in (10), where $(\det \psi_1(\xi))^{-\mathrm{sgn}(q_1)/2}$ is interpreted as:

$$(\det \psi_1(\xi))^{-\mathrm{sgn}(q_1)/2} := \begin{cases} 1 & (q_1 = 0), \\ (\det \psi_1(\xi))^{-1/2} & (q_1 > 0). \end{cases}$$

When $q_1 = 0$, we have:

$$\begin{aligned} \varphi_{\mathcal{V}}(\xi) &= \int_0^\infty \int_{\mathcal{P}_{\mathcal{V}'}} e^{-x_{11}\xi_{11}} e^{-(x'|\xi')} dx_{11}\, dx' \\ &= \xi_{11}^{-1} \varphi_{\mathcal{V}'}(\xi'), \end{aligned}$$

which means (24).

When $q_1 > 0$, the Euclidean measure dx equals $2^{q_1/2} x_{11}^{q_1}\, dx_{11} d\tilde{u} d\tilde{x}'$ by the change of variables in (6). Indeed, the coefficient $2^{q_1/2}$ comes from the normalization of the inner product on $\mathcal{W} \simeq \mathbb{R}^{q_1}$ regarded as a subspace of $\mathcal{Z}_{\mathcal{V}}$. Then, we have by (15):

$$\begin{aligned} \varphi_{\mathcal{V}}(\xi) = \int_0^\infty \int_{\mathbb{R}^{q_1}} \int_{\mathcal{P}_{\mathcal{V}'}} & e^{-x_{11}(\xi_{11} - {}^t v \psi(\xi')^{-1} v)} e^{-x_{11} {}^t(\tilde{u} + \psi(\xi')^{-1} v) \psi(\xi') (\tilde{u} + \psi(\xi')^{-1} v)} e^{-(\tilde{x}'|\xi')} \\ & \times 2^{q_1/2} x_{11}^{q_1}\, dx_{11} d\tilde{u} d\tilde{x}'. \end{aligned}$$

By the Gaussian integral formula, we have:

$$\int_{\mathbb{R}^{q_1}} e^{-x_{11} {}^t(\tilde{u} + \psi(\xi')^{-1} v) \psi(\xi') (\tilde{u} + \psi(\xi')^{-1} v)} d\tilde{u} = \pi^{q_1/2} x_{11}^{-q_1/2} (\det \psi(\xi'))^{-1/2}.$$

Therefore, we get:

$$\varphi_{\mathcal{V}}(\xi) = (2\pi)^{q_1/2}(\det\psi(\xi'))^{-1/2}\int_0^\infty e^{-x_{11}(\xi_{11}-{}^{t}v\psi(\xi')^{-1}v)}x_{11}^{q_1/2}\,dx_{11}\int_{\mathcal{P}_{\mathcal{V}'}} e^{-(\tilde{x}'|\xi')}\,d\tilde{x}'$$
$$= (2\pi)^{q_1/2}(\det\psi_1(\xi))^{-1/2}\Gamma(1+\frac{q_1}{2})(\xi_{11}-{}^{t}v\psi(\xi')^{-1}v)^{-1-q_k/2}\varphi_{\mathcal{V}'}(\xi'),$$

which together with (18) leads us to (24). □

Example 2. *Let* $\mathcal{V} = \{\mathcal{V}_{lk}\}_{1\le k<l\le 3}$ *be as in Example 1. For:*

$$\xi = \begin{pmatrix} \xi_1 & 0 & \xi_4 & 0 \\ 0 & \xi_1 & 0 & \xi_5 \\ \xi_4 & 0 & \xi_2 & \xi_6 \\ 0 & \xi_5 & \xi_6 & \xi_3 \end{pmatrix} \in \mathcal{Z}_{\mathcal{V}}, \tag{25}$$

we have:

$$\phi_1(\xi) = \begin{pmatrix} \xi_1 & \xi_4 & \xi_5 \\ \xi_4 & \xi_2 & 0 \\ \xi_5 & 0 & \xi_3 \end{pmatrix}, \quad \phi_2(\xi) = \begin{pmatrix} \xi_2 & \xi_6 \\ \xi_6 & \xi_3 \end{pmatrix}, \quad \phi_3(\xi) = \xi_3,$$
$$\psi_1(\xi) = \begin{pmatrix} \xi_2 & 0 \\ 0 & \xi_3 \end{pmatrix}, \quad \psi_2(\xi) = \xi_3.$$

The cone $\mathcal{P}_{\mathcal{V}}^*$ *is described as:*

$$\mathcal{P}_{\mathcal{V}}^* = \left\{ \xi \in \mathcal{Z}_{\mathcal{V}};\ \begin{vmatrix} \xi_1 & \xi_4 & \xi_5 \\ \xi_4 & \xi_2 & 0 \\ \xi_5 & 0 & \xi_3 \end{vmatrix} > 0,\ \begin{vmatrix} \xi_2 & \xi_6 \\ \xi_6 & \xi_3 \end{vmatrix} > 0,\ \xi_3 > 0 \right\},$$

and its Koszul–Vinberg characteristic function $\varphi_{\mathcal{V}}$ *is expressed as:*

$$\varphi_{\mathcal{V}}(\xi) = C_{\mathcal{V}} \left\{ \begin{vmatrix} \xi_1 & \xi_4 & \xi_5 \\ \xi_4 & \xi_2 & 0 \\ \xi_5 & 0 & \xi_3 \end{vmatrix} / (\xi_2\xi_3) \right\}^{-2} \left\{ \begin{vmatrix} \xi_2 & \xi_6 \\ \xi_6 & \xi_3 \end{vmatrix} / \xi_3 \right\}^{-3/2} \xi_3^{-1} \cdot (\xi_2\xi_3)^{-1/2}(\xi_3)^{-1/2}$$
$$= C_{\mathcal{V}} \begin{vmatrix} \xi_1 & \xi_4 & \xi_5 \\ \xi_4 & \xi_2 & 0 \\ \xi_5 & 0 & \xi_3 \end{vmatrix}^{-2} \begin{vmatrix} \xi_2 & \xi_6 \\ \xi_6 & \xi_3 \end{vmatrix}^{-3/2} \xi_2^{3/2}\xi_3^{3/2},$$

where $C_{\mathcal{V}} = (2\pi)^{3/2}\Gamma(2)\Gamma(3/2)\Gamma(1) = \sqrt{2}\pi^2$.

Suppose that the cone $\mathcal{P}_{\mathcal{V}}$ *is homogeneous. Then,* $\mathcal{P}_{\mathcal{V}}^*$*, as well as* $\mathcal{P}_{\mathcal{V}}$*, is a homogeneous cone of rank 3, so that the Koszul–Vinberg characteristic function of* $\mathcal{P}_{\mathcal{V}}^*$ *has at most three irreducible factors (see [8]). However, we have seen that there are four irreducible factors in the function* $\varphi_{\mathcal{V}}$*. Therefore, we conclude that neither* $\mathcal{P}_{\mathcal{V}}$*, nor* $\mathcal{P}_{\mathcal{V}}^*$ *is homogeneous.*

4. Γ-Type Integral Formulas

For an $n \times n$ matrix $A = (A_{ij})$ and $1 \le m \le n$, we denote by $A^{[m]}$ the upper-left $m \times m$ submatrix $(A_{ij})_{i,j \le m}$ of A. Put $M_k := \sum_{i=1}^{k} n_k \ (k = 1, \ldots, r)$. For $\underline{s} = (s_1, \ldots, s_r) \in \mathbb{C}^r$, we define functions $\Delta_{\underline{s}}^{\mathcal{V}}$ on $\mathcal{P}_{\mathcal{V}}$ and $\delta_{\underline{s}}^{\mathcal{V}}$ on $\mathcal{P}_{\mathcal{V}}^*$ respectively by:

$$\begin{aligned}\Delta_{\underline{s}}^{\mathcal{V}}(x) &:= (\det x^{[M_1]})^{s_1/n_1} \prod_{k=2}^{r} \Big(\frac{\det x^{[M_k]}}{\det x^{[M_{k-1}]}} \Big)^{s_k/n_k} \\ &= (\det x)^{s_r/n_r} \prod_{k=1}^{r-1} (\det x^{[M_k]})^{s_k/n_k - s_{k-1}/n_{k-1}} \qquad (x \in \mathcal{P}_{\mathcal{V}}),\end{aligned} \tag{26}$$

$$\begin{aligned}\delta_{\underline{s}}^{\mathcal{V}}(\xi) &:= \prod_{k=1}^{r} (\phi_k(\xi)^{-1})_{11}^{-s_k} \\ &= \prod_{q_k=0} \xi_{kk}^{s_k} \prod_{q_k>0} (\xi_{kk} - {}^t v_k \psi_k(\xi)^{-1} v_k)^{s_k} \qquad (\xi \in \mathcal{P}_{\mathcal{V}}^*).\end{aligned} \tag{27}$$

Recall (22) for the second equality of (27).

For $\underline{a} = (a_1, \ldots, a_r) \in \mathbb{R}_{>0}^r$, let $D_{\underline{a}}$ denote the diagonal matrix defined by:

$$D_{\underline{a}} := \begin{pmatrix} a_1 I_{n_1} & & & \\ & a_2 I_{n_2} & & \\ & & \ddots & \\ & & & a_r I_{n_r} \end{pmatrix} \in GL(n, \mathbb{R}).$$

Then, the linear map $\mathcal{Z}_{\mathcal{V}} \ni x \mapsto D_a x D_a \in \mathcal{Z}_{\mathcal{V}}$ preserves both $\mathcal{P}_{\mathcal{V}}$ and $\mathcal{P}_{\mathcal{V}}^*$, and we have:

$$\Delta_{\underline{s}}^{\mathcal{V}}(D_a x D_a) = (\prod_{k=1}^{r} a_k^{2s_k}) \Delta_{\underline{s}}^{\mathcal{V}}(x) \qquad (x \in \mathcal{P}_{\mathcal{V}}), \tag{28}$$

$$\delta_{\underline{s}}^{\mathcal{V}}(D_a \xi D_a) = (\prod_{k=1}^{r} a_k^{2s_k}) \delta_{\underline{s}}^{\mathcal{V}}(\xi) \qquad (\xi \in \mathcal{P}_{\mathcal{V}}). \tag{29}$$

Assume $q_1 > 0$. For $B \in \mathcal{W}$, we denote by τ_B the linear transform on $\mathcal{Z}_{\mathcal{V}}$ given by:

$$\begin{aligned}\tau_B x &:= \begin{pmatrix} I_{n_1} & \\ B & I_{n-n_1} \end{pmatrix} \begin{pmatrix} x_{11} I_{n_1} & {}^t U \\ U & x' \end{pmatrix} \begin{pmatrix} I_{n_1} & {}^t B \\ & I_{n-n_1} \end{pmatrix} \\ &= \begin{pmatrix} x_{11} I_{n_1} & {}^t U + x_{11} {}^t B \\ U + x_{11} B & x' + U {}^t B + B {}^t U + x_{11} B {}^t B \end{pmatrix},\end{aligned}$$

where $x \in \mathcal{Z}_{\mathcal{V}}$ is as in (3). Indeed, since:

$$U {}^t B + B {}^t U = (U + B) {}^t (U + B) - U {}^t U - B {}^t B \in \mathcal{Z}_{\mathcal{V}'},$$

the matrix $\tau_B x$ belongs to $\mathcal{Z}_{\mathcal{V}}$. Clearly, τ_B preserves $\mathcal{P}_{\mathcal{V}}$, and we have:

$$\Delta_{\underline{s}}^{\mathcal{V}}(\tau_B x) = \Delta_{\underline{s}}^{\mathcal{V}}(x) \qquad (x \in \mathcal{P}_{\mathcal{V}}). \tag{30}$$

The formula (5) is rewritten as:

$$\tau_{-x_{11}^{-1} U}(x) = \begin{pmatrix} x_{11} I_{n_1} & \\ & x' - x_{11}^{-1} U {}^t U \end{pmatrix},$$

which together with (30) tells us that:

$$\Delta_{\underline{s}}^{\mathcal{V}}(x) = x_{11}^{s_1}\Delta_{\underline{s}'}^{\mathcal{V}'}(x' - x_{11}^{-1}U^{\mathrm{t}}U), \tag{31}$$

where $\underline{s}' := (s_2, \ldots, s_r) \in \mathbb{C}^{r-1}$.

Let us consider the adjoint map $\tau_B^* : \mathcal{Z}_{\mathcal{V}} \to \mathcal{Z}_{\mathcal{V}}$ of τ_B with respect to the standard inner product. Let $b \in \mathbb{R}^{q_1}$ be the vector corresponding to $B \in \mathcal{W}$. For $x \in \mathcal{Z}_{\mathcal{V}}$ and $\xi \in \mathcal{Z}_{\mathcal{V}}$ as in (3) and (10), respectively, we observe that:

$$\begin{aligned}(\tau_B x|\xi) &= x_{11}\xi_{11} + 2^{\mathrm{t}}(u + x_{11}b)v + (x' + U^{\mathrm{t}}B + B^{\mathrm{t}}U + x_{11}B^{\mathrm{t}}B|\xi') \\ &= x_{11}(\xi_{11} + 2^{\mathrm{t}}bv + {}^{\mathrm{t}}b\psi(\xi')b) + 2^{\mathrm{t}}u(v + \psi(\xi')b) + (x'|\xi').\end{aligned}$$

Thus, if we write:

$$\iota(\xi_{11}, v, \xi') := \begin{pmatrix} \xi_{11}I_{n_1} & {}^{\mathrm{t}}V \\ V & \xi' \end{pmatrix},$$

we have:

$$\tau_B^*\iota(\xi_{11}, v, \xi') = \iota(\xi_{11} + 2^{\mathrm{t}}bv + {}^{\mathrm{t}}b\psi(\xi')b, v + \psi(\xi')b, \xi'). \tag{32}$$

Furthermore, we see from (12) that $\phi_1(\tau_B^*\iota(\xi_{11}, v, \xi'))$ equals:

$$\begin{pmatrix} \xi_{11} + 2^{\mathrm{t}}bv + {}^{\mathrm{t}}b\psi(\xi')b & {}^{\mathrm{t}}v + {}^{\mathrm{t}}b\psi(\xi') \\ v + \psi(\xi')b & \psi(\xi') \end{pmatrix} = \begin{pmatrix} 1 & {}^{\mathrm{t}}b \\ & I_{q_1} \end{pmatrix}\begin{pmatrix} \xi_{11} & {}^{\mathrm{t}}v \\ v & \psi(\xi') \end{pmatrix}\begin{pmatrix} 1 & \\ b & I_{q_1} \end{pmatrix},$$

so that we get for $\xi = \iota(\xi_{11}, v, \xi')$:

$$\phi_1(\tau_B^*\xi) = \begin{pmatrix} 1 & {}^{\mathrm{t}}b \\ & I_{q_1} \end{pmatrix}\phi_1(\xi)\begin{pmatrix} 1 & \\ b & I_{q_1} \end{pmatrix}.$$

Therefore:

$$(\phi_1(\tau_B^*\xi)^{-1})_{11} = (\phi_1(\xi)^{-1})_{11}.$$

On the other hand, we have for $\xi = \iota(\xi_{11}, v, \xi') \in \mathcal{P}_{\mathcal{V}}^*$:

$$\delta_{\underline{s}}^{\mathcal{V}}(\xi) = (\phi_1(\xi)^{-1})_{11}^{-s_1}\,\delta_{\underline{s}'}^{\mathcal{V}'}(\xi'). \tag{33}$$

Thus, we conclude that:

$$\delta_{\underline{s}}^{\mathcal{V}}(\tau_B^*\xi) = \delta_{\underline{s}}^{\mathcal{V}}(\xi). \tag{34}$$

Theorem 3. *When $\Re s_k > -1 - q_k/2$ for $k = 1, \ldots, r$, one has:*

$$\int_{\mathcal{P}_{\mathcal{V}}} e^{-(x|\xi)}\Delta_{\underline{s}}^{\mathcal{V}}(x)\,dx = C_{\mathcal{V}}^{-1}\gamma_{\mathcal{V}}(\underline{s})\,\delta_{-\underline{s}}^{\mathcal{V}}(\xi)\varphi_{\mathcal{V}}(\xi), \tag{35}$$

where $\gamma_{\mathcal{V}}(\underline{s}) := (2\pi)^{(N-r)/2}\prod_{k=1}^{r}\Gamma(s_k + 1 + \frac{q_k}{2})$.

Proof. Recalling Theorem 2, we rewrite the right-hand side of (35) as:

$$(2\pi)^{(N-r)/2}\prod_{k=1}^{r}\Gamma(s_k + 1 + \frac{q_k}{2})\prod_{k=1}^{r}(\phi_k(\xi)^{-1})_{11}^{s_k+1+q_k/2}\prod_{q_k>0}(\det\psi_k(\xi))^{-1/2},$$

which is similar to the right-hand side of (23). Thus, the proof is parallel to Theorem 2. Namely, by induction on the rank, it suffices to show that:

$$\begin{aligned}&\int_{\mathcal{P}_\mathcal{V}} e^{-(x|\xi)}\Delta_{\underline{s}}^{\mathcal{V}}(x)\,dx\\&=(2\pi)^{q_1/2}\Gamma(s_1+1+\frac{q_1}{2})(\phi_1(\xi)^{-1})_{11}^{s_1+1+q_1/2}(\det\psi_1(\xi))^{-\mathrm{sgn}(q_1)/2}\\&\quad\times\int_{\mathcal{P}_{\mathcal{V}'}} e^{-(x'|\xi)}\Delta_{\underline{s}'}^{\mathcal{V}'}(x')\,dx'\end{aligned}\tag{36}$$

thanks to (33).

When $q_1=0$, we have $(x|\xi)=x_{11}\xi_{11}+(x'|\xi')$ and $\Delta_{\underline{s}}^{\mathcal{V}}(x)=x_{11}^{s_1}\Delta_{\underline{s}}^{\mathcal{V}'}(x')$. Thus:

$$\int_{\mathcal{P}_\mathcal{V}} e^{-(x|\xi)}\Delta_{\underline{s}}^{\mathcal{V}}(x)\,dx=\int_0^\infty e^{-x_{11}\xi_{11}}x_{11}^{s_1}\,dx_{11}\times\int_{\mathcal{P}_{\mathcal{V}'}} e^{-(x'|\xi)}\Delta_{\underline{s}'}^{\mathcal{V}'}(x')\,dx'.$$

Since $\int_0^\infty e^{-x_{11}\xi_{11}}x_{11}^{s_1}\,dx_{11}=\Gamma(s_1+1)\xi_{11}^{-s_1-1}$, we get (36).

When $q_1>1$, we use the change of variable (6). Since $\tilde{x}'=x'-x_{11}^{-1}U{}^tU$, we have $\Delta_{\underline{s}}^{\mathcal{V}}(x)=x_{11}^{s_1}\Delta_{\underline{s}'}^{\mathcal{V}'}(\tilde{x}')$ by (31). Therefore, by the same Gaussian integral formula as in the proof of Theorem 2, the integral $\int_{\mathcal{P}_\mathcal{V}} e^{-(x|\xi)}\Delta_{\underline{s}}^{\mathcal{V}}(x)\,dx$ equals:

$$\begin{aligned}&\int_0^\infty\int_{\mathcal{W}}\int_{\mathcal{P}_{\mathcal{V}'}} e^{-x_{11}(\xi_{11}-{}^tv\psi(\xi')^{-1}v)}e^{-x_{11}{}^t(\tilde{u}+\psi(\xi')^{-1}v)\psi(\xi')(\tilde{u}+\psi(\xi')^{-1}v)}e^{-(\tilde{x}'|\xi')}x_{11}^{s_1}\Delta_{\underline{s}'}^{\mathcal{V}'}(\tilde{x}')\\&\qquad\times 2^{q_1/2}x_{11}^{q_1}\,dx_{11}d\tilde{u}d\tilde{x}'\\&=(2\pi)^{q_1/2}(\det\psi(\xi))^{-1/2}\int_0^\infty e^{-x_{11}(\xi_{11}-{}^tv\psi(\xi')^{-1}v)}x_{11}^{s_1+q_1/2}\,dx_{11}\\&\quad\times\int_{\mathcal{P}_{\mathcal{V}'}} e^{-(\tilde{x}'|\xi')}\Delta_{\underline{s}'}^{\mathcal{V}'}(\tilde{x}')\,d\tilde{x}'\\&=(2\pi)^{q_1/2}(\det\psi(\xi))^{-1/2}\Gamma(s_k+1+\frac{q_1}{2})(\xi_{11}-{}^tv\psi(\xi')^{-1}v)^{-s_k-1-q_k/2}\\&\quad\times\int_{\mathcal{P}_{\mathcal{V}'}} e^{-(\tilde{x}'|\xi')}\Delta_{\underline{s}'}^{\mathcal{V}'}(\tilde{x}')\,d\tilde{x}'.\end{aligned}$$

Hence, we get (36) by (18). □

We shall obtain an integral formula over $\mathcal{P}_\mathcal{V}^*$ as follows.

Theorem 4. *When $\Re s_k>q_k/2$ for $k=1,\ldots,r$, one has:*

$$\int_{\mathcal{P}_\mathcal{V}^*} e^{-(x|\xi)}\delta_{\underline{s}}^{\mathcal{V}}(\xi)\,\varphi_\mathcal{V}(\xi)\,d\xi=C_\mathcal{V}\Gamma_\mathcal{V}(\underline{s})\Delta_{-\underline{s}}^{\mathcal{V}}(x)\qquad(x\in\mathcal{P}_\mathcal{V}),\tag{37}$$

where $\Gamma_\mathcal{V}(\underline{s}):=(2\pi)^{(N-r)/2}\prod_{k=1}^r\Gamma(s_k-q_k/2)$.

Proof. Using (24), (31) and (33), we rewrite (37) as:

$$\begin{aligned}&\int_{\mathcal{P}_\mathcal{V}^*} e^{-(x|\xi)}(\phi_1(\xi)^{-1})_{11}^{-s_1+1+q_1/2}(\det\psi_1(\xi))^{-\mathrm{sgn}(q_1)/2}\delta_{\underline{s}'}^{\mathcal{V}'}(\xi')\,\varphi_{\mathcal{V}'}(\xi')\,d\xi\\&=C_{\mathcal{V}'}(2\pi)^{q_1/2}\Gamma(s_1-q_1/2)\Gamma_{\mathcal{V}'}(\underline{s}')\,x_{11}^{-s_1}\Delta_{-\underline{s}'}^{\mathcal{V}'}(\tilde{x}'),\end{aligned}\tag{38}$$

where:

$$\tilde{x}':=\begin{cases}x' & (q_1=0),\\ x'-x_{11}^{-1}U{}^tU & (q_1>0).\end{cases}$$

Therefore, by induction on the rank, it suffices to show that the left-hand side of (38) equals:

$$(2\pi)^{q_1/2}\Gamma(s_1 - q_1/2)x_{11}^{-s_1} \int_{\mathcal{P}^*_{\mathcal{V}'}} e^{-(\tilde{x}'|\xi')} \delta^{\mathcal{V}'}_{\underline{s}'}(\xi')\, \varphi_{\mathcal{V}'}(\xi')\, d\xi'. \tag{39}$$

When $q_1 = 0$, since $d\xi = d\xi_{11} d\xi'$, the left-hand side of (38) equals:

$$\int_0^\infty e^{-x_{11}\xi_{11}} \xi_{11}^{s_1-1}\, d\xi_{11} \int_{\mathcal{P}^*_{\mathcal{V}'}} e^{-(x'|\xi')} \delta^{\mathcal{V}'}_{\underline{s}'}(\xi')\, \varphi_{\mathcal{V}'}(\xi')\, d\xi',$$

which coincides with (39) in this case.

Assume $q_1 > 0$. Keeping (16) and (18) in mind, we put $\tilde{\xi}_{11} := \xi_{11} - {}^t v \psi(\xi')^{-1} v = (\phi_1(\xi)^{-1})_{11}^{-1} > 0$. By the change of variables $\xi = \iota(\tilde{\xi}_{11} + {}^t v\psi(\xi')^{-1} v,\ v,\ \xi')$, we have $d\xi = 2^{q_1/2}\, d\tilde{\xi}_{11} dv d\xi'$. On the other hand, we observe:

$$\begin{aligned}(x|\xi) &= x_{11}(\tilde{\xi}_{11} + {}^t v\psi(\xi')^{-1} v) + 2{}^t uv + (x'|\xi') \\ &= x_{11}\tilde{\xi}_{11} + x_{11}{}^t(v + x_{11}^{-1}\psi(\xi')u)\psi(\xi')^{-1}(v + x_{11}^{-1}\psi(\xi')u) + (x - x_{11}^{-1}U{}^tU|\xi').\end{aligned}$$

Thus, the left-hand side of (39) equals:

$$\begin{aligned}\int_0^\infty \int_{\mathbb{R}^{q_1}} \int_{\mathcal{P}^*_{\mathcal{V}'}} & e^{-x_{11}\tilde{\xi}_{11}} e^{-x_{11}{}^t(v + x_{11}^{-1}\psi(\xi')u)\psi(\xi')^{-1}(v + x_{11}^{-1}\psi(\xi')u)} e^{-(x - x_{11}^{-1}U{}^tU|\xi')} \\ & \times \tilde{\xi}_{11}^{s_1-1-q_1/2} (\det \psi(\xi'))^{-1/2} \delta^{\mathcal{V}'}_{\underline{s}'}(\xi')\, \varphi_{\mathcal{V}'}(\xi')\, 2^{q_1/2}\, d\tilde{\xi}_{11} dv d\xi'.\end{aligned} \tag{40}$$

By the Gaussian integral formula, we have:

$$\int_{\mathbb{R}^{q_1}} e^{-x_{11}{}^t(v + x_{11}^{-1}\psi(\xi')u)\psi(\xi')^{-1}(v + x_{11}^{-1}\psi(\xi')u)}\, dv = \pi^{q_1/2} x_{11}^{-q_1/2} (\det \psi(\xi'))^{1/2},$$

so that (40) equals:

$$(2\pi)^{q_1/2} x_{11}^{-q_1/2} \int_0^\infty e^{-x_{11}\tilde{\xi}_{11}} \tilde{\xi}_{11}^{s_1-1-q_1/2}\, d\tilde{\xi}_{11} \int_{\mathcal{P}^*_{\mathcal{V}'}} e^{-(x - x_{11}^{-1}U{}^tU|\xi')} \delta^{\mathcal{V}'}_{\underline{s}'}(\xi')\, \varphi_{\mathcal{V}'}(\xi') d\xi',$$

which coincides with (39) because: $\int_0^\infty e^{-x_{11}\tilde{\xi}_{11}} \tilde{\xi}_{11}^{s_1-1-q_1/2}\, d\tilde{\xi}_{11} = \Gamma(s_1 - q_1/2) x_{11}^{-s_1+q_1/2}$. □

Example 3. *Let $\mathcal{Z}_{\mathcal{V}}$ be as in Example 1, and let $x \in \mathcal{P}_{\mathcal{V}}$ and $\xi \in \mathcal{P}^*_{\mathcal{V}}$ be as in (2) and (25), respectively. Then, we have for $\underline{s} = (s_1, s_2, s_3) \in \mathbb{C}^3$,*

$$\begin{aligned}\Delta^{\mathcal{V}}_{\underline{s}}(x) &= (x_{11}^2)^{s_1/2 - s_2} \begin{vmatrix} x_1 & 0 & x_4 \\ 0 & x_1 & 0 \\ x_4 & 0 & x_2 \end{vmatrix}^{s_2 - s_3} (\det x)^{s_3} \\ &= x_{11}^{s_1 - s_2 - s_3} \begin{vmatrix} x_1 & x_4 \\ x_4 & x_2 \end{vmatrix}^{s_2 - s_3} (\det x)^{s_3},\end{aligned}$$

and:

$$\begin{aligned}\delta^{\mathcal{V}}_{\underline{s}}(\xi) &= (\xi_1 - \frac{\xi_4^2}{\xi_2} - \frac{\xi_5^2}{\xi_3})^{s_1} (\xi_2 - \frac{\xi_6^2}{\xi_3})^{s_2} \xi_3^{s_3} \\ &= \begin{vmatrix} \xi_1 & \xi_4 & \xi_5 \\ \xi_4 & \xi_2 & 0 \\ \xi_5 & 0 & \xi_3 \end{vmatrix}^{s_1} \begin{vmatrix} \xi_2 & \xi_6 \\ \xi_6 & \xi_3 \end{vmatrix}^{s_2} \xi_2^{-s_1} \xi_3^{s_3 - s_1 - s_2}.\end{aligned}$$

When $\Re s_1 > -2, \Re s_2 > -3/2$ and $\Re s_3 > -1$, the integral formula (35) holds with:

$$\gamma_{\mathcal{V}}(\underline{s}) = (2\pi)^{3/2}\Gamma(s_1+2)\Gamma(s_2+3/2)\Gamma(s_3+1).$$

Furthermore, when $\Re s_1 > 1$, $\Re s_2 > 1/2$ and $\Re s_3 > 0$, the integral formula (37) holds with:

$$\Gamma_{\mathcal{V}}(\underline{s}) = (2\pi)^{3/2}\Gamma(s_1-1)\Gamma(s_2-1/2)\Gamma(s_3).$$

5. Multiplicative Legendre Transform of Generalized Power Functions

For $\underline{s} \in \mathbb{R}^r_{>0}$, we see that $\log \Delta_{-\underline{s}}$ is a strictly convex function on the cone $\mathcal{P}_{\mathcal{V}}$. In fact, $\Delta_{-\underline{s}}$ is defined naturally on $\mathcal{P}_n$ as a product of powers of principal minors, and it is well known that such $\log \Delta_{-\underline{s}}$ is strictly convex on the whole $\mathcal{P}_n$. In this section, we shall show that $\log \Delta^{\mathcal{V}}_{-\underline{s}}$ and $\log \delta^{\mathcal{V}}_{-\underline{s}}$ are related by the Fenchel–Legendre transform.

For $x \in \mathcal{P}_{\mathcal{V}}$, we denote by $\mathcal{I}^{\mathcal{V}}_{\underline{s}}(x)$ the minus gradient $-\nabla \log \Delta_{-\underline{s}}(x)$ at x with respect to the inner product. Namely, $\mathcal{I}^{\mathcal{V}}_{\underline{s}}(x)$ is an element of $\mathcal{Z}_{\mathcal{V}}$ for which:

$$(\mathcal{I}^{\mathcal{V}}_{\underline{s}}(x)|y) = -\left(\frac{d}{dt}\right)_{t=0} \log \Delta_{-\underline{s}}(x+ty) \qquad (y \in \mathcal{Z}_{\mathcal{V}}).$$

Similarly, $\mathcal{J}^{\mathcal{V}}_{\underline{s}}(\xi) := -\nabla \log \delta_{-\underline{s}}(\xi)$ is defined for $\xi \in \mathcal{P}^*_{\mathcal{V}}$. If $q_1 > 0$, then for any $B \in \mathcal{W}$, we have:

$$\mathcal{I}^{\mathcal{V}}_{\underline{s}} \circ \tau_B = \tau^*_B \circ \mathcal{I}^{\mathcal{V}}_{\underline{s}}, \tag{41}$$
$$\mathcal{J}^{\mathcal{V}}_{\underline{s}} \circ \tau^*_B = \tau_B \circ \mathcal{J}^{\mathcal{V}}_{\underline{s}} \tag{42}$$

owing to (30) and (34), respectively.

Theorem 5. *For any $\underline{s} \in \mathbb{R}^r_{>0}$, the map $\mathcal{I}^{\mathcal{V}}_{\underline{s}} : \mathcal{P}_{\mathcal{V}} \to \mathcal{Z}_{\mathcal{V}}$ gives a diffeomorphism from $\mathcal{P}_{\mathcal{V}}$ onto $\mathcal{P}^*_{\mathcal{V}}$, and $\mathcal{J}^{\mathcal{V}}_{\underline{s}}$ gives the inverse map.*

Proof. We shall prove the statement by induction on the rank. When $r = 1$, we have $\mathcal{I}^{\mathcal{V}}_{\underline{s}}(x_{11}I_{n_1}) = \frac{s_1}{x_{11}}I_{n_1} = \mathcal{J}^{\mathcal{V}}_{\underline{s}}(x_{11}I_{n_1})$ for $x_{11} > 0$. Thus, the statement is true in this case.

When $r > 1$, assume that the statement holds for the system of rank $r-1$. Let $\mathcal{Z}^0_{\mathcal{V}}$ be the subspace of $\mathcal{Z}_{\mathcal{V}}$ defined by:

$$\mathcal{Z}^0_{\mathcal{V}} := \left\{ \begin{pmatrix} x_{11}I_{n_1} & 0 \\ 0 & x' \end{pmatrix} ; x_{11} \in \mathbb{R},\ x' \in \mathcal{Z}_{\mathcal{V}'} \right\}.$$

By direct computation with (31) and (33), we have:

$$\mathcal{I}^{\mathcal{V}}_{\underline{s}} \begin{pmatrix} x_{11}I_{n_1} & 0 \\ 0 & x' \end{pmatrix} = \begin{pmatrix} \frac{s_1}{x_{11}}I_{n_1} & 0 \\ 0 & \mathcal{I}^{\mathcal{V}'}_{\underline{s}'}(x') \end{pmatrix}, \tag{43}$$
$$\mathcal{J}^{\mathcal{V}}_{\underline{s}} \begin{pmatrix} \xi_{11}I_{n_1} & 0 \\ 0 & \xi' \end{pmatrix} = \begin{pmatrix} \frac{s_1}{\xi_{11}}I_{n_1} & 0 \\ 0 & \mathcal{J}^{\mathcal{V}'}_{\underline{s}'}(\xi') \end{pmatrix} \tag{44}$$

for $x_{11}, \xi_{11} > 0$, $x' \in \mathcal{P}_{\mathcal{V}'}$ and $\xi' \in \mathcal{P}^*_{\mathcal{V}'}$. By the induction hypothesis, we see that $\mathcal{I}^{\mathcal{V}}_{\underline{s}} : \mathcal{P}_{\mathcal{V}} \cap \mathcal{Z}^0_{\mathcal{V}} \to \mathcal{P}^*_{\mathcal{V}} \cap \mathcal{Z}^0_{\mathcal{V}}$ is bijective with the inverse map $\mathcal{J}^{\mathcal{V}}_{\underline{s}} : \mathcal{P}^*_{\mathcal{V}} \cap \mathcal{Z}^0_{\mathcal{V}} \to \mathcal{P}_{\mathcal{V}} \cap \mathcal{Z}^0_{\mathcal{V}}$.

If $q_1 = 0$, the statement holds because $\mathcal{Z}_{\mathcal{V}} = \mathcal{Z}^0_{\mathcal{V}}$. Assume $q_1 > 0$. Lemma 1 (ii) tells us that, for $x \in \mathcal{P}_{\mathcal{V}}$, there exist unique $x^0 \in \mathcal{Z}^0_{\mathcal{V}} \cap \mathcal{P}_{\mathcal{V}}$ and $B \in \mathcal{W}$ for which $x = \tau_B x^0$. Similarly, we see from (32) that, for $\xi \in \mathcal{P}^*_{\mathcal{V}}$, there exist unique $\xi^0 \in \mathcal{Z}^0_{\mathcal{V}} \cap \mathcal{P}^*_{\mathcal{V}}$ and $C \in \mathcal{W}$ for which $\xi = \tau^*_C \xi^0$. Therefore, we deduce from (41) and (42) that $\mathcal{I}^{\mathcal{V}}_{\underline{s}} : \mathcal{P}_{\mathcal{V}} \mapsto \mathcal{P}^*_{\mathcal{V}}$ is a bijection with $\mathcal{J}^{\mathcal{V}}_{\underline{s}}$ the inverse map. □

Proposition 1. *Let $\underline{s} \in \mathbb{R}^r_{>0}$. For $\xi \in \mathcal{P}^*_{\mathcal{V}}$, one has:*

$$\Delta_{-\underline{s}}(\mathcal{J}_{\underline{s}}(\xi))^{-1} = (\prod_{k=1}^{r} s_k^{s_k})\delta_{-\underline{s}}(\xi). \tag{45}$$

Proof. We prove the statement by induction on the rank. When $r = 1$, the equality (45) is verified directly. Indeed, the left-hand side of (45) is computed as $(\frac{s_1}{\xi_{11}})^{s_1} = s_1^{s_1}\xi_{11}^{-s_1}$.

When $r > 1$, assume that (45) holds for a system of rank $r-1$. We deduce from (31), (33), (43), (44) and the induction hypothesis that (45) holds for $\xi \in \mathcal{P}^*_{\mathcal{V}} \cap \mathcal{Z}^0_{\mathcal{V}}$. Therefore, (45) holds for all $\xi \in \mathcal{P}^*_{\mathcal{V}}$ by (30), (34) and (42). □

In general, for a non-zero function f, the function $\frac{1}{f\circ(\nabla \log f)^{-1}}$ is called the multiplicative Legendre transform of f. Thanks to Theorem 5 and Proposition 1, we see that the multiplicative Legendre transform of $\Delta_{-\underline{s}}(x)$ is equal to $\delta_{-\underline{s}}(-\xi)$ on $-\mathcal{P}^*_{\mathcal{V}}$ up to constant multiple. As a corollary, we arrive at the following result.

Theorem 6. *The Fenchel–Legendre transform of the convex function $\log \Delta_{-\underline{s}}$ on $\mathcal{P}_{\mathcal{V}}$ is equal to the function $\log \delta_{-\underline{s}}(-\xi)$ of $\xi \in -\mathcal{P}^*$ up to constant addition.*

6. Application to Statistics and Optimization

Take $\underline{s} \in \mathbb{R}^r$ for which $s_k > q_k/2$ $(k = 1, \dots, r)$. We define a measure $\rho^{\mathcal{V}}_{\underline{s}}$ on $\mathcal{P}^*_{\mathcal{V}}$ by:

$$\rho^{\mathcal{V}}_{\underline{s}}(d\xi) := C_{\mathcal{V}}^{-1}\Gamma_{\mathcal{V}}(\underline{s})^{-1}\delta^{\mathcal{V}}_{\underline{s}}(\xi)\varphi_{\mathcal{V}}(\xi)\,d\xi \qquad (\xi \in \mathcal{P}^*_{\mathcal{V}}). \tag{46}$$

Theorem 4 states that:

$$\int_{\mathcal{P}^*_{\mathcal{V}}} e^{-(x|\xi)}\rho^{\mathcal{V}}_{\underline{s}}(d\xi) = \Delta^{\mathcal{V}}_{-\underline{s}}(x) \qquad (x \in \mathcal{P}_{\mathcal{V}}).$$

Then, we obtain the natural exponential family generated by $\rho^{\mathcal{V}}_{\underline{s}}$, that is a family $\{\mu^{\mathcal{V}}_{\underline{s},x}\}_{x\in\mathcal{P}_{\mathcal{V}}}$ of probability measures on $\mathcal{P}^*_{\mathcal{V}}$ given by:

$$\mu^{\mathcal{V}}_{\underline{s},x}(d\xi) := \Delta^{\mathcal{V}}_{\underline{s}}(x)e^{-(x|\xi)}\rho^{\mathcal{V}}_{\underline{s}}(d\xi).$$

In particular, when $\underline{s} = (n_1\alpha, n_2\alpha, \dots, n_r\alpha)$ for sufficiently large α, we have $\mu^{\mathcal{V}}_{\underline{s},x}(d\xi) = (\det x)^{\alpha}e^{-(x|\xi)}\rho^{\mathcal{V}}_{\underline{s}}(d\xi)$. We call $\mu^{\mathcal{V}}_{\underline{s},x}$ the Wishart distributions on $\mathcal{P}^*_{\mathcal{V}}$ in general.

From a sample $\xi_0 \in \mathcal{P}^*_{\mathcal{V}}$, let us estimate the parameter $x \in \mathcal{P}_{\mathcal{V}}$ in such a way that the likelihood function $\Delta^{\mathcal{V}}_{\underline{s}}(x)e^{-(x|\xi)}$ attains its maximum at the estimator x_0. Then, we have the likelihood equation $\xi_0 = \mathcal{I}^{\mathcal{V}}_{\underline{s}}(x_0)$, whereas Theorem 5 gives a unique solution by $x_0 = \mathcal{J}^{\mathcal{V}}_{\underline{s}}(\xi_0)$.

The same argument leads us to the following result in semidefinite programming. For a fixed $\xi_0 \in \mathcal{P}^*_{\mathcal{V}}$ and $\alpha > 0$, a unique solution x_0 of the minimization problem of $(x|\xi_0) - \alpha \log \det x$ subject to $x \in \mathcal{P}_{\mathcal{V}} = \mathcal{Z}_{\mathcal{V}} \cap \mathcal{P}_n$ is given by $x_0 = \mathcal{J}^{\mathcal{V}}_{\underline{s}}(\xi_0)$, where $\underline{s} = (n_1\alpha, \dots, n_r\alpha)$. Note that $\mathcal{J}^{\mathcal{V}}_{\underline{s}}$ is a rational map because $\delta^{\mathcal{V}}_{\underline{s}}$ is a product of powers of rational functions.

7. Special Cases

7.1. Matrix Realization of Homogeneous Cones

Let us assume that the system $\mathcal{V} = \{\mathcal{V}_{lk}\}_{1\le k<l\le r}$ satisfies not only the conditions (V1) and (V2), but also the following:

(V3) $A \in \mathcal{V}_{lk}$, $B \in \mathcal{V}_{kj} \Rightarrow AB \in \mathcal{V}_{lj}$ $\quad (1 \le j < k < l \le r)$.

Then, the set $H_{\mathcal{V}}$ of lower triangular matrices T of the form:

$$T = \begin{pmatrix} T_{11} & & & \\ T_{21} & T_{22} & & \\ \vdots & & \ddots & \\ T_{r1} & T_{r2} & \dots & T_{rr} \end{pmatrix}$$

becomes a linear Lie group, and $H_{\mathcal{V}}$ acts on the space $\mathcal{Z}_{\mathcal{V}}$ by $\rho(T)x := Tx\,{}^{t}T$ $(T \in H_{\mathcal{V}},\, x \in \mathcal{Z}_{\mathcal{V}})$. The group $H_{\mathcal{V}}$ acts on the cone $\mathcal{P}_{\mathcal{V}}$ simply transitively by this action ρ, so that $\mathcal{P}_{\mathcal{V}}$ is a homogeneous cone. Moreover, it is shown in [15] that every homogeneous cone is linearly isomorphic to such $\mathcal{P}_{\mathcal{V}}$ (see also [18]).

Let $\mathcal{V}^0 = \{\mathcal{V}^0_{lk}\}_{1 \le k < l \le 3}$ be the system given by $\mathcal{V}^0_{21} = \{0\}$ and $\mathcal{V}^0_{lk} = \mathbb{R}$ $((l,k) \ne (2,1))$. Then:

$$\mathcal{Z}_{\mathcal{V}^0} = \left\{ \begin{pmatrix} x_1 & 0 & x_4 \\ 0 & x_2 & x_5 \\ x_4 & x_5 & x_3 \end{pmatrix} ; x_1, \dots, x_5 \in \mathbb{R} \right\},$$

and $\mathcal{P}_{\mathcal{V}^0} := \mathcal{Z}_{\mathcal{V}^0} \cap \mathcal{P}_3$ is homogeneous because (V1)–(V3) are satisfied in this case. On the other hand, let $\mathcal{V}^1 = \{\mathcal{V}^1_{lk}\}_{1 \le k < l \le 3}$ be the system given by $\mathcal{V}^1_{31} = \{0\}$ and $\mathcal{V}^1_{lk} = \mathbb{R}$ $((l,k) \ne (3,1))$. Then:

$$\mathcal{Z}_{\mathcal{V}^1} = \left\{ \begin{pmatrix} x_1 & x_4 & 0 \\ x_4 & x_2 & x_5 \\ 0 & x_5 & x_3 \end{pmatrix} ; x_1, \dots, x_5 \in \mathbb{R} \right\}.$$

Note that $\mathcal{V}^1$ satisfies only (V1) and (V2), but $\mathcal{P}_{\mathcal{V}^1}$ is homogeneous because $\mathcal{P}_{\mathcal{V}^1}$ is isomorphic to the homogeneous cone $\mathcal{P}_{\mathcal{V}^0}$ via the map:

$$\mathcal{P}_{\mathcal{V}^1} \ni \begin{pmatrix} x_1 & x_4 & 0 \\ x_4 & x_2 & x_5 \\ 0 & x_5 & x_3 \end{pmatrix} \mapsto \begin{pmatrix} 1 & 0 & 0 \\ 0 & 0 & 1 \\ 0 & 1 & 0 \end{pmatrix} \begin{pmatrix} x_1 & x_4 & 0 \\ x_4 & x_2 & x_5 \\ 0 & x_5 & x_3 \end{pmatrix} \begin{pmatrix} 1 & 0 & 0 \\ 0 & 0 & 1 \\ 0 & 1 & 0 \end{pmatrix} = \begin{pmatrix} x_1 & 0 & x_4 \\ 0 & x_3 & x_5 \\ x_4 & x_5 & x_2 \end{pmatrix} \in \mathcal{P}_{\mathcal{V}^0}.$$

This example tells us that our matrix realization of a convex cone is not unique and that the condition (V3) is merely a sufficient condition for the homogeneity of the cone.

Many ideas in this work are inspired by the theory of homogeneous cones. The notion of generalized power functions, as well as the Γ-type integral formulas are due to Gindikin [8] (see also [23]). The Wishart distributions for homogeneous cones are studied in [17,21,24,25].

7.2. Cones Associated with Chordal Graphs

If $n_1 = n_2 = \cdots = n_r = 1$, then $\mathcal{V}_{lk}$ equals either $\mathbb{R}$ or $\{0\}$. In this case, $\mathcal{Z}_{\mathcal{V}}$ is the space of symmetric matrices with prescribed zero components. Such a space is described by using an undirected graph in the graphical model theory.

Let us recall some notion in the graph theory. Let G be a graph and V_G the set of vertices of G. We assume that G has no multiple edge, that is, for any two vertices $i, j \in V_G$, either there is one edge connecting them or there is no edge between them. These relations of the vertices i and j are denoted by $i \sim j$ and $i \not\sim j$, respectively. Assume further that G has no loop, which means that $i \not\sim i$ for $i \in V_G$. We define the edge set $E_G \subset V_G \times V_G$ by:

$$E_G := \{\, (i,j) \in V_G \times V_G \,;\, i \sim j \,\}.$$

Since V_G and E_G have all of the information of G, the graph G is often identified with the pair (V_G, E_G). For a non-empty subset V' of V_G, put $E' := E_G \cap (V' \times V')$. The graph $G' := (V', E')$ is called an induced subgraph of G. The graph G is said to be chordal or decomposable if G contains no cycle of length greater than three as an induced subgraph, and said to be A_4-free if G contains no A_4 graph $\bullet - \bullet - \bullet - \bullet$ as an induced subgraph. Let $\preceq$ be a total order on the vertex set V_G, and for $i \in V_G$, put $V_G^{[i]} := \{ j \in V_G\,;\, i \sim j \text{ and } i \preceq j \} \subset V_G$. Then, $\preceq$ is said to be an eliminating order on the graph G if the induced subgraph with the vertex set $V_G^{[i]}$ is complete for each $i \in V_G$. It is known that there exists an eliminating order on G if and only if the graph G is chordal.

Let us identify the vertex set V_G with $\{1,2,\dots,r\}$. Let $\mathcal{Z}_G$ be the space of symmetric matrices $x = (x_{ij}) \in \mathrm{Sym}(r,\mathbb{R})$, such that, if $i \neq j$ and $i \not\sim j$, then $x_{ij} = 0$. Define $\mathcal{P}_G := \mathcal{Z}_G \cap \mathcal{P}_r$. We can show ([11] (Theorem 2.2), [26]) that the cone $\mathcal{P}_G$ is homogeneous if and only if the graph G is chordal and A_4-free. On the other hand, it is known in the graphical model theory as well as the sparse matrix linear algebra that even though $\mathcal{P}_G$ is not homogeneous, various formulas still hold for $\mathcal{P}_G$ if G is chordal.

The cone $\mathcal{P}_G$ is expressed as $\mathcal{P}_{\mathcal{V}}$ with $n_1 = n_2 = \cdots = n_r = 1$ and:

$$\mathcal{V}_{ji} = \begin{cases} \mathbb{R} & (j \sim i), \\ \{0\} & (j \not\sim i). \end{cases}$$

Then, the condition (V2) means exactly that the order $\leq$ is an eliminating order on G. Therefore, any cone $\mathcal{P}_G$ with chordal G can be treated as $\mathcal{P}_{\mathcal{V}}$ in our framework. Most of the integral formulas for $\mathcal{P}_G$ in [11,27] can be deduced from Theorems 3 and 4, while the Wishart distribution is a central object in the theory of graphical model. In [28], the analysis for generalized power functions associated with all eliminating orders is discussed for a specific graph $A_n : \bullet - \bullet - \cdots - \bullet$ by direct computations.

Acknowledgments: The author would like to express sincere gratitude to Piotr Graczyk and Yoshihiko Konno for stimulating discussions about this subject. He is also grateful to Frédéric Barbaresco for his interest in and encouragement of this work. He thanks to anonymous referees for valuable comments, which were helpful for the improvement of the present paper. This work was supported by JSPS KAKENHI Grant Number 16K05174.

References

1. Koszul, J.L. Ouverts convexes homogènes des espaces affines. *Math. Z.* **1962**, *79*, 254–259.
2. Vinberg, E.B. The theory of convex homogeneous cones. *Trans. Moscow Math. Soc.* **1963**, *12*, 340–403.
3. Vey, J. Sur les automorphismes affines des ouverts convexes saillants. *Annali della Scuola Normale Superiore di Pisa* **1970**, *24*, 641–665.
4. Shima, H. *The Geometry of Hessian Structures*; World Scientific: Hackensack, NJ, USA, 2007.
5. Nesterov, Y.; Nemirovskii, A. *Interior-Point Polynomial Algorithms in Convex Programming*; Society for Industrial and Applied Mathematics: Philadelphia, PA, USA, 1994.
6. Barbaresco, F. Koszul information geometry and Souriau geometric temperature/capacity of Lie group thermodynamics. *Entropy* **2014**, *16*, 4521–4565.
7. Barbaresco, F. Symplectic structure of information geometry: Fisher etric and Euler-Poincaré equation of Souriau Lie group thermodynamics. In *Geometric Science of Information*; Nielsen, F., Barbaresco, F., Eds.; (Lecture Notes in Computer Science); Springer International Publishing: Basel, Switzerland, 2015; Volume 9389, pp. 529–540.
8. Gindikin, S.G. Analysis in homogeneous domains. *Russ. Math. Surv.* **1964**, *19*, 1–89.
9. Truong, V.A.; Tunçel, L. Geometry of homogeneous convex cones, duality mapping, and optimal self-concordant barriers. *Math. Program.* **2004**, *100*, 295–316.
10. Tunçel, L.; Xu, S. On homogeneous convex cones, the Caratheodory number, and the duality mapping. *Math. Oper. Res.* **2001**, *26*, 234–247.
11. Letac, G.; Massam, H. Wishart distributions for decomposable graphs. *Ann. Stat.* **2007**, *35*, 1278–1323.
12. Rothaus, O.S. The construction of homogeneous convex cones. *Ann. Math.* **1966**, *83*, 358–376.
13. Xu, Y.-C. *Theory of Complex Homogeneous Bounded Domains*; Kluwer: Dordrecht, The Netherlands, 2005.

14. Chua, C.B. Relating homogeneous cones and positive definite cones via T-algebras. *SIAM J. Optim.* **2003**, *14*, 500–506.
15. Ishi, H. On symplectic representations of normal j-algebras and their application to Xu's realizations of Siegel domains. *Differ. Geom. Appl.* **2006**, *24*, 588–612.
16. Yamasaki, T.; Nomura, T. Realization of homogeneous cones through oriented graphs. *Kyushu J. Math.* **2015**, *69*, 11–48.
17. Ishi, H. Homogeneous cones and their applications to statistics. In *Modern Methods of Multivariate Statistics*; Graczyk, P., Hassairi, A, Eds.; Hermann: Paris, France, 2014; pp. 135–154.
18. Ishi, H. Matrix realization of homogeneous cones. In *Geometric Science of Information*; Nielsen, F., Barbaresco, F., Eds.; (Lecture Notes in Computer Science); Springer International Publishing: Basel, Switzerland, 2015; Volume 9389, pp. 248–256.
19. Lauritzen, S.L. *Graphical Models*; Clarendon Press: Oxford, UK, 1996.
20. Faraut, J.; Korányi, A. *Analysis on Symmetric Cones*; Clarendon Press: Oxford, UK; 1994.
21. Graczyk, P.; Ishi, H. Riesz measures and Wishart laws associated with quadratic maps. *J. Math. Soc. Jpn.* **2014**, *66*, 317–348.
22. Shima, H. Homogeneous Hessian manifolds. *Ann. Inst. Fourier* **1980**, *30*, 91–128.
23. Güler, O.; Tunçel, L. Characterization of the barrier parameter of homogeneous convex cones. *Math. Program. A* **1988**, *81*, 55–76.
24. Andersson, S.A.; Wojnar, G.G. Wishart distributions on homogeneous cones. *J. Theor. Probab.* **2004**, *17*, 781–818.
25. Graczyk, P.; Ishi, H.; Kołodziejek, B. Wishart exponential families and variance function on homogeneous cones. *HAL* **2016**, submitted for publication.
26. Ishi, H. On a class of homogeneous cones consisting of real symmetric matrices. *Josai Math. Monogr.* **2013**, *6*, 71–80.
27. Roverato, A. Cholesky decomposition of a hyper inverse Wishart matrix. *Biometrika* **2000**, *87*, 99–112.
28. Graczyk, P.; Ishi, H.; Mamane, S. Wishart exponential families on cones related to A_n graphs. *HAL* **2016**, submitted for publication.

4

The Information Geometry of Sparse Goodness-of-Fit Testing

Paul Marriott [1,*], Radka Sabolová [2], Germain Van Bever [3] and Frank Critchley [2]

1. Department of Statistics and Actuarial Science, University of Waterloo, 200 University Avenue West, Waterloo, ON N2L 3G1, Canada
2. School of Mathematics and Statistics, The Open University, Walton Hall, Milton Keynes, Buckinghamshire MK7 6AA, UK; radka.sabolova@open.ac.uk (R.S.); f.critchley@open.ac.uk (F.C.)
3. Department of Mathematics & ECARES, Université libre de Bruxelles, Avenue F.D. Roosevelt 42, 1050 Brussels, Belgium; gvbever@ulb.ac.be

* Correspondence: pmarriot@uwaterloo.ca.

Academic Editors: Frédéric Barbaresco and Frank Nielsen

Abstract: This paper takes an information-geometric approach to the challenging issue of goodness-of-fit testing in the high dimensional, low sample size context where—potentially—boundary effects dominate. The main contributions of this paper are threefold: first, we present and prove two new theorems on the behaviour of commonly used test statistics in this context; second, we investigate—in the novel environment of the extended multinomial model—the links between information geometry-based divergences and standard goodness-of-fit statistics, allowing us to formalise relationships which have been missing in the literature; finally, we use simulation studies to validate and illustrate our theoretical results and to explore currently open research questions about the way that discretisation effects can dominate sampling distributions near the boundary. Novelly accommodating these discretisation effects contrasts sharply with the essentially continuous approach of skewness and other corrections flowing from standard higher-order asymptotic analysis.

Keywords: extended multinomial models; goodness-of-fit testing; information geometry

1. Introduction

We start by emphasising the threefold achievements of this paper, spelled out in detail in terms of the paper's section structure below. First, we present and prove two new theorems on the behaviour of some standard goodness-of-fit statistics in the high dimensional, low sample size context, focusing on behaviour "near the boundary" of the extended multinomial family. We also comment on the methods of proof which allow explicit calculations of higher order moments in this context. Second, working again explicitly in the extended multinomial context, we fill a hole in the literature by linking information-geometric-based divergences and standard goodness-of-fit statistics. Finally, we use simulation studies to explore discretisation effects that can dominate sampling distributions "near the boundary". Indeed, we illustrate and explore how—in the high dimensional, low sample size context—all distributions are affected by boundary effects. We also use these simulation results to explore currently open research questions. As can be seen, the overarching theme is the importance of working in the geometry of the extended exponential family [1], rather than the traditional manifold-based structure of information geometry.

In more detail, the paper extends and builds on the results of [2], and we use notation and definitions consistently across these two papers. Both papers investigate the issue of goodness-of-fit testing in the high dimensional sparse extended multinomial context, using the tools of Computational Information Geometry (CIG) [1].

Section 2 gives formal proofs of two results, Theorems 1 and 2, which were announced in [2]. These results explore the sampling performance of standard goodness-of-fit statistics—Wald, Pearson's χ^2, score and deviance—in the sparse setting. In particular, they look at the case where the data generation process is "close to the boundary" of the parameter space where one or more cell probabilities vanish. This complements results in much of the literature, where the centre of the parameter space—i.e., the uniform distribution—is often the focus of attention.

Section 3 starts with a review of the links between Information Geometry (IG) [3] and goodness-of-fit testing. In particular, it looks at the power family of Cressie and Read [4,5] in terms of the geometric theory of divergences. In the case of regular exponential families, these links have been well-explored in the literature [6], as has the corresponding sampling behaviour [7]. What is novel here is the exploration of the geometry with respect to the closure of the exponential family; i.e., the extended multinomial model—a key tool in CIG. We illustrate how the boundary can dominate the statistical properties in ways that are surprising compared to standard—and even high-order—analyses, which are asymptotic in sample size.

Through simulation experiments, Section 4 explores the consequences of working in the sparse multinomial setting, with the design of the numerical experiments being inspired by the information geometry.

2. Sampling Distributions in the Sparse Case

One of the first major impacts that information geometry had on statistical practice was through the geometric analysis of higher order asymptotic theory (e.g., [8,9]). Geometric interpretations and invariant expressions of terms in the higher order corrections to approximations of sampling distributions are a good example, [8] (Chapter 4). Geometric terms are used to correct for skewness and other higher order moment (cumulant) issues in the sampling distributions. However, these correction terms grow very large near the boundary [1,10]. Since this region plays a key role in modelling in the sparse setting—the maximum likelihood estimator (MLE) often being on the boundary—extensions to the classical theory are needed. This paper, together with [2], start such a development. This work is related to similar ideas in categorical, (hierarchical) log–linear, and graphical models [1,11–13]. As stated in [13], "their statistical properties under sparse settings are still very poorly understood. As a result, analysis of such data remains exceptionally difficult".

In this section we show why the Wald—equivalently, the Pearson χ^2 and score statistics—are unworkable when near the boundary of the extended multinomial model, but that the deviance has a simple, accurate, and tractable sampling distribution—even for moderate sample sizes. We also show how the higher moments of the deviance are easily computable, in principle allowing for higher order adjustments. However, we also make some observations about the appropriateness of these classical adjustments in Section 4.

First, we define some notation, consistent with that of [2]. With i ranging over $\{0, 1, ..., k\}$, let $n = (n_i) \sim$ Multinomial $(N, (\pi_i))$, where here each $\pi_i > 0$. In this context, the Wald, Pearson's χ^2, and score statistics all coincide, their common value, W, being

$$W := \sum_{i=0}^{k} \frac{(\pi_i - n_i/N)^2}{\pi_i} \equiv \frac{1}{N^2} \sum_{i=0}^{k} \frac{n_i^2}{\pi_i} - 1.$$

Defining $\pi^{(\alpha)} := \sum_i \pi_i^\alpha$, we note the inequality, for each $m \geq 1$,

$$\pi^{(-m)} - (k+1)^{m+1} \geq 0,$$

in which equality holds if and only if $\pi_i \equiv 1/(k+1)$—i.e., iff (π_i) is uniform. We then have the following theorem, which establishes that the statistic W is unworkable as $\pi_{\min} := \min(\pi_i) \to 0$ for fixed k and N.

Theorem 1. *For $k > 1$ and $N \geq 6$, the first three moments of W are:*

$$E(W) = \frac{k}{N}, \quad Var(W) = \frac{\left\{\pi^{(-1)} - (k+1)^2\right\} + 2k(N-1)}{N^3}$$

and $E[\{W - E(W)\}^3]$ given by

$$\frac{\left\{\pi^{(-2)} - (k+1)^3\right\} - (3k + 25 - 22N)\left\{\pi^{(-1)} - (k+1)^2\right\} + g(k,N)}{N^5},$$

where $g(k,N) = 4(N-1)k(k+2N-5) > 0$.

In particular, for fixed k and N, as $\pi_{\min} \to 0$

$$Var(W) \to \infty \text{ and } \gamma(W) \to +\infty,$$

where $\gamma(W) := E[\{W - E(W)\}^3]/\{Var(W)\}^{3/2}$.

A detailed proof is found in Appendix A, and we give here an outline of its important features. The machinery developed is capable of delivering much more than a proof of Theorem 1. As indicated there, it provides a generic way to explicitly compute arbitrary moments or mixed moments of multinomial counts, and could in principle be implemented by computer algebra. Overall, there are four stages. First, a key recurrence relation is established; secondly, it is exploited to deliver moments of a single cell count. Third, mixed moments of any order are derived from those of lower order, exploiting a certain functional dependence. Finally, results are combined to find the first three moments of W, higher moments being similarly obtainable.

The practical implication of Theorem 1 is that standard first (and higher-order) asymptotic approximations to the sampling distribution of the Wald, χ^2, and score statistics break down when the data generation process is "close to" the boundary, where at least one cell probability is zero. This result is qualitatively similar to results in [10], which shows how asymptotic approximations to the distribution of the maximum likelihood estimate fail; for example, in the case of logistic regression, when the boundary is close in terms of distances as defined by the Fisher information.

Unlike statistics considered in Theorem 1, the deviance has a workable distribution in the same limit: that is, for fixed N and k as we approach the boundary of the probability simplex. In sharp contrast to that theorem, we see the very stable and workable behaviour of the k-asymptotic approximation to the distribution of the deviance, in which the number of cells increases without limit.

Define the deviance D via

$$\begin{aligned} D/2 &= \sum\nolimits_{\{0 \leq i \leq k : n_i > 0\}} n_i \log(n_i/N) - \sum_{i=0}^{k} n_i \log(\pi_i) \\ &= \sum\nolimits_{\{0 \leq i \leq k : n_i > 0\}} n_i \log(n_i/\mu_i), \end{aligned}$$

where $\mu_i := E(n_i) = N\pi_i$. We will exploit the characterisation that the multinomial random vector (n_i) has the same distribution as a vector of independent Poisson random variables conditioned on their sum. Specifically, let the elements of (n_i^*) be *independently* distributed as Poisson $Po(\mu_i)$. Then, $N^* := \sum_{i=0}^{k} n_i^* \sim Po(N)$, while $(n_i) := (n_i^* | N^* = N) \sim$ Multinomial$(N, (\pi_i))$. Define the vector

$$S^* := \begin{pmatrix} N^* \\ D^*/2 \end{pmatrix} = \sum_{i=0}^{k} \begin{pmatrix} n_i^* \\ n_i^* \log(n_i^*/\mu_i) \end{pmatrix},$$

where D^* is defined implicitly and $0\log 0 := 0$. The terms ν, τ, and ρ are defined by the first two moments of S^* via the vectors

$$\begin{pmatrix} N \\ \nu \end{pmatrix} := E(S^*) = \begin{pmatrix} N \\ \sum_{i=0}^{k} E(n_i^* \log{(n_i^*/\mu_i)}) \end{pmatrix}, \tag{1}$$

$$\begin{pmatrix} N & \rho\tau\sqrt{N} \\ \cdot & \tau^2 \end{pmatrix} := Cov(S^*) = \begin{pmatrix} N & \sum_{i=0}^{k} C_i \\ \cdot & \sum_{i=0}^{k} V_i \end{pmatrix}, \tag{2}$$

where $C_i := Cov(n_i^*, n_i^* \log(n_i^*/\mu_i))$ and $V_i := Var(n_i^* \log(n_i^*/\mu_i))$.

Theorem 2. *Each of the terms ν, τ, and ρ remains bounded as $\pi_{\min} \to 0$.*

We start with some preliminary remarks. We use the following notation: $\mathcal{N} := \{1, 2, ...\}$ denotes the natural numbers, while $\mathcal{N}_0 := \{0\} \cup \mathcal{N}$. Throughout, $X \sim Po(\mu)$ denotes a Poisson random variable having positive mean μ—that is, X is discrete with support $\mathcal{N}_0$ and probability mass function $p : \mathcal{N}_0 \to (0, 1)$ given by:

$$p(x) := e^{-\mu}\mu^x/x! \ (\mu > 0). \tag{3}$$

Putting:

$$\forall m \in \mathcal{N}_0, \ F^{[m]}(\mu) := \Pr(X \leq m) = \textstyle\sum_{x=0}^{m} p(x) \in (0, 1), \tag{4}$$

for given μ, $\{1 - F^{[m]}(\mu)\}$ is strictly decreasing with m, vanishing as $m \to \infty$. For all $(x, m) \in \mathcal{N}_0^2$, we define $x_{(m)}$ by:

$$x_{(0)} := 1; \ x_{(m)} := x(x-1)...(x-(m-1)) \ (m \in \mathcal{N}) \tag{5}$$

so that, if $x \geq m$, $x_{(m)} = x!/(x-m)!$.

The set $\mathcal{A}_0$ comprises all functions $a_0 : (0, \infty) \to R$ such that, as $\xi \to 0_+$:

(i) $a_0(\xi)$ tends to an infinite limit $a_0(0_+) \in \{-\infty, +\infty\}$, while: (ii) $\xi a_0(\xi) \to 0$.

Of particular interest here, by l'Hôspital's rule,

$$\forall m \in \mathcal{N}, \ (\log)^m \in \mathcal{A}_0, \tag{6}$$

where $(\log)^m : \xi \to (\log \xi)^m$ $(\xi > 0)$. For each $a_0 \in \mathcal{A}_0$, $\overline{a_0}$ denotes its continuous extension from $(0, \infty)$ to $[0, \infty)$—that is: $\overline{a_0}(0) := a_0(0_+)$; $\overline{a_0}(\xi) := a_0(\xi)$ $(\xi > 0)$—while, appealing to continuity, we also define $0\overline{a_0}(0) := 0$. Overall, denoting the extended reals by $\overline{R} := R \cup \{-\infty\} \cup \{+\infty\}$, and putting

$$\mathcal{A} := \{a : \mathcal{N}_0 \to \overline{R} \text{ such that } 0a(0) = 0\}$$

we have that $\mathcal{A}$ contains the disjoint union:

$$\{\text{all functions } a : \mathcal{N}_0 \to R\} \cup \{\overline{a_0}|_{\mathcal{N}_0} : a_0 \in \mathcal{A}_0\}.$$

We refer to $\overline{a_0}|_{\mathcal{N}_0}$ as *the member of $\mathcal{A}$ based on $a_0 \in \mathcal{A}_0$*.

We make repeated use of two simple facts. First:

$$\forall x \in \mathcal{N}_0, \ 0 \leq \log(x+1) \leq x, \tag{7}$$

equality holding in both places if, and only if, $x = 0$. Second, (3) and (5) give:

$$\forall (x, m) \in \mathcal{N}_0^2 \text{ with } x \geq m, \ x_{(m)}p(x) = \mu^m p(x-m) \tag{8}$$

so that, by definition of $\mathcal{A}$:

$$\forall m \in \mathcal{N}_0, \forall a \in \mathcal{A},\ E(X_{(m)}a(X)) = \mu^m E(a(X+m)), \tag{9}$$

equality holding trivially when $m = 0$. In particular, taking $a = 1 \in \mathcal{A}$—that is, $a(x) = 1$ $(x \in \mathcal{N}_0)$—(9) recovers, at once, the Poisson factorial moments:

$$\forall m \in \mathcal{N}_0,\ E(X_{(m)}) = \mu^m$$

whence, in further particular, we also recover:

$$E(X) = \mu,\ E(X^2) = \mu^2 + \mu \text{ and } E(X^3) = \mu^3 + 3\mu^2 + \mu. \tag{10}$$

We are ready now to prove Theorem 2.

Proof of Theorem 2. In view of (1) and (2), it suffices to show that the first two moments of S^* remain bounded as $\pi_{\min} \to 0$. By the Cauchy–Schwarz inequality, this in turn is a direct consequence of the following result. □

Lemma 1. *Let $X \sim Po(\mu)$ $(\mu > 0)$, and put $X_\mu := X\log(X/\mu)$, with $0\log 0 := 0$. Then, there exist $b^{(1)}, b^{(2)} : (0,\infty) \to (0,\infty)$ such that:*

(a) $0 \le E(X_\mu) \le b^{(1)}(\mu)$ and $0 \le E(X_\mu^2) \le b^{(2)}(\mu)$, while:

(b) for $i = 1,2:$ $b^{(i)}(\mu) \to 0$ as $\mu \to 0_+$.

Proof. By (6), $a_0^{(1)}(\xi) := \log(\xi/\mu) \in \mathcal{A}_0$. Taking $m = 1$ and $a \in \mathcal{A}$ based on $a_0^{(1)}$ in (9), and using (7), gives at once the stated bounds on $E(X_\mu)$ with $b^{(1)}(\mu) = \mu(\mu - \log\mu)$, which does indeed tend to 0 as $\mu \to 0_+$.

Further, let $a_0^{(2)}(\xi) := \xi(\log(\xi/\mu))^2$. Taking $m = 1$ and a as the restriction of $a_0^{(2)}$ to $\mathcal{N}_0$ in (9) gives $E(X_\mu^2) = \mu E(a^{(2)}(X+1))$. Noting that

$$\{x \in \mathcal{N}_0 : \log((x+1)/\mu) < 0\} = \begin{cases} \varnothing & (\mu \le 1) \\ \{0, \ldots, \overline{\mu} - 2\} & (\mu > 1) \end{cases},$$

in which $\overline{\mu}$ denotes the smallest integer greater than or equal to μ, and putting

$$B(\mu) := \begin{cases} 0 & (\mu \le 1) \\ \mu\sum_{x=0}^{\overline{\mu}-2} a^{(2)}(x+1)p(x) & (\mu > 1) \end{cases},$$

(7), (10), and l'Hôpital's rule give the stated bounds on $E(X_\mu^2)$, with

$$\begin{aligned} b^{(2)}(\mu) &= B(\mu) + \mu\sum_{x=0}^{\infty}(x+1)(x - \log\mu)^2 p(x) \\ &= B(\mu) + \mu E\{X^3 + X^2(1 - 2\log\mu) + X((\log\mu)^2 - 2\log\mu) + (\log\mu)^2\} \\ &= B(\mu) + \mu^4 + 4\mu^3 + 2\mu^2 + \mu(\log\mu)^2 + (\mu\log\mu)^2 - 2\mu(\mu+2)(\mu\log\mu) \end{aligned}$$

which, indeed, tends to 0 as $\mu \to 0_+$. □

As a result of Theorem 2, the distribution of the deviance is stable in this limit. Further, as noted in [2], each of ν, τ, and ρ can be easily and accurately approximated by standard truncate and bound methods in the limit as $\pi_{\min} \to 0$. These are detailed in Appendix B.

3. Divergences and Goodness-of-Fit

The emphasis of this section is the importance of the boundary of the extended multinomial when understanding the links between information geometric divergences and families of goodness-of-fit statistics. For completeness, a set of well-known results linking the Power-Divergence family and information geometry in the manifold sense are surveyed in Sections 3.1–3.3. The extension to the extended multinomial family is discussed in Section 3.4, where we make clear how the global behaviour of divergences is dominated by boundary effects. This complements the usual local analysis, which links divergences with the Fisher information, [8]. Perhaps the key point is that, since counts in the data can be zero, information geometric structures should also allow probabilities to be zero. Hence, closures of exponential families seem to be the correct geometric object to work on.

3.1. The Power-Divergence Family

The results of Section 2 concern the boundary behaviour of two important members of a rich class of goodness-of-fit statistics. An important unifying framework which encompasses these and other important statistics can be found in [5] (page 16) with the so-called Power-Divergence statistics. These are defined, for $-\infty < \lambda < \infty$, by

$$2NI^{\lambda}\left(\frac{n}{N} : \pi\right) := \frac{2}{\lambda(\lambda+1)} \sum_{i=0}^{k} n_i \left[\left(\frac{n_i}{N\pi_i}\right)^{\lambda} - 1\right], \tag{11}$$

with the cases $\lambda = -1, 0$ being defined by taking the appropriate limit to give

$$\lim_{\lambda \to -1} 2NI^{\lambda}\left(\frac{n}{N} : \pi\right) = 2\sum_{i=0}^{k} N\pi_i \log\left(N\pi_i / n_i\right), \ \lim_{\lambda \to 0} 2NI^{\lambda}\left(\frac{n}{N} : \pi\right) = 2\sum_{i=0}^{k} n_i \log\left(n_i / N\pi_i\right).$$

Important special cases are shown in Table 1 (whose first column is described below in Section 3.3), and we also note the case $\lambda = 2/3$, which Read and Cressie recommend [5] (page 79) as a reasonably robust statistic with an easily calculable critical value for small N. In a sense, it lies "between" the Pearson χ^2 and deviance statistics, which we compared in Section 2.

Table 1. Special cases of the Power-Divergence statistics.

$\alpha := 1 + 2\lambda$	λ	Formula	Name
3	1	$\sum_{i=0}^{k} \frac{(n_i - N\pi_i)^2}{N\pi_i}$	Pearson χ^2
7/3	2/3	$\frac{9}{5}\sum_{i=0}^{k} n_i \left[\left(\frac{n_i}{N\pi_i}\right)^{\frac{2}{3}} - 1\right]$	Read–Cressie
1	0	$2\sum_{i=0}^{k} n_i \log\left(n_i / N\pi_i\right)$	Twice log-likelihood (deviance)
0	$-\frac{1}{2}$	$4\sum_{i=0}^{k} \left(\sqrt{n_i} - \sqrt{N\pi_i}\right)^2$	Freeman–Tukey or Hellinger
-1	-1	$2\sum_{i=0}^{k} N\pi_i \log\left(N\pi_i / n_i\right)$	Twice modified log-likelihood
-3	-2	$\sum_{i=0}^{k} \frac{(n_i - N\pi_i)^2}{n_i}$	Neyman χ^2

This paper is primarily concerned with the sparse case where many of the n_i counts are zero, and we are also interested in letting probabilities, π_i, becoming arbitrarily small, or even zero.

3.2. Literature Review

Before we look at this, we briefly review the literature on the geometry of goodness-of-fit statistics. A good source for the historical developments (in the discrete context) can be found in [5] (pages 131–153) and [7]. Important examples include the analysis of contingency tables, log-linear, and discrete graphical models. Testing is often used to check the consistency of a parametric model with given data, and to check dependency assumptions such as independence between categorical variables. However, we note an important caveat: as pointed out by [14,15], the fact that a parametric model "passes" a goodness-of-fit test only weakly constrains the resulting inference. The essential point here is that goodness-of-fit is a necessary, but not sufficient, condition for model choice, since—in general—many models will be empirically supported. This issue has recently been explored geometrically in [16] using CIG.

There have been many possible test statistics proposed for goodness-of-fit testing, and one of the attractions of the Power-Divergence family, defined in (11), is that the most important ones are included in the family and indexed by a single scalar λ. Of course, when there is a choice of test statistic, different inferences can result from different choices. One of the main themes of [5] is to give the analyst insight about selecting a particular λ. Key considerations for making the selection of λ include the tractability of the sampling distribution, its power against important alternatives, and interpretation when hypotheses are rejected.

The first order, asymptotic in N, χ^2-sampling distribution for all members of the Power-Divergence family, which is appropriate when all observed counts are "large enough", is the most commonly used tool, and a very attractive feature of the family. However, this can fail badly in the "sparse" case and when the model is close to the boundary. Elementary, moment based corrections, to improve small sample performance, are discussed in [5] (Chapter 5). More formal asymptotic approaches to these issues include the doubly asymptotic, in N and k, approach of [17], discussed in Section 2 and similar normal approximation ideas in [18]. See also [19]. Extensive simulation experiments have been undertaken to learn in practice what 'large enough' means, see [5,20,21].

When there are nuisance parameters to be estimated (as is common), [22] points out that it is the sampling distribution *conditional* upon these estimates which needs to be approximated, and proposes higher order methods based on the Edgeworth expansion. Simulation approaches are often used in the conditional context due to the common intractability of the conditional distribution [23,24], and importance sampling methods play an important role—see [25–27]. Other approaches used to investigate the sampling distribution include jackknifing [28], the Chen–Stein method [29], and detailed asymptotic analysis in [30–32].

In very high dimensional model spaces, considerations of the power of tests rarely generates uniformly best procedures but, we feel, geometry can be an important tool in understanding the choices that need to be made. Further, [5], states the situation is "complicated", showing this through simulation experiments. One of the reasons for Read and Cressie's preferred choice of $\lambda = 2/3$ is its good power against some important types of alternative–the so-called bump or dip cases–as well as the relative tractability of its sampling distribution under the null. Other considerations about power can be found in [33] which looks specifically at mixture model based alternatives.

3.3. Links with Information Geometry

At the time that the Power-Divergence family was being examined, there was a parallel development in Information Geometry; oddly, however, it seemed to have taken some time before the links between the two areas were fully recognised. A good treatment of these links can be found in [6] (Chapter 9). Since it is important to understand the extreme values of divergence functions, considerations of convexity can clearly play an important role. The general class of Bregman divergences, [6,34] (page 240), and [35] (page 13) is very useful here. For each Bregman divergence, there will exist affine parameters of the exponential family in which the divergence function is convex. In the class of product Poisson models—which are the key building blocks of log–linear models—all members of the Power-Divergence family have the Bregman property. These are then α-divergences, capable of generating the complete Information Geometry of the model [35], with the link between α and λ given in Table 1. The α-representation highlights the duality properties, which are a cornerstone of Information Geometry, but which is rather hidden in the λ representation. The Bregman divergence representation for the Poisson is given in Table 2. The divergence parameter—in which we have convexity—is shown for each λ, as is the so-called potential function, which generates the complete information geometry for these models.

Table 2. Power-Divergence in the Poisson model with mean μ, where $\lambda^* = 1 - \lambda$.

λ	α	Divergence $D_\lambda(\mu_1, \mu_2)$	Divergence Parameter ξ	Potential
-1	-1	$\mu_1 - \mu_2 - \mu_2(\log(\mu_1) - \log(\mu_2))$	$\xi = \log(\mu)$	$\exp(\xi)$
0	1	$\mu_2 - \mu_1 - \mu_1(\log(\mu_2) - \log(\mu_1))$	$\xi = \mu$	$\xi\log(\xi) - \xi$
$\lambda \neq 0, -1$	$\alpha \neq \pm 1$	$\frac{\left(\lambda^*\mu_1 - \lambda^*\mu_2 - \mu_2\left(\left(\frac{\mu_1}{\mu_2}\right)^{\lambda^*} - 1\right)\right)}{\lambda^*(1-\lambda^*)}$	$\xi = \frac{1}{\lambda^*}\mu^{\lambda^*}$	$\frac{(\lambda^*\xi)^{1/\lambda^*}}{1-\lambda^*}$

3.4. Extended Multinomial Case

In this paper, we are focusing on the class of log–linear models where the multinomial is the underlying class of distributions; that is, we condition on the sample size, N, being fixed in the product Poisson space. In particular, we focus on extended multinomials, which includes the closure of the multinomials, so we have a boundary. Due to the conditioning (which induces curvature), only the cases where $\lambda = 0, -1$ remain Bregman divergences, but all are still divergences in the sense of being Csiszár f-divergences [36,37].

The closure of an exponential family (e.g., [11,38–40]), and its application in the theory of log–linear models has been explored in [12,13,41,42]. The key here is understanding the limiting behaviour in the natural—$\alpha = 1$ in the sense of [8]—parameter space. This can be done by considering the polar dual [43], or, alternatively, the directions of recession—[12] or [42]. The boundary polytope determines key statistical properties of the model, including the behaviour of the sampling distribution of (functions of) the MLE and the shape of level sets of divergence functions.

Figures 1 and 2 show level sets of the $\alpha = \pm 1$ Power-Divergences in the $(+1)$-affine and (-1)-affine parameters (Panels (a) and (b), respectively) for the $k = 2$ extended multinomial model. The boundary polytope in this case is a simple triangle "at infinity", and the shape of this is strongly reflected in the behaviour of the level sets. In Figure 1, we show—in the simplex $\left\{(\pi_0, \pi_1, \pi_2) | \sum_{i=0}^{2} \pi_i = 1, \pi_i \geq 0\right\}$—the level sets of the $\alpha = -1$ divergence, which, in the Csiszár f-divergence form, is

$$K(\pi^0, \pi) := \sum_{i=0}^{2} \log\left(\frac{\pi_i^0}{\pi_i}\right)\pi_i^0.$$

The figures show how in Panel (a), the directions of recession dominate the shape of level sets, and in Panel (b) the duals of these directions (i.e., the vertices of the simplex) each have different maximal behaviour. The lack of convexity of the level sets in Panel (a) corresponds to the fact that the natural parameters are not the affine divergence parameters for this divergence, so we do not expect convex behaviour. In Panel (b), we do get non-convex level sets, as expected.

Figure 2 shows the same story, but this time for the dual divergence,

$$K^*(\pi, \pi^0) := K(\pi^0, \pi).$$

Now, the affine divergence parameters are shown in Panel (a), the natural parameters. We see that in the limit the shape of the divergence is converging to that of the polar of the boundary polytope. In general, local behaviour is quadratic, but boundary behaviour is polygonal.

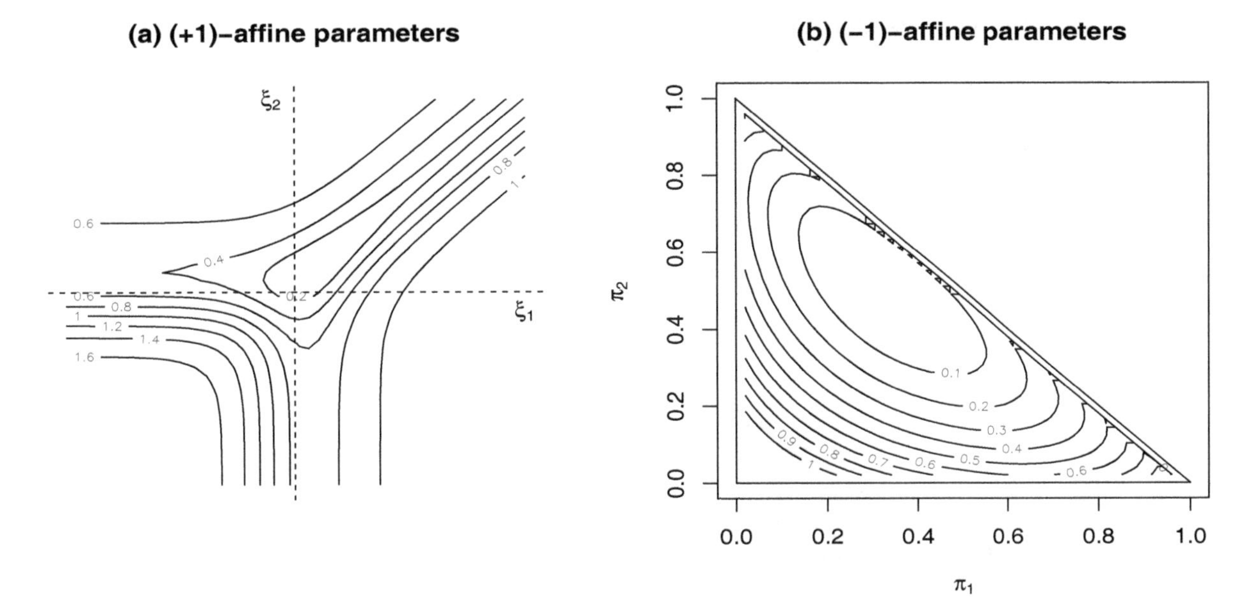

Figure 1. Level sets of $K(\pi^0, \pi)$, for fixed $\pi^0 = (\frac{1}{6}, \frac{2}{6}, \frac{3}{6})$ in: (**a**) the natural parameters, and (**b**) the mean parameters.

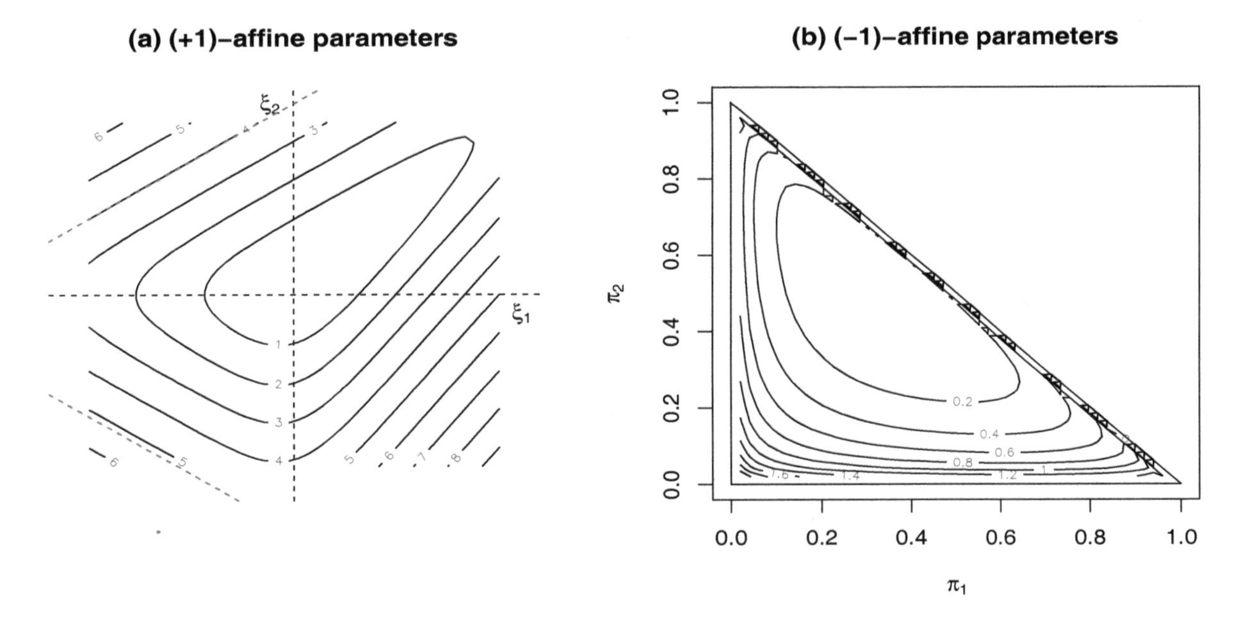

Figure 2. Level sets of $K^*(\pi^0, \pi)$, for fixed $\pi^0 = (\frac{1}{6}, \frac{2}{6}, \frac{3}{6})$ in: (**a**) the natural parameters, and (**b**) the mean parameters.

4. Simulation Studies

In this section, we undertake simulation studies to numerically explore what has been discussed above. Separate sub-sections address three general topics—focusing on one particular instance of each, as follows:

1. The transition as (N, k) varies between discrete and continuous features of the sampling distributions of goodness-of-fit statistics—focusing on the behaviour of the deviance at the uniform discrete distribution;
2. The comparative behaviour of a range of Power-Divergence statistics—focusing on the relative stability of their sampling distributions near the boundary;
3. The lack of uniformity—across the parameter space—of the finite sample adequacy of standard asymptotic sampling distributions, focusing on testing independence in 2×2 contingency tables.

For each topic, the results presented invite further investigation.

4.1. Transition Between Discrete and Continuous Features of Sampling Distributions

Earlier work [2] used the decomposition:

$$D^*/2 = \sum_{\{0 \le i \le k : n_i^* > 0\}} n_i^* \log(n_i^*/\mu_i) = \Gamma^* + \Delta^*,$$

$$\Gamma^* := \sum_{i=0}^{k} \alpha_i n_i^* \text{ and } \Delta^* := \sum_{\{0 \le i \le k : n_i^* > 1\}} n_i^* \log n_i^* \ge 0, \text{ where } \alpha_i := -\log \mu_i,$$

to show that a particularly bad case for the adequacy of any continuous approximation to the sampling distribution of the deviance $D := D^*|(N^* = N)$ is the uniform discrete distribution: $\pi_i = 1/(k+1)$. In this case, the Γ^* term contributes a constant to the deviance, while the Δ^* term has no contributions from cells with 0 or 1 observations—these being in the vast majority in the $N << k$ situation considered here. In other words, *all* of the variability in D comes from that between the $n_i \log n_i$ values for the (relatively rare) cell counts above 1. This gives rise to a discreteness phenomenon termed "granularity" in [2], whose meaning was conveyed graphically there in the case $N = 30$ and $k = 200$. Work by Holst [19] predicts that continuous (indeed, normal) approximations will improve with larger values of N/k, as is intuitive. Remarkably, simply doubling the sample size to $N = 60$ was shown in [2] to be sufficient to give a good enough approximation for most goodness-of-fit testing purposes. In other words, N being 30% of $k = 200$ was found to be good enough for practical purposes.

Here, we illustrate the role of k-asymptotics (Section 2) in this transition between discrete and continuous features by repeating the above analyses for different values of k. Figures 3 and 4 (where $k = 100$ while $N = 20$ and 40, respectively) are qualitatively the same as those presented in [2]. The difference here is that the smaller value of k means that a higher value of N/k (40%) is needed in Figure 4 to adequately remove the granularity evident in Figure 3. For $k = 400$, the figures with $N = 50$ and $N = 100$ (omitted here for brevity) are, again, qualitatively the same as in [2]—the larger value of k needing only a smaller value of N/k (25%) for practical purposes. Note the QQ-plots used in these two figures are relative to normal quantiles.

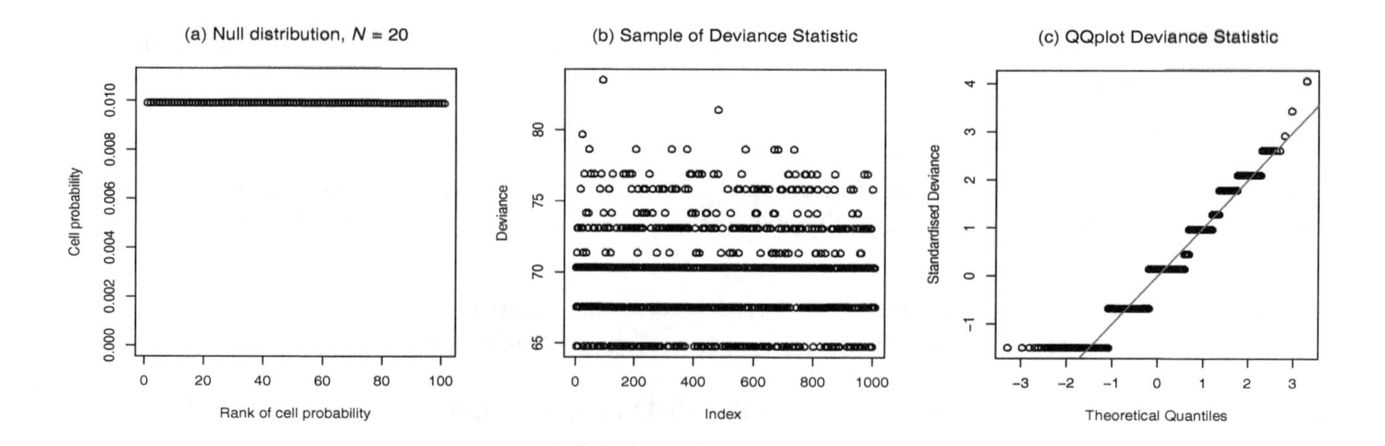

Figure 3. $k = 100$, $N = 20$.

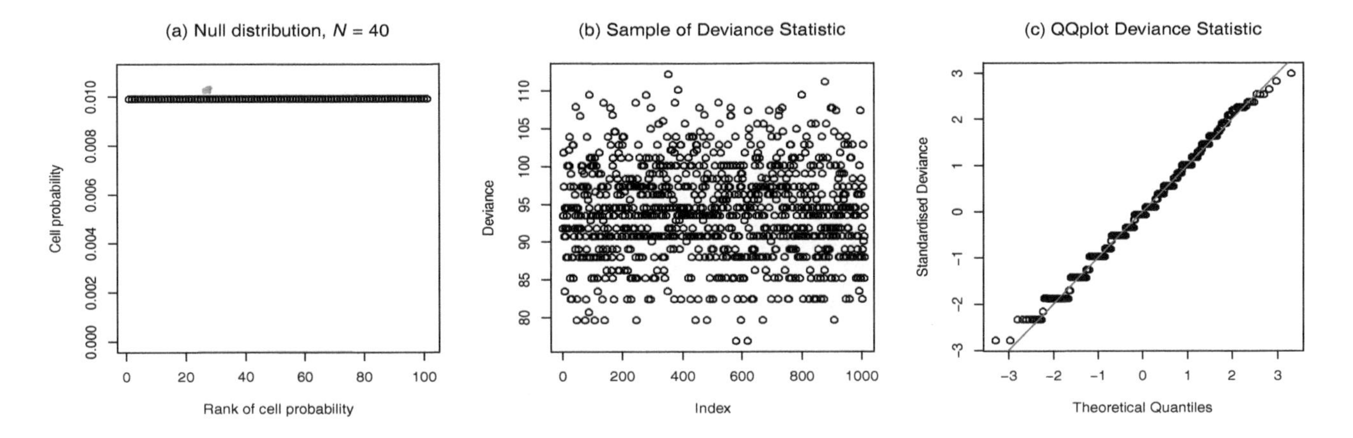

Figure 4. $k = 100$, $N = 40$.

The results of this section show the universality of boundary effects. The simulations of Figures 3 and 4 are undertaken under the uniform model, which might be felt to be far from the boundary. In fact, the results show that in the high dimensional, low sample size case, all distributions are "close to" the boundary, and that discretisation effects can dominate.

4.2. Comparative Behaviour of Power-Divergence Statistics near the Boundary

Here we study the relative stability—near the boundary of the simplex—of the sampling distributions of a range of Power-Divergence statistics indexed by Amari's parameter α. Figure 5 shows histograms for six different values of α, $N = 50$, $k = 200$, and exponentially decreasing values of $\{\pi_i\}$, as plotted in Figure 6. In it, red lines depict kernel density estimates using the bandwidth suggested in [44].

These sampling distributions differ markedly. The instability for $\alpha = 3$ expected from Theorem 1 is clearly visible: very large values contribute to high variance and skewness. Analogous instability features (albeit at a lower level) remain with the Cressie–Read recommended value $\alpha = 7/3$. In contrast (as expected from the discussion around Theorem 2), the distribution of the deviance ($\alpha = 1$) is stable and roughly normal. Lower values of α retain these same features.

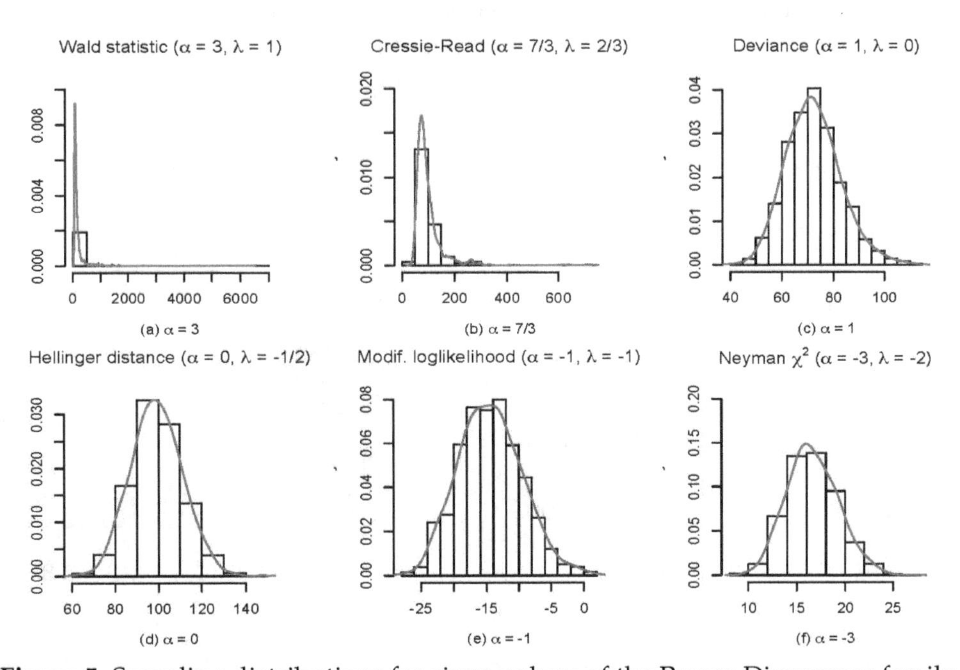

Figure 5. Sampling distributions for six members of the Power-Divergence family.

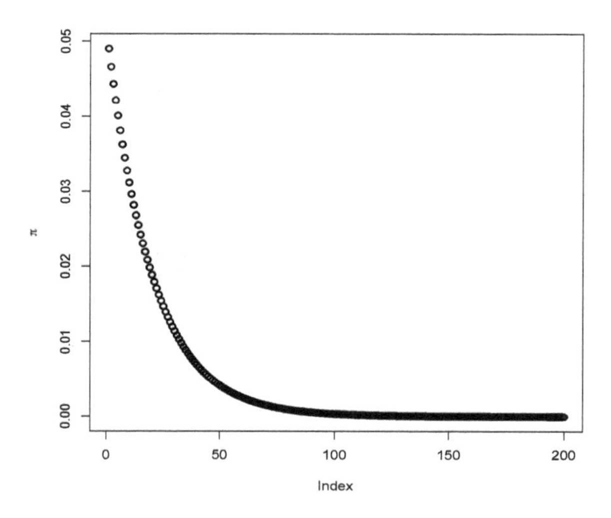

Figure 6. Exponentially decreasing values of π_i.

4.3. Variation in Finite Sample Adequacy of Asymptotic Distributions across the Parameter Space

Pearson's χ^2 statistic ($\alpha = 3$) is widely used to test independence in contingency tables, a standard rule-of-thumb for its validity being that each expected cell frequency should be at least 5. For illustrative purposes, we consider 2×2 contingency tables, the relevant N-asymptotic null distribution being χ_1^2. We assess the adequacy of this asymptotic approximation by comparing nominal and actual significance levels of this test, based on 10,000 replications. Particular interest lies in how these actual levels vary across different data generation processes within the same null hypothesis of independence.

Figures 7 and 8 show the actual level of the Pearson χ^2 test for nominal levels 0.1 and 0.05 for sample sizes $N = 20$ and $N = 50$, with π_r and π_c denoting row and column probabilities, respectively. The above general rule applies only at the central black dot in Figure 7, and inside the closed black curved region in Figure 8. The actual level was computed for all pairs of values of π_r and π_c, averaged using the symmetry of the parameter space, and smoothed using the kernel smoother for irregular 2D data (implemented in the package *fields* in R). In each case, the white tone contains the nominal level, while red tones correspond to liberal and blue tones to conservative actual levels.

The finite sample adequacy of this standard asymptotic test clearly varies across the parameter space. In particular, its nominal and actual levels agree well at some parameter values outside the standard rule-of-thumb region; and, conversely, disagree somewhat at other parameter values inside it. Intriguingly, the agreement between nominal and actual levels does not improve everywhere with sample size. Overall, the clear patterns evident in this lack of uniformity invite further theoretical investigation.

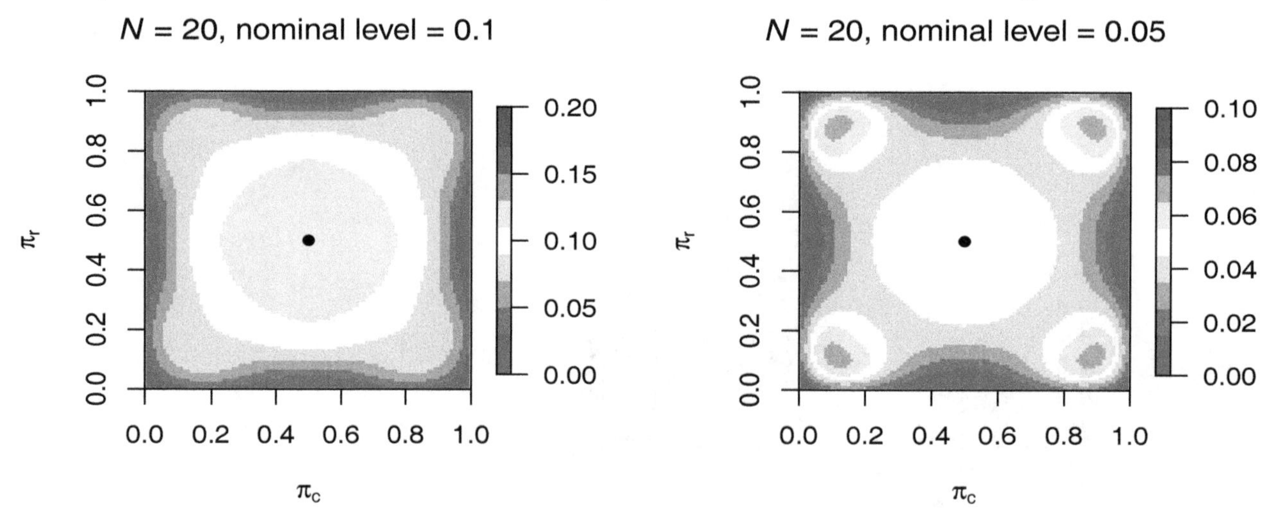

Figure 7. Heatmap of the actual level of the test for $N = 20$ at nominal levels 0.1 and 0.05; the standard rule-of-thumb (where expected counts are greater than 5) applies only at the black dot.

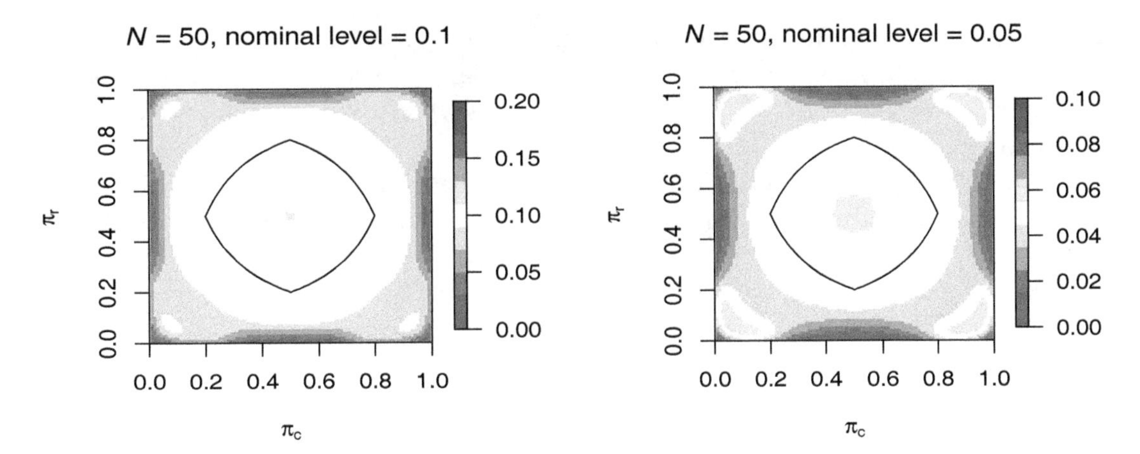

Figure 8. Heatmap of the actual level of the test for $N = 50$ at nominal levels 0.1 and 0.05; the standard rule-of-thumb (where expected counts are greater than 5) applies inside the closed black curved region.

5. Discussion

This paper has illustrated the key importance of working with the boundary of the closure of exponential families when studying goodness-of-fit testing in the high dimensional, low sample size context. Some of this work is new (Section 2), while some uses the structure of extended exponential families to add insight to standard results in the literature (Section 3). The last section, Section 4, uses simulation studies to start to explore open questions in this area.

One open question—related to the results of Theorems 1 and 2—is to see if a unified theory, for all values of α, and over large classes of extended exponential families, can be developed.

Acknowledgments: The authors would like to thank the EPSRC for the support of grant number EP/L010429/1. Germain Van Bever would also like to thank FRS-FNRS for its support through the grant FC84444. We would also like to thank the referees for very helpful comments.

Author Contributions: All four authors made critical contributions to the paper. R.S. made key contribution to, especially, Section 4. P.M. and F.C. provided the overall structure and key content details of the paper. G.V.B. provided invaluable suggestions throughout.

Appendix A. Proof of Theorem 1

We start by noting an important recurrence relation which will be exploited in the computations below. By definition, for any $t := (t_i) \in R^{k+1}$, $n = (n_i)$ has moment generating function

$$M(t;N) := E\{\exp(t^T n)\} = [m(t)]^N$$

with $m(t) = \sum_{i=0}^k a_i$ and $a_i = a_i(t_i) = \pi_i e^{t_i}$. Putting

$$f_{N,i}(t;r) := N_{(r)}\,[m(t)]^{N-r}\,a_i^r \quad (0 \le r \le N),$$

where

$$N_{(r)} := {}^N P_r = \begin{cases} 1 & \text{if } r = 0 \\ N(N-1)...(N-(r-1)) & \text{if } r \in \{1,...,N\} \end{cases},$$

we have

$$M(t;N) = f_{N,i}(t;0) \quad (0 \le i \le k) \tag{A1}$$

and the recurrence relation:

$$\frac{\partial f_{N,i}(t;r)}{\partial t_i} = f_{N,i}(t;r+1) + r f_{N,i}(t;r) \quad (0 \le i \le k;\ 0 \le r < N). \tag{A2}$$

When there is no risk of confusion, we may abbreviate $M(t;N)$ to M and $f_{N,i}(t;r)$ to $f_N(r)$, or even to $f(r)$—so that (A1) becomes $M = f(0)$. Again, we may write $\partial^r M(t;N)/\partial t_i^r$ as M_r, $\partial^{r+s} M(t;N)/\partial t_i^r \partial t_j^s$ as $M_{r,s}$ and $\partial^{r+s+u} M(t;N)/\partial t_i^r \partial t_j^s \partial t_l^u$ as $M_{r,s,u}$, with similar conventions for higher order mixed derivatives.

We can now use this to explicitly calculate low order moments of the count vectors. Using $E(n_i^r) = \partial^r M(t;N)/\partial t_i^r|_{t=0}$, the first N moments of n_i now follow from (A1) and repeated use of (A2), noting that $m(0) = 1$ and $a_i(0) = \pi_i$.

In particular, the first 6 moments of each n_i can be obtained as follows, where $N \geq 6$ is assumed. Using (A1) and (A2), we have

$$\begin{aligned}
M_1 &= f(1)\\
M_2 &= f(2) + f(1)\\
M_3 &= f(3) + 2f(2) + f(2) + f(1) = f(3) + 3f(2) + f(1)\\
M_4 &= f(4) + 6f(3) + 7f(2) + f(1)\\
M_5 &= f(5) + 10f(4) + 25f(3) + 15f(2) + f(1)\\
M_6 &= f(6) + 15f(5) + 65f(4) + 90f(3) + 31f(2) + f(1).
\end{aligned}$$

Substituting in, we have

$$\begin{aligned}
E(n_i) &= N\pi_i\\
E(n_i^2) &= N_{(2)}\pi_i^2 + N\pi_i\\
E(n_i^3) &= N_{(3)}\pi_i^3 + 3N_{(2)}\pi_i^2 + N\pi_i\\
E(n_i^4) &= N_{(4)}\pi_i^4 + 6N_{(3)}\pi_i^3 + 7N_{(2)}\pi_i^2 + N\pi_i\\
E(n_i^5) &= N_{(5)}\pi_i^5 + 10N_{(4)}\pi_i^4 + 25N_{(3)}\pi_i^3 + 15N_{(2)}\pi_i^2 + N\pi_i\\
E(n_i^6) &= N_{(6)}\pi_i^6 + 15N_{(5)}\pi_i^5 + 65N_{(4)}\pi_i^4 + 90N_{(3)}\pi_i^3 + 31N_{(2)}\pi_i^2 + N\pi_i.
\end{aligned}$$

This can be formalised in the following Lemma

Lemma A1. *The integer coefficients in any expansion*

$$M_r = \sum_{s=1}^{r} c_r(s) f(s) \quad (1 \leq r \leq N)$$

can be computed using $c_r(1) = c_r(r) = 1$ together, for $r \geq 3$, with the update:

$$c_r(s) = c_{r-1}(s-1) + s c_{r-1}(s) \quad (1 < s < r).$$

We note that if M_r is required for $r > N$, we may repeatedly differentiate

$$M_N = \sum_{s=1}^{N} c_N(s) f(s)$$

w.r.t. t_i, noting that $f(N) = N! a_i^N$ no longer depends on $m(t)$ so that, for all $h > 0$, $\partial^h f(N)/\partial t_i^h = N^h f(N)$.

Mixed moments of any order can be derived from those of lower order, exploiting the fact that a_i depends on t only via t_i. We illustrate this by deriving those required for the second and third moments of W.

First consider the mixed moments required for the second moment of W. Of course, $Var(W) = 0$ if $k = 0$. Otherwise, $k > 0$, and computing $Var(W)$ requires $E(n_i^2 n_j^2)$ for $i \neq j$. We find this as follows, assuming $N \geq 4$.

The relation $M_2 = f(2) + f(1)$ established above gives

$$\partial^2 M/\partial t_j^2 = N_{(2)} a_j^2 f_{N-2}(0) + N a_j f_{N-1}(0). \tag{A3}$$

Repeated use of (A3) now gives

$$M_{2,2} = N_{(4)} a_i^2 a_j^2 f_{N-4}(0) + N_{(3)} a_i a_j (a_i + a_j) f_{N-3}(0) + N_{(2)} a_i a_j f_{N-2}(0) \tag{A4}$$

so that

$$E(n_i^2 n_j^2) = N_{(4)} \pi_i^2 \pi_j^2 + N_{(3)} \pi_i \pi_j (\pi_i + \pi_j) + N_{(2)} \pi_i \pi_j.$$

We further look at the mixed moments needed for the third moment of W. For the skewness of W, we need $E(n_i^2 n_j^4)$ for $i \neq j$ and, when $k > 1$, $E(n_i^2 n_j^2 n_l^2)$ for i, j, l distinct. We find these similarly, as follows, assuming $k > 1$ and $N \geq 6$.

Equation (A4) above gives

$$\partial^2 M/\partial t_j^2 \partial t_l^2 = N_{(4)} a_j^2 a_l^2 f_{N-4}(0) + N_{(3)} a_j a_l (a_j + a_l) f_{N-3}(0) + N_{(2)} a_j a_l f_{N-2}(0)$$

from which, using (A3) repeatedly, we have

$$\begin{aligned} M_{2,2,2} &= a_j^2 a_l^2 \{N_{(6)} a_i^2 f_{N-6}(0) + N_{(5)} a_i f_{N-5}(0)\} + a_j a_l (a_j + a_l) \{N_{(5)} a_i^2 f_{N-5}(0) + N_{(4)} a_i f_{N-4}(0)\} + \\ &\quad a_j a_l \{N_{(4)} a_i^2 f_{N-4}(0) + N_{(3)} a_i f_{N-3}(0)\} \\ &= N_{(6)} a_i^2 a_j^2 a_l^2 f_{N-6}(0) + N_{(5)} a_i a_j a_l \{a_i a_j + a_j a_l + a_l a_i\} f_{N-5}(0) + N_{(4)} a_i a_j a_l \{a_i + a_j + a_l\} f_{N-4}(0) + \\ &\quad N_{(3)} a_i a_j a_l f_{N-3}(0) \end{aligned}$$

so that $E(n_i^2 n_j^2 n_l^2)$ equals

$$N_{(6)} \pi_i^2 \pi_j^2 \pi_l^2 + N_{(5)} \pi_i \pi_j \pi_l \{\pi_i \pi_j + \pi_j \pi_l + \pi_l \pi_i\} + N_{(4)} \pi_i \pi_j \pi_l \{\pi_i + \pi_j + \pi_l\} + N_{(3)} \pi_i \pi_j \pi_l.$$

Finally, the relation $M_4 = f(4) + 6f(3) + 7f(2) + f(1)$ established above gives

$$\partial^4 M/\partial t_j^4 = N_{(4)} a_j^4 f_{N-4}(0) + 6N_{(3)} a_j^3 f_{N-3}(0) + 7N_{(2)} a_j^2 f_{N-2}(0) + N a_j f_{N-1}(0)$$

so that, again using (A3) repeatedly, yields

$$E(n_i^2 n_j^4) = N_{(6)} \pi_i^2 \pi_j^4 + N_{(5)} \pi_i \pi_j^3 (6\pi_i + \pi_j) + N_{(4)} \pi_i \pi_j^2 (7\pi_i + 6\pi_j) + N_{(3)} \pi_i \pi_j (\pi_i + 7\pi_j) + N_{(2)} \pi_i \pi_j.$$

Combining above results, we obtain here the first three moments of W. Higher moments may be found similarly.

We first look at $E(W)$. We have $W = \frac{1}{N^2} \sum_{i=0}^{k} \frac{n_i^2}{\pi_i} - 1$ and $E(n_i^2) = N_{(2)} \pi_i^2 + N\pi_i$, so that

$$E(W) = \frac{N_{(2)}}{N^2} + \frac{(k+1)}{N} - 1 = \frac{k}{N}.$$

The variance is computed by recalling that $N^2(W+1) = \sum_i \frac{n_i^2}{\pi_i}$, while $E(W) = \frac{k}{N}$,

$$Var(W) = Var(W+1) = \frac{A^{(2)}}{N^4} - \left(\frac{k}{N} + 1\right)^2,$$

where

$$A^{(2)} := N^4 E\{(W+1)^2\} = \sum_i \frac{E(n_i^4)}{\pi_i^2} + \sum\sum_{i \neq j} \frac{E(n_i^2 n_j^2)}{\pi_i \pi_j}.$$

Using expressions for $E(n_i^4)$ and $E(n_i^2 n_j^2)$ established above, and putting

$$\pi^{(\alpha)} := \sum_i \pi_i^{\alpha},$$

we have

$$\begin{aligned}\sum_i \frac{E(n_i^4)}{\pi_i^2} &= \sum_i \{N_{(4)}\pi_i^2 + 6N_{(3)}\pi_i + 7N_{(2)} + N\pi_i^{-1}\} \\ &= N_{(4)}\pi^{(2)} + 6N_{(3)} + 7N_{(2)}(k+1) + N\pi^{(-1)}\end{aligned}$$

and

$$\begin{aligned}\sum\sum_{i\neq j} \frac{E(n_i^2 n_j^2)}{\pi_i \pi_j} &= \sum_{i\neq j}\{N_{(4)}\pi_i\pi_j + N_{(3)}(\pi_i + \pi_j) + N_{(2)}\} \\ &= N_{(4)}(1-\pi^{(2)}) + 2N_{(3)}k + N_{(2)}k(k+1),\end{aligned}$$

so that

$$A^{(2)} = N_{(4)} + 2N_{(3)}(k+3) + N_{(2)}(k+1)(k+7) + N\pi^{(-1)},$$

whence

$$\begin{aligned}Var(W) &= \frac{N_{(4)} + 2N_{(3)}(k+3) + N_{(2)}(k+1)(k+7) + N\pi^{(-1)}}{N^4} - \left(1 + \frac{k}{N}\right)^2 \\ &= \frac{\left\{\pi^{(-1)} - (k+1)^2\right\} + 2k(N-1)}{N^3}, \quad \text{after some simplification.}\end{aligned}$$

Note that $Var(W)$ depends on (π_i) *only* via $\pi^{(-1)}$ while, by strict convexity of $x \to 1/x$ $(x > 0)$,

$$\pi^{(-1)} \geq (k+1)^2, \text{ equality holding iff } \pi_i \overset{i}{\equiv} 1/(k+1).$$

Thus, for given k and N, $Var(W)$ is strictly increasing as (π_i) departs from uniformity, tending to ∞ as one or more $\pi_i \to 0_+$.

Finally, for these calculations, we look at $E[\{W - E(W)\}^3]$. Recalling again that $N^2(W+1) = \sum_i \frac{n_i^2}{\pi_i}$,

$$\begin{aligned}E[\{W - E(W)\}^3] &= E[\{(W+1) - E(W+1)\}^3] \\ &= N^{-6}A^{(3)} - 3\,Var(W)(E(W)+1) - (E(W)+1)^3,\end{aligned}$$

where $A^{(3)} := N^6 E\{(W+1)^3\}$ is given by

$$A^{(3)} = \sum_i \frac{E(n_i^6)}{\pi_i^3} + 3\sum\sum_{i\neq j} \frac{E(n_i^2 n_j^4)}{\pi_i \pi_j^2} + \sum\sum\sum_{i,j,l \text{ distinct}} \frac{E(n_i^2 n_j^2 n_l^2)}{\pi_i \pi_j \pi_l}.$$

Given that

$$E(W) = k/N \text{ and } Var(W) = \frac{\left\{\pi^{(-1)} - (k+1)^2\right\} + 2k(N-1)}{N^3},$$

it suffices to find $A^{(3)}$.

Using expressions for $E(n_i^6)$, $E(n_i^2 n_j^2 n_l^2)$, and $E(n_i^2 n_j^4)$ established above, we have

$$\sum_i \frac{E(n_i^6)}{\pi_i^3} = N_{(6)}\pi^{(3)} + 15N_{(5)}\pi^{(2)} + 65N_{(4)} + 90N_{(3)}(k+1) + 31N_{(2)}\pi^{(-1)} + N\pi^{(-2)}$$

$$\sum\sum_{i \neq j} \frac{E(n_i^2 n_j^4)}{\pi_i \pi_j^2} = N_{(6)}\pi_i\pi_j^2 + N_{(5)}\pi_j(6\pi_i + \pi_j) + N_{(4)}(7\pi_i + 6\pi_j) + N_{(3)}(\pi_i/\pi_j + 7) + N_{(2)}\pi_j^{-1}$$
$$= N_{(6)}\{\pi^{(2)} - \pi^{(3)}\} + N_{(5)}\{6 + (k-6)\pi^{(2)}\} +$$
$$13N_{(4)}k + N_{(3)}\{\pi^{(-1)} + (7k-1)(k+1)\} + N_{(2)}k\pi^{(-1)}$$

and

$$\sum\sum\sum_{i,j,l \text{ distinct}} \frac{E(n_i^2 n_j^2 n_l^2)}{\pi_i \pi_j \pi_l} = N_{(6)}\{1 + 2\pi^{(3)} - 3\pi^{(2)}\} + 3N_{(5)}(k-1)\{1 - \pi^{(2)}\} +$$
$$3N_{(4)}k(k-1) + N_{(3)}k(k^2-1)$$

so that, after some simplification,

$$A^{(3)} = N_{(6)} + 3N_{(5)}(k+5) + N_{(4)}\{3k(k+12) + 65\} +$$
$$N_{(3)}\{k^3 + 21k^2 + 107k + 87\} + 3N_{(3)}\pi^{(-1)} + N_{(2)}(31 + 3k)\pi^{(-1)} + N\pi^{(-2)}.$$

Substituting in and simplifying, we find $E[\{W - E(W)\}^3]$ to be:

$$\frac{\left\{\pi^{(-2)} - (k+1)^3\right\} - (3k + 25 - 22N)\left\{\pi^{(-1)} - (k+1)^2\right\} + g(k,N)}{N^5},$$

where

$$g(k,N) = 4(N-1)k(k + 2N - 5) > 0.$$

Note that $E[\{W - E(W)\}^3]$ depends on (π_i) *only* via $\pi^{(-1)}$ and the *larger* quantity $\pi^{(-2)}$. In particular, for given k and N, the skewness of W tends to $+\infty$ as one or more $\pi_i \to 0_+$.

Appendix B. Truncate and Bound Approximations

In the notation of Lemma 1, it suffices to find truncate and bound approximations for each of $E(X_\mu)$, $E(X.X_\mu)$, and $E(X_\mu^2)$.

For all r, s in $\mathcal{N}$, define $h_{r,s}(\mu) := E\{(\log(X + r))^s\}$. Appropriate choices of $m \in \mathcal{N}_0$ and $a \in \mathcal{A}$ in (9), together with (10), give:

$$E(X_\mu) = \mu h_{1,1}(\mu) - \mu \log \mu,$$
$$E(X.X_\mu) = \{\mu^2 h_{2,1}(\mu) + \mu h_{1,1}(\mu)\} - (\mu^2 + \mu)\log \mu, \text{ and:}$$
$$E(X_\mu^2) = \mu^2 h_{2,2}(\mu) + \mu h_{1,2}(\mu) + (\mu^2 + \mu)(\log \mu)^2 - 2\log \mu\{\mu^2 h_{2,1}(\mu) + \mu h_{1,1}(\mu)\},$$

so that it suffices to truncate and bound $h_{r,s}(\mu)$ for $r, s \in \{1, 2\}$.

For all r, s in $\mathcal{N}$, and for all $m \in \mathcal{N}_0$, we write:

$$h_{r,s}(\mu) = h_{r,s}^{[m]}(\mu) + \varepsilon_{r,s}^{[m]}(\mu)$$

in which:

$$h_{r,s}^{[m]}(\mu) := \sum_{x=0}^{m}\{(\log(x + r))^s\}p(x) \text{ and } \varepsilon_{r,s}^{[m]}(\mu) := \sum_{x=m+1}^{\infty}\{(\log(x + r))^s\}p(x).$$

Using again (7), the "error term" $\varepsilon_{r,s}^{[m]}(\mu)$ has lower and upper bounds:

$$0 < \varepsilon_{r,s}^{[m]}(\mu) < \bar{\varepsilon}_{r,s}^{[m]}(\mu) := \sum_{x=m+1}^{\infty}(x + (r-1))^s p(x).$$

Restricting attention now to $r, s \in \{1,2\}$, as we may, and requiring $m \geq s$ so that $F^{[m-s]}(\mu)$ given by (4) is defined, (8) gives:

$$\bar{\varepsilon}_{1,1}^{[m]}(\mu) = \sum_{x=m+1}^{\infty} x p(x) = \mu \sum_{x=m}^{\infty} p(x) = \mu\{1 - F^{[m-1]}(\mu)\},$$

$$\bar{\varepsilon}_{2,1}^{[m]}(\mu) = \sum_{x=m+1}^{\infty} (x+1) p(x) = \bar{\varepsilon}_{1,1}^{[m]}(\mu) + \{1 - F^{[m]}(\mu)\},$$

$$\begin{aligned} \bar{\varepsilon}_{1,2}^{[m]}(\mu) &= \sum_{x=m+1}^{\infty} x^2 p(x) = \sum_{x=m+1}^{\infty} \{x(x-1) + x\} p(x) \\ &= \mu^2\{1 - F^{[m-2]}(\mu)\} + \bar{\varepsilon}_{1,1}^{[m]}(\mu) \end{aligned}$$

and:

$$\begin{aligned} \bar{\varepsilon}_{2,2}^{[m]}(\mu) &= \sum_{x=m+1}^{\infty} (x+1)^2 p(x) = \sum_{x=m+1}^{\infty} \{x^2 + (x+1) + x\} p(x) \\ &= \bar{\varepsilon}_{1,2}^{[m]}(\mu) + \bar{\varepsilon}_{2,1}^{[m]}(\mu) + \bar{\varepsilon}_{1,1}^{[m]}(\mu). \end{aligned}$$

Accordingly, for given μ, each $\bar{\varepsilon}_{r,s}^{[m]}(\mu)$ decreases strictly to zero with m providing—to any desired accuracy—truncate and bound approximations for each of ν, τ, and ρ. In this connection, we note that the upper tail probabilities involved here can be bounded by standard Chernoff arguments.

References

1. Critchley, F.; Marriott, P. Computational Information Geometry in Statistics: Theory and practice. *Entropy* **2014**, *16*, 2454–2471.
2. Marriott, P.; Sabolova, R.; Van Bever, G.; Critchley, F. Geometry of goodness-of-fit testing in high dimensional low sample size modelling. In *Geometric Science of Information: Second International Conference, GSI 2015, Palaiseau, France, October 28–30, 2015, Proceedings*; Nielsen, F., Barbaresco, F., Eds.; Springer: Berlin, Germany, 2015; pp. 569–576.
3. Amari, S.-I.; Nagaoka, H. *Methods of Information Geometry*; Translations of Mathematical Monographs; American Mathematical Society: Providence, RI, USA, 2000.
4. Cressie, N.; Read, T.R.C. Multinomial goodness-of-fit tests. *J. R. Stat. Soc. B* **1984**, *46*, 440–464.
5. Read, T.R.C.; Cressie, N.A.C. *Goodness-of-Fit Statistics for Discrete Multivariate Data*; Springer: New York, NY, USA, 1988.
6. Kass, R.E.; Vos, P.W. *Geometrical Foundations of Asymptotic Inference*; John Wiley & Sons, Inc.: Hoboken, NJ, USA, 1997.
7. Agresti, A. *Categorical Data Analysis*, 3rd ed.; Wiley: Hoboken, NJ, USA, 2013.
8. Amari, S.-I. Differential-geometrical methods in statistics. In *Lecture Notes in Statistics*; Springer: New York, NY, USA, 1985; Volume 28.
9. Barndorff-Nielsen, O.E.; Cox, D.R. *Asymptotic Techniques for Use in Statistics*; Chapman & Hall: London, UK, 1989.
10. Anaya-Izquierdo, K.; Critchley, F.; Marriott, P. When are first-order asymptotics adequate? A diagnostic. *STAT* **2014**, *3*, 17–22.
11. Lauritzen, S.L. *Graphical Models*; Clarendon Press: Oxford, UK, 1996.
12. Geyer, C.J. Likelihood inference in exponential families and directions of recession. *Electron. J. Stat.* **2009**, *3*, 259–289.
13. Fienberg, S.E.; Rinaldo, A. Maximum likelihood estimation in log-linear models. *Ann. Stat.* **2012**, *40*, 996–1023.
14. Eguchi, S.; Copas, J. Local model uncertainty and incomplete-data bias. *J. R. Stat. Soc. B* **2005**, *67*, 1–37.
15. Copas, J.; Eguchi, S. Likelihood for statistically equivalent models. *J. R. Stat. Soc. B* **2010**, *72*, 193–217.
16. Anaya-Izquierdo, K.; Critchley, F.; Marriott, P.; Vos, P. On the geometric interplay between goodness-of-fit and estimation: Illustrative examples. In *Computational Information Geometry: For Image and Signal Processing*; Lecture Notes in Computer Science (LNCS); Nielsen, F., Dodson, K., Critchley, F., Eds.; Springer: Berlin, Germany, 2016.

17. Morris, C. Central limit theorems for multinomial sums. *Ann. Stat.* **1975**, *3*, 165–188.
18. Osius, G.; Rojek, D. Normal goodness-of-fit tests for multinomial models with large degrees of freedom. *JASA* **1992**, *87*, 1145–1152.
19. Holst, L. Asymptotic normality and efficiency for certain goodness-of-fit tests. *Biometrika* **1972**, *59*, 137–145.
20. Koehler, K.J.; Larntz, K. An empirical investigation of goodness-of-fit statistics for sparse multinomials. *JASA* **1980**, *75*, 336–344.
21. Koehler, K.J. Goodness-of-fit tests for log-linear models in sparse contingency tables. *JASA* **1986**, 81, 483–493.
22. McCullagh, P. The conditional distribution of goodness-of-fit statistics for discrete data. *JASA* **1986**, *81*, 104–107.
23. Forster, J.J.; McDonald, J.W.; Smith, P.W.F. Monte Carlo exact conditional tests for log-linear and logistic models. *J. R. Stat. Soc. B* **1996**, *58*, 445–453.
24. Kim, D.; Agresti, A. Nearly exact tests of conditional independence and marginal homogeneity for sparse contingency tables. *Comput. Stat. Data Anal.* **1997**, *24*, 89–104.
25. Booth, J.G.; Butler, R.W. An importance sampling algorithm for exact conditional tests in log-linear models. *Biometrika* **1999**, *86*, 321–332.
26. Caffo, B.S.; Booth, J.G. Monte Carlo conditional inference for log-linear and logistic models: A survey of current methodology. *Stat. Methods Med. Res.* **2003**, *12*, 109–123.
27. Lloyd, C.J. Computing highly accurate or exact P-values using importance sampling. *Comput. Stat. Data Anal.* **2012**, *56*, 1784–1794.
28. Simonoff, J.S. Jackknifing and bootstrapping goodness-of-fit statistics in sparse multinomials. *JASA* **1986**, *81*, 1005–1011.
29. Gaunt, R.E.; Pickett, A.; Reinert, G. Chi-square approximation by Stein's method with application to Pearson's statistic. *arXiv* **2015**, arXiv:1507.01707.
30. Fan, J.; Hung, H.-N.; Wong, W.-H. Geometric understanding of likelihood ratio statistics. *JASA* **2000**, *95*, 836–841.
31. Ulyanov, V.V.; Zubov, V.N. Refinement on the convergence of one family of goodness-of-fit statistics to chi-squared distribution. *Hiroshima Math. J.* **2009**, *39*, 133–161.
32. Asylbekov, Z.A.; Zubov, V.N.; Ulyanov, V.V. On approximating some statistics of goodness-of-fit tests in the case of three-dimensional discrete data. *Sib. Math. J.* **2011**, *52*, 571–584.
33. Zelterman, D. Goodness-of-fit tests for large sparse multinomial distributions. *JASA* **1987**, *82*, 624–629.
34. Bregman, L.M. The relaxation method of finding the common point of convex sets and its application to the solution of problems in convex programming. *USSR Comp. Math. Math.* **1967**, *7*, 200–217.
35. Amari, S.-I. *Information Geometry and Its Applications*; Springer: Tokyo, Japan, 2015.
36. Csiszár, I. On topological properties of f-divergences. *Stud. Sci. Math. Hung.* **1967**, *2*, 329–339.
37. Csiszár, I. Information measures: A critical survey. In *Transactions of the Seventh Prague Conference on Information Theory, Statistical Decision Functions, Random Processes and of the 1974 European Meeting of Statisticians*; Kozesnik, J., Ed.; Springer: Houten, The Netherlands, 1977; Volume B, pp. 73–86.
38. Barndorff-Nielsen, O. *Information and Exponential Families in Statistical Theory*; John Wiley & Sons, Ltd.: Chichester, UK, 1978.
39. Brown, L.D. *Fundamentals of Statistical Exponential Families with Applications in Statistical Decision Theory*; Lecture Notes-Monograph Series; Integrated Media Systems (IMS): Hayward, CA, USA, 1986; Volume 9.
40. Csiszár, I.; Matúš, F. Closures of exponential families. *Ann. Probab.* **2005**, *33*, 582–600.
41. Eriksson, N.; Fienberg, S.E.; Rinaldo, A.; Sullivant, S. Polyhedral conditions for the nonexistence of the MLE for hierarchical log-linear models. *J. Symb. Comput.* **2006**, *41*, 222–233.
42. Rinaldo, A.; Feinberg, S.; Zhou, Y. On the geometry of discrete exponential families with applications to exponential random graph models. *Electron. J. Stat.* **2009**, *3*, 446–484.
43. Critchley, F.; Marriott, P. Computing with Fisher geodesics and extended exponential families. *Stat. Comput.* **2016**, *26*, 325–332.
44. Sheather, S.J.; Jones, M.C. A reliable data-based bandwidth selection method for kernel density estimation. *J. R. Stat. Soc. B* **1991**, *53*, 683–690.

5

Anisotropically Weighted and Nonholonomically Constrained Evolutions on Manifolds

Stefan Sommer

Department of Computer Science, University of Copenhagen, DK-2100 Copenhagen E, Denmark; sommer@di.ku.dk

† This paper is an extended version of our paper published in the 2nd Conference on Geometric Science of Information, Paris, France, 28–30 October 2015.

Academic Editors: Frédéric Barbaresco and Frank Nielsen

Abstract: We present evolution equations for a family of paths that results from anisotropically weighting curve energies in non-linear statistics of manifold valued data. This situation arises when performing inference on data that have non-trivial covariance and are anisotropic distributed. The family can be interpreted as most probable paths for a driving semi-martingale that through stochastic development is mapped to the manifold. We discuss how the paths are projections of geodesics for a sub-Riemannian metric on the frame bundle of the manifold, and how the curvature of the underlying connection appears in the sub-Riemannian Hamilton–Jacobi equations. Evolution equations for both metric and cometric formulations of the sub-Riemannian metric are derived. We furthermore show how rank-deficient metrics can be mixed with an underlying Riemannian metric, and we relate the paths to geodesics and polynomials in Riemannian geometry. Examples from the family of paths are visualized on embedded surfaces, and we explore computational representations on finite dimensional landmark manifolds with geometry induced from right-invariant metrics on diffeomorphism groups.

Keywords: sub-Riemannian geometry; geodesics; most probable paths; stochastic development; non-linear data analysis; statistics

---

1. Introduction

When manifold valued data have non-trivial covariance (i.e., when *anisotropy* asserts higher variance in some directions than others), non-zero curvature necessitates special care when generalizing Euclidean space normal distributions to manifold valued distributions: in the Euclidean situation, normal distributions can be seen as transition distributions of diffusion processes, but on the manifold, holonomy makes transport of covariance path-dependent in the presence of curvature, preventing a global notion of a spatially constant covariance matrix. To handle this, in the diffusion principal component analysis (PCA) framework [1], and with the class of anisotropic normal distributions on manifolds defined in [2,3], data on non-linear manifolds are modelled as being distributed according to transition distributions of anisotropic diffusion processes that are mapped from Euclidean space to the manifold by stochastic development (see [4]). The construction is connected to a non-bracket-generating sub-Riemannian metric on the bundle of linear frames of the manifold, the frame bundle, and the requirement that covariance stays covariantly constant gives a nonholonomically constrained system.

Velocity vectors and geodesic distances are conventionally used for estimation and statistics in Riemannian manifolds; for example, for estimation of the Frechét mean [5], for Principal Geodesic Analysis [6], and for tangent space statistics [7]. In contrast to this, anisotropy as modelled with anisotropic normal distributions makes a distance for a sub-Riemannian metric the natural vehicle for estimation and statistics. This metric naturally accounts for anisotropy in a similar way as the precision matrix weights the inner product in the negative log-likelihood of a Euclidean normal distribution. The connection between the weighted distance and statistics of manifold valued data was presented in [2], and the underlying sub-Riemannian and fiber-bundle geometry, together with properties of the generated densities, was further explored in [3]. The fundamental idea is to perform statistics on manifolds by maximum likelihood (ML) instead of parametric constructions that use, for example, approximating geodesic subspaces; by defining natural families of probability distributions (in this case using diffusion processes), ML parameter estimates give a coherent way to statistically model non-linear data. The anisotropically weighted distance and the resulting family of extremal paths arises in this situation when the diffusion processes have non-isotropic covariance (i.e., when the distribution is not generated from a standard Brownian motion).

In this paper, we focus on the family of *most probable paths* for the semi-martingales that drives the stochastic development, and in turn the manifold valued anisotropic stochastic processes. Such paths, as exemplified in Figure 1, extremize the anisotropically weighted action functional. We present derivations of evolution equations for the paths from different viewpoints, and we discuss the role of frames as representing either metrics or cometrics. In the derivation, we explicitly see the influence of the connection and its curvature. We then turn to the relation between the sub-Riemannian metric and the Sasaki–Mok metric on the frame bundle, and we develop a construction that allows the sub-Riemannian metric to be defined as a sum of a rank-deficient generator and an underlying Riemannian metric. Finally, we relate the paths to geodesics and polynomials in Riemannian geometry, and we explore computational representations on different manifolds including a specific case: the finite dimensional manifolds arising in the Large Deformation Diffeomorphic Metric Mapping (LDDMM) [8] landmark matching problem. The paper ends with a discussion concerning statistical implications, open questions, and concluding remarks.

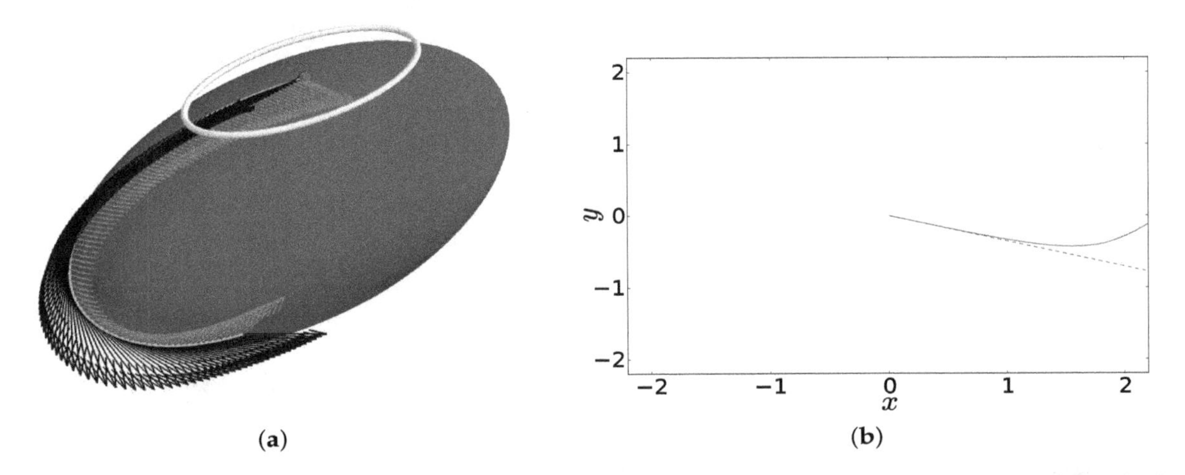

Figure 1. (**a**) A *most probable path* (MPP) for a driving Euclidean Brownian motion on an ellipsoid. The gray ellipsis over the starting point (red dot) indicates the covariance of the anisotropic diffusion. A frame u_t (black/gray vectors) representing the square root covariance is parallel transported along the curve, enabling the anisotropic weighting with the precision matrix in the action functional. With isotropic covariance, normal MPPs are Riemannian geodesics. In general situations, such as the displayed anisotropic case, the family of MPPs is much larger; (**b**) The corresponding anti-development in $\mathbb{R}^2$ (red line) of the MPP. Compare with the anti-development of a Riemannian geodesic with same initial velocity (blue dotted line). The frames $u_t \in \mathrm{GL}(\mathbb{R}^2, T_{x_t}M)$ provide local frame coordinates for each time t.

Background

Generalizing common statistical tools for performing inference on Euclidean space data to manifold valued data has been the subject of extensive work (e.g., [9]). Perhaps most fundamental is the notion of Frechét or Karcher means [5,10], defined as minimizers of the square Riemannian distance. Generalizations of the Euclidean principal component analysis procedure to manifolds are particularly relevant for data exhibiting anisotropy. Approaches include principal geodesic analysis (PGA, [6]), geodesic PCA (GPCA, [11]), principal nested spheres (PNS, [12]), barycentric subspace analysis (BSA, [13]), and horizontal component analysis (HCA, [14]). Common to these constructions are explicit representations of approximating low-dimensional subspaces. The fundamental challenge here is that the notion of Euclidean linear subspace on which PCA relies has no direct analogue in non-linear spaces.

A different approach taken by diffusion PCA (DPCA, [1,2]) and probabilistic PGA [15] is to base the PCA problem on a maximum likelihood fit of normal distributions to data. In Euclidean space, this approach was first introduced with probabilistic PCA [16]. In DPCA, the process of stochastic development [4] is used to define a class of anisotropic distributions that generalizes the family of Euclidean space normal distributions to the manifold context. DPCA is then a simple maximum likelihood fit in this family of distributions mimicking the Euclidean probabilistic PCA. The approach transfers the geometric complexities of defining subspaces common in the approaches listed above to the problem of defining a geometrically natural notion of normal distributions.

In Euclidean space, squared distances $\|x - x_0\|^2$ between observations x and the mean x_0 are affinely related to the negative log-likelihood of a normal distribution $\mathcal{N}(x_0, \mathrm{Id})$. This makes an ML fit of the mean such as performed in probabilistic PCA equivalent to minimizing squared distances. On a manifold, distances $d_g(x, x_0)^2$ coming from a Riemannian metric g are equivalent to tangent space distances $\|\mathrm{Log}_{x_0} x\|^2$ when mapping data from M to $T_{x_0}M$ using the inverse exponential map Log_{x_0}. Assuming $\mathrm{Log}_{x_0} x$ are distributed according to a normal distribution in the linear space $T_{x_0}M$, this restores the equivalence with a maximum likelihood fit. Let $\{e_1, \ldots, e_d\}$ be the standard basis for $\mathbb{R}^d$. If $u : \mathbb{R}^d \to T_{x_0}M$ is a linear invertible map with $ue_1, \ldots, ue_d$ orthonormal with respect to g, the normal distribution in $T_{x_0}M$ can be defined as $u\mathcal{N}(0, \mathrm{Id})$ (see Figure 2).

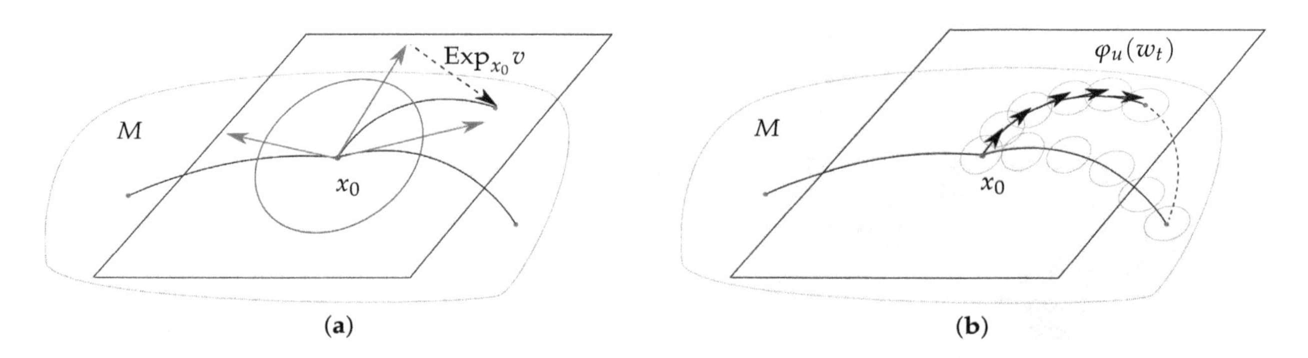

Figure 2. (**a**) Normal distributions $u\mathcal{N}(0, \mathrm{Id})$ in the tangent space $T_{x_0}M$ with covariance uu^T (blue ellipsis) can be mapped to the manifold by applying the exponential map Exp_{x_0} to sampled vectors $v \in T_{x_0}M$ (red vectors). This effectively linearises the geometry around x_0; (**b**) The stochastic development map φ_u maps $\mathbb{R}^d$ valued paths w_t to M by transporting the covariance in each step (blue ellipses) giving a covariance u_t along the entire sample path. The approach does not linearise around a single point. Holonomy of the connection implies that the covariance "rotates" around closed loops—an effect which can be illustrated by continuing the transport along the loop created by the dashed path. The anisotropic metric g_{FM} weights step lengths by the transported covariance at each time t.

The map u can be represented as a point in the frame bundle FM of M. When the orthonormal requirement on u is relaxed so that $u\mathcal{N}(0, \mathrm{Id})$ is a normal distribution in $T_{x_0}M$ with anisotropic covariance, the negative log-likelihood in $T_{x_0}M$ is related to $(u^{-1}\mathrm{Log}_{x_0} x)^T(u^{-1}\mathrm{Log}_{x_0} x)$ in the same way as the precision matrix Σ^{-1} is related to the negative log-likelihood $(x - x_0)^T\Sigma^{-1}(x - x_0)$ in

Euclidean space. The distance is thus weighted by the anisotropy of u, and u can be interpreted as a square root covariance matrix $\Sigma^{1/2}$.

However, the above approach does not specify how u changes when moving away from the base point x_0. The use of $\text{Log}_{x_0} x$ effectively linearises the geometry around x_0, but a geometrically natural way to relate u at points nearby to x_0 will be to parallel transport it, equivalently specifying that u when transported does not change as measured from the curved geometry. This constraint is *nonholonomic*, and it implies that any path from x_0 to x carries with it a parallel transport of u lifting paths from M to paths in the frame bundle FM. It therefore becomes natural to equip FM with a form of metric that encodes the anisotropy represented by u. The result is the sub-Riemannian metric on FM defined below that weights infinitesimal movements on M using the parallel transport of the frame u. Optimal paths for this metric are sub-Riemannian geodesics giving the family of *most probable paths* for the driving process that this paper concerns. Figure 1 shows one such path for an anisotropic normal distribution with M an ellipsoid embedded in $\mathbb{R}^3$.

2. Frame Bundles, Stochastic Development, and Anisotropic Diffusions

Let M be a finite dimensional manifold of dimension d with connection $\mathcal{C}$, and let x_0 be a fixed point in M. When a Riemannian metric is present, and $\mathcal{C}$ is its Levi–Civita connection, we denote the metric g_R. For a given interval $[0, T]$, we let $W(M)$ denote the Wiener space of continuous paths in M starting at x_0. Similarly, $W(\mathbb{R}^d)$ is the Wiener space of paths in $\mathbb{R}^d$. We let $H(\mathbb{R}^d)$ denote the subspace of $W(\mathbb{R}^d)$ of finite energy paths.

Let now $u = (u_1, \ldots, u_d)$ be a frame for T_xM, $x \in M$; i.e., $u_1, \ldots, u_d$ is an ordered set of linearly independent vectors in T_xM with $\text{span}\{u_1, \ldots, u_d\} = T_xM$. We can regard the frame as an isomorphism $u : \mathbb{R}^d \to T_xM$ with $u(e_i) = u_i$, where $e_1, \ldots, e_d$ denotes the standard basis in $\mathbb{R}^d$. Stochastic development (e.g., [4]) provides an invertible map φ_u from $W(\mathbb{R}^d)$ to $W(M)$. Through φ_u, Euclidean semi-martingales map to stochastic processes on M. When M is Riemannian and u orthonormal, the result is the Eells–Elworthy–Malliavin construction of Brownian motion [17]. We here outline the geometry behind development, stochastic development, the connection, and curvature, focusing in particular on frame bundle geometry.

2.1. The Frame Bundle

For each point $x \in M$, let F_xM be the set of frames for T_xM (i.e., the set of ordered bases for T_xM). The set $\{F_xM\}_{x \in M}$ can be given a natural differential structure as a fiber bundle on M called the frame bundle FM. It can equivalently be defined as the principal bundle $\text{GL}(\mathbb{R}^d, TM)$. We let the map $\pi : FM \to M$ denote the canonical projection. The kernel of $\pi_* : TFM \to TM$ is the sub-bundle of TFM that consists of vectors tangent to the fibers $\pi^{-1}(x)$. It is denoted the vertical subspace VFM. We will often work in a local trivialization $u = (x, u_1, \ldots, u_d) \in FM$, where $x = \pi(u) \in M$ denotes the base point, and for each $i = 1, \ldots, d$, $u_i \in T_xM$ is the ith frame vector. For $v \in T_xM$ and $u \in FM$ with $\pi(u) = x$, the vector $u^{-1}v \in \mathbb{R}^d$ expresses v in components in terms of the frame u. We will denote the vector $u^{-1}v$ frame coordinates of v.

For a differentiable curve x_t in M with $x = x_0$, a frame u for $T_{x_0}M$ can be parallel transported along x_t by parallel transporting each vector in the frame, thus giving a path $u_t \in FM$. Such paths are called horizontal, and have zero acceleration in the sense $\mathcal{C}(\dot{u}_{i,t}) = 0$. For each $x \in M$, their derivatives form a d-dimensional subspace of the $d + d^2$-dimensional tangent space T_uFM. This horizontal subspace HFM and the vertical subspace VFM together split the tangent bundle of FM (i.e., $TFM = HFM \oplus VFM$). The split induces a map $\pi_* : HFM \to TM$, see Figure 3. For fixed $u \in FM$, the restriction $\pi_*|_{H_uFM} : H_uFM \to T_xM$ is an isomorphism. Its inverse is called the horizontal lift and is denoted h_u in the following. Using h_u, horizontal vector fields H_e on FM are defined for vectors $e \in \mathbb{R}^d$ by $H_e(u) = h_u(ue)$. In particular, the standard basis $(e_1, \ldots, e_d)$ on $\mathbb{R}^d$ gives d globally defined horizontal vector fields $H_i \in HFM$, $i = 1, \ldots, d$ by $H_i = H_{e_i}$. Intuitively, the fields $H_i(u)$ model infinitesimal transformations in M of x_0 in direction $u_i = ue_i$ with corresponding infinitesimal parallel transport of the vectors $u_1, \ldots, u_d$ of the frame along the direction u_i. A *horizontal lift* of a differentiable curve $x_t \in M$ is a curve in FM tangent to HFM that projects to x_t. Horizontal lifts are unique up to the choice of initial frame u_0.

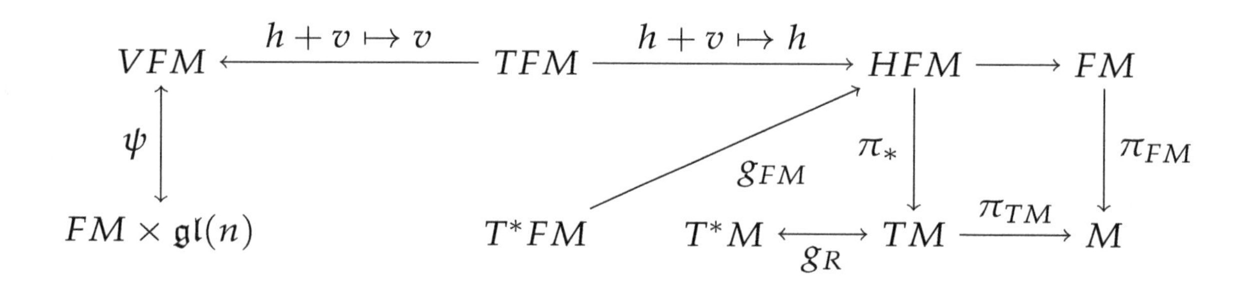

Figure 3. Relations between the manifold, frame bundle, the horizontal distribution HFM, the vertical bundle VFM, a Riemannian metric g_R, and the sub-Riemannian metric g_{FM}, defined below. The connection C provides the splitting $TFM = HFM \oplus VFM$. The restrictions $\pi_*|_{H_uM}$ are invertible maps $H_uM \to T_{\pi(u)}M$ with inverse h_u, the horizontal lift. Correspondingly, the vertical bundle VFM is isomorphic to the trivial bundle $FM \times \mathfrak{gl}(n)$. The metric $g_{FM} : T^*FM \to TFM$ has an image in the subspace HFM.

2.2. Development and Stochastic Development

Let x_t be a differentiable curve on M and u_t a horizontal lift. If s_t is a curve in $\mathbb{R}^d$ with components s_t^i such that $\dot{x}_t = H_i(u)s_t^i$, x_t is said to be a development of s_t. Correspondingly, s_t is the anti-development of x_t. For each t, the vector s_t contains frame coordinates of $\dot{x}_t$ as defined above. Similarly, let W_t be an $\mathbb{R}^d$ valued Brownian motion so that sample paths $W_t(\omega) \in W(\mathbb{R}^d)$. A solution to the stochastic differential equation $dU_t = \sum_{i=1}^d H_i(U_t) \circ dW_t^i$ in FM is called a stochastic development of W_t in FM. The solution projects to a stochastic development $X_t = \pi(U_t)$ in M. We call the process W_t in $\mathbb{R}^d$, that through φ maps to X_t, the driving process of X_t. Let $\varphi_u : W(\mathbb{R}^d) \to W(M)$ be the map that for fixed u sends a path in $\mathbb{R}^d$ to its development on M. Its inverse φ_u^{-1} is the anti-development in $\mathbb{R}^d$ of paths on M given u.

Equivalent to the fact that normal distributions $\mathcal{N}(0, \Sigma)$ in $\mathbb{R}^d$ can be obtained as the transition distributions of diffusion processes $\Sigma^{1/2}W_t$ stopped at time $t = 1$, a general class of distributions on the manifold M can be defined by stochastic development of processes W_t, resulting in M-valued random variables $X = X_1$. This family of distributions on M introduced in [2] is denoted *anisotropic normal distributions*. The stochastic development by construction ensures that U_t is horizontal, and the frames are thus parallel transported along the stochastic displacements. The effect is that the frames stay covariantly constant, thus resembling the Euclidean situation where $\Sigma^{1/2}$ is spatially constant and therefore does not change as W_t evolves. Thus, as further discussed in Section 3.2, the covariance is kept constant at each of the infinitesimal stochastic displacements. The existence of a smooth density for the target process X_t and small time asymptotics are discussed in [3].

Stochastic development gives a map $\int_{\text{Diff}} : FM \to \text{Prob}(M)$ to the space of probability distributions on M. For each point $u \in FM$, the map sends a Brownian motion in $\mathbb{R}^d$ to a distribution μ_u by stochastic development of the process U_t in FM, starting at u and letting μ_u be the distribution of $X = \pi(U_1)$. The pair (x, u), $x = \pi(u)$ is analogous to the parameters (μ, Σ) for a Euclidean normal distribution: the point $x \in M$ represents the starting point of the diffusion, and the frame u represents a square root $\Sigma^{1/2}$ of the covariance Σ. In the general situation where μ_u has smooth density, the construction can be used to fit the parameters u to data by maximum likelihood. As an example, diffusion PCA fits distributions obtained through $\int_{\text{Diff}}$ by maximum likelihood to observed samples in M; i.e., it optimizes for the most likely parameters $u = (x, u_1, \ldots, u_d)$ for the anisotropic diffusion process, giving a fit to the data of the manifold generalization of the Euclidean normal distribution.

2.3. Adapted Coordinates

For concrete expressions of the geometric constructions related to frame bundles, and for computational purposes, it is useful to apply coordinates that are adapted to the horizontal bundle HFM and the vertical bundle VFM together with their duals H^*FM and V^*FM. The notation below follows the notation used in, for example, [18]. Let $z = (u, \xi)$ be a local trivialization of T^*FM, and let (x^i, u^i_α) be coordinates on FM with u^i_α satisfying $u_\alpha = u^i_\alpha \partial_{x^i}$ for each $\alpha = 1, \ldots, d$.

To find a basis that is adapted to the horizontal distribution, define the d linearly independent vector fields $D_j = \partial_{x^j} - \Gamma^{h_\gamma}_j \partial_{u^h_\gamma}$ where $\Gamma^{h_\gamma}_j = \Gamma^h{}_{ji} u^i_\gamma$ is the contraction of the Christoffel symbols $\Gamma^h{}_{ij}$ for the connection $\mathcal{C}$ with u^i_α. We denote this adapted frame D. The vertical distribution is correspondingly spanned by $D_{j_\beta} = \partial_{u^j_\beta}$. The vectors $D^h = dx^h$, and $D^{h_\gamma} = \Gamma^{h_\gamma}_j dx^j + du^h_\gamma$ constitutes a dual coframe D^*. The map $\pi_* : HFM \to TM$ is in coordinates of the adapted frame $\pi_*(w^j D_j) = w^j \partial_{x^j}$. Correspondingly, the horizontal lift h_u is $h_u(w^j \partial_{x^j}) = w^j D_j$. The map $u : \mathbb{R}^d \to T_x M$ is given by the matrix $[u^i_\alpha]$ so that $uv = u^i_\alpha v^\alpha \partial_{x^i} = u_\alpha v^\alpha$.

Switching between standard coordinates and the adapted frame and coframes can be expressed in terms of the component matrices A below the frame and coframe induced by the coordinates (x^i, u^i_α) and the adapted frame D and coframe D^*. We have

$$_{(\partial_{x^i}, \partial_{u^i_\alpha})}A_D = \begin{bmatrix} I & 0 \\ -\Gamma & I \end{bmatrix} \text{ with inverse } {}_D A_{(\partial_{x^i}, \partial_{u^i_\alpha})} = \begin{bmatrix} I & 0 \\ \Gamma & I \end{bmatrix}$$

writing Γ for the matrix $[\Gamma^{h_\gamma}_j]$. Similarly, the component matrices of the dual frame D^* are

$$_{(\partial_{x^i}, \partial_{u^i_\alpha})^*}A_{D^*} = \begin{bmatrix} I & \Gamma^T \\ 0 & I \end{bmatrix} \text{ and } {}_{D^*}A_{(\partial_{x^i}, \partial_{u^i_\alpha})^*} = \begin{bmatrix} I & -\Gamma^T \\ 0 & I \end{bmatrix} .$$

2.4. Connection and Curvature

The TM valued connection $\mathcal{C} : TM \times TM \to TM$ lifts to a principal connection $TFM \times TFM \to VFM$ on the principal bundle FM. $\mathcal{C}$ can then be identified with the $\mathfrak{gl}(n)$-valued connection form ω on TFM. The identification occurs by the isomorphism ψ between $FM \times \mathfrak{gl}(n)$ and VFM given by $\psi(u, v) = \frac{d}{dt} u \exp(tv)|_{t=0}$ (e.g., [19,20]).

The map ψ is equivariant with respect to the $\mathrm{GL}(n)$ action $g \mapsto ug^{-1}$ on FM. In order to explicitly see the connection between the usual covariant derivative $\nabla : \Gamma(TM) \times \Gamma(TM) \to \Gamma(TM)$ on M determined by $\mathcal{C}$ and $\mathcal{C}$ regarded as a connection on the principal bundle FM, following [19], we let $s : M \to TM$ be a local vector field on M; equivalently, $s \in \Gamma(TM)$ is a local section of TM. s determines a map $s^{FM} : FM \to \mathbb{R}^d$ by $s^{FM}(u) = u^{-1} s(\pi(u))$; i.e., it gives the coordinates of $s(x)$ in the frame u at x. The pushforward $(s^{FM})_* : TFM \to \mathbb{R}^d$ has in its ith component the exterior derivative $d(s^{FM})^i$. Let now $w(x)$ be a local section of FM. The composition $w \circ (s^{FM})_* \circ h_w : TM \to TM$ is identical to the covariant derivative $\nabla_\cdot s : TM \to TM$. The construction is independent of the choice of w because of the $\mathrm{GL}(n)$-equivariance of s^{FM}. The connection form ω can be expressed as the matrix $(s^{FM}_1 \circ h_w, \ldots, s^{FM}_d \circ h_w)$ when letting $s^{FM}_i(u) = e_i$.

The identification becomes particularly simple if the covariant derivative is taken along a curve x_t on which w_t is the horizontal lift. In this case, we can let $s_t = w_{t,i} s^i_t$. Then, $s^{FM}(w_t) = (s^1_t, \ldots, s^d_t)^T$, and

$$w_t^{-1} \nabla_{\dot{x}_t} s = (s^F M)_*(h_{w_t}(\dot{x}_t)) = \tfrac{d}{dt}(s^1_t, \ldots, s^d_t)^T ; \tag{1}$$

i.e., the covariant derivative takes the form of the standard derivative applied to the frame coordinates s^i_t.

The curvature tensor $R \in \mathcal{T}_1^3(M)$ gives the $\mathfrak{gl}(n)$-valued curvature form $\Omega : TFM \times TFM \to \mathfrak{gl}(n)$ on TFM by

$$\Omega(v_u, w_u) = u^{-1}R(\pi_*(v_u), \pi_*(w_u))u \ , \ v_u, w_v \in TFM \ .$$

Note that $\Omega(v_u, w_u) = \Omega(h_u(\pi_*(v_u)), h_u(\pi_*(w_u)))$, which we can use to write the curvature R as the $\mathfrak{gl}(n)$-valued map $R_u : T^2(T_{\pi(u)}M) \to \mathfrak{gl}(n)$, $(v,w) \mapsto \Omega(h_u(\pi_*(v_u)), h_u(\pi_*(w_u)))$ for fixed u. In coordinates, the curvature is

$$R_{ijk}{}^{s} = \Gamma^{l}{}_{ik}\Gamma^{s}{}_{jl} - \Gamma^{l}{}_{jk}\Gamma^{s}{}_{il} + \Gamma^{s}{}_{ik;j} - \Gamma^{s}{}_{jk;i}$$

where $\Gamma^{s}{}_{ik;j} = \partial_{x^j}\Gamma^{s}{}_{ik}$.

Let $x_{t,s}$ be a family of paths in M, and let $u_{t,s} \in \pi^{-1}(x_{t,s})$ be horizontal lifts of $x_{t,s}$ for each fixed s. Write $\dot{x}_{t,s} = \partial_t x_{t,s}$ and $\dot{u}_{t,s} = \partial_t u_{t,s}$. The s-derivative of $u_{t,s}$ can be regarded a pushforward of the horizontal lift and is the curve in TFM

$$\begin{aligned} \partial_s u_{t,s} &= \psi(u_{t,s}, \psi_{u_{0,s}}^{-1}(\mathcal{C}(\partial_s u_{0,s})) + \int_0^s \Omega(\dot{u}_{r,s}, \partial_s u_{r,s})dr) + h_{u_{t,s}}(\partial_s x_{t,s}) \\ &= \psi(u_{t,s}, \psi_{0,s}^{-1}(\mathcal{C}(\partial_s u_{0,s})) + \int_0^s R_{u_{r,s}}(\dot{x}_{r,s}, \partial_s x_{r,s})dr) + h_{u_{t,s}}(\partial_s x_{t,s}) \ . \end{aligned} \tag{2}$$

This follows from the structure equation $d\omega = -\omega \wedge \omega + \Omega$ (e.g., [21]). Note that the curve depends on the vertical variation $\mathcal{C}(\partial_s u_{0,s})$ at only one point along the curve. The remaining terms depend on the horizontal variation or, equivalently, $\partial_s x_{t,s}$. The t-derivative of $\partial_s u_{t,s}$ is the curve in $TTFM$ satisfying

$$\begin{aligned} \partial_s h_{u_{t,s}}(\dot{x}_{t,s}) &= \psi(u_{t,s}, R_{u_{t,s}}(\dot{x}_{t,s}, \partial_s x_{t,s})) + \partial_t \psi(u_{t,s}, \psi_{0,s}^{-1}(\mathcal{C}(\partial_s u_{0,s}))) + \partial_t(h_{u_{t,s}}(\partial_s x_{t,s})) \\ &= \psi(u_{t,s}, R_{u_{t,s}}(\dot{x}_{t,s}, \partial_s x_{t,s})) + \partial_t \psi(u_{t,s}, \psi_{0,s}^{-1}(\mathcal{C}(\partial_s u_{0,s}))) \\ &\quad + h_{u_{t,s}}(\partial_t \partial_s x_{t,s}) + (\partial_t h_{u_{t,s}})(\partial_s x_{t,s}). \end{aligned} \tag{3}$$

Here, the first and third term in the last expression are identified with elements of $T_{\partial_s u_{t,s}}TFM$ by the natural mapping $T_{u_{t,s}}FM \to T_{\partial_s u_{t,s}}TFM$. When $\mathcal{C}(\partial_s u_{0,s})$ is zero, the relation reflects the property that the curvature arises when computing brackets between horizontal vector fields. Note that the first term of (3) has values in VFM, while the third term has values in HFM.

3. The Anisotropically Weighted Metric

For a Euclidean driftless diffusion process with spatially constant stochastic generator Σ, the log-probability of a sample path can formally be written

$$\ln \tilde{p}_\Sigma(x_t) \propto -\int_0^1 \|\dot{x}_t\|_\Sigma^2 dt + c_\Sigma \tag{4}$$

with the norm $\|\cdot\|_\Sigma$ given by the inner product $\langle v, w\rangle_\Sigma = \left\langle \Sigma^{-1/2}v, \Sigma^{-1/2}w \right\rangle = v\Sigma^{-1}w$; i.e., the inner product weighted by the precision matrix Σ^{-1}. Though only formal, as the sample paths are almost surely nowhere differentiable, the interpretation can be given a precise meaning by taking limits of piecewise linear curves [21]. Turning to the manifold situation with the processes mapped to M by stochastic development, the probability of observing a differentiable path can either be given a precise meaning in the manifold by taking limits of small tubes around the curve, or in $\mathbb{R}^d$ by considering infinitesimal tubes around the anti-development of the curves. With the former formulation, a scalar curvature correction term must be added to (4), giving the Onsager–Machlup function [22]. The latter formulation corresponds to defining a notion of path density for the driving $\mathbb{R}^d$-valued process W_t. When M is Riemannian and Σ unitary, taking the maximum of (4) gives geodesics as most probable paths for the driving process.

Let now u_t be a path in FM, and choose a local trivialization $u_t = (x_t, u_{1,t}, \ldots, u_{d,t})$ such that the matrix $[u^i_{\alpha,t}]$ represents the square root covariance matrix $\Sigma^{1/2}$ at x_t. Since u_t being a frame defines an invertible map $\mathbb{R}^d \to T_{x_t}M$, the norm $\|\cdot\|_\Sigma$ above has a direct analogue in the norm $\|\cdot\|_{u_t}$ defined by the inner product

$$\langle v, w \rangle_{u_t} = \left\langle u_t^{-1} v, u_t^{-1} w \right\rangle_{\mathbb{R}^d} \tag{5}$$

for vectors $v, w \in T_{x_t}M$. The transport of the frame along paths in effect defines a transport of inner product along sample paths: the paths carry with them the inner product weighted by the precision matrix, which in turn is a transport of the square root covariance u_0 at x_0.

The inner product can equivalently be defined as a metric $g_u : T^*_x M \to T_x M$. Again using that u can be considered a map $\mathbb{R}^d \to T_x M$, g_u is defined by $\xi \mapsto u(\xi \circ u)^\sharp$, where $\sharp$ is the standard identification $(\mathbb{R}^d)^* \to \mathbb{R}^d$. The sequence of mappings defining g_u is illustrated below:

$$\begin{array}{ccccccc} T^*_x M & \to & (\mathbb{R}^d)^* & \to & \mathbb{R}^d & \to & T_x M \\ \xi & \mapsto & \xi \circ u & \mapsto & (\xi \circ u)^\sharp & \mapsto & u(\xi \circ u)^\sharp . \end{array} \tag{6}$$

This definition uses the $\mathbb{R}^d$ inner product in the definition of $\sharp$. Its inverse gives the cometric $g_u^{-1} : T_x M \to T^*_x M$; i.e., $v \mapsto (u^{-1}v)^\flat \circ u^{-1}$.

$$\begin{array}{ccccccc} T_x M & \to & \mathbb{R}^d & \to & (\mathbb{R}^d)^* & \to & T^*_x M \\ v & \mapsto & u^{-1} v & \mapsto & (u^{-1})^\flat & \mapsto & (u^{-1})^\flat \circ u^{-1} . \end{array} \tag{7}$$

3.1. Sub-Riemannian Metric on the Horizontal Distribution

We now lift the path-dependent metric defined above to a sub-Riemannian metric on HFM. For any $w, \tilde{w} \in H_u FM$, the lift of (5) by π_* is the inner product

$$\langle w, \tilde{w} \rangle = \left\langle u^{-1} \pi_* w, u^{-1} \pi_* \tilde{w} \right\rangle_{\mathbb{R}^d} .$$

The inner product induces a sub-Riemannian metric $g_{FM} : TFM^* \to HFM \subset TFM$ by

$$\langle w, g_{FM}(\xi) \rangle = (\xi | w) \,, \ \forall w \in H_u FM \tag{8}$$

with $(\xi|w)$ denoting the evaluation $\xi(w)$ for the covector $\xi \in T^*FM$. The metric g_{FM} gives FM a non-bracket-generating sub-Riemannian structure [23] on FM (see also Figure 3). It is equivalent to the lift

$$\xi \mapsto h_u(g_u(\xi \circ h_u)) \,, \ \xi \in T_u FM \tag{9}$$

of the metric g_u above. In frame coordinates, the metric takes the form

$$u^{-1} \pi_* g_{FM}(\xi) = \begin{pmatrix} \xi(H_1(u)) \\ \vdots \\ \xi(H_d(u)) \end{pmatrix} . \tag{10}$$

In terms of the adapted coordinates for TFM described in Section 2.3, with $w = w^j D_j$ and $\tilde{w} = \tilde{w}^j D_j$, we have

$$\begin{aligned} \langle w, \tilde{w} \rangle &= \left\langle w^i D_i, \tilde{w}^j D_j \right\rangle = \left\langle u^{-1} w^i \partial_{x^i}, u^{-1} \tilde{w}^j \partial_{x^j} \right\rangle \\ &= \left\langle w^i u^\alpha_i, \tilde{w}^j u^\alpha_j \right\rangle_{\mathbb{R}^d} = \delta_{\alpha\beta} w^i u^\alpha_i \tilde{w}^j u^\beta_j = W_{ij} w^i \tilde{w}^j \end{aligned}$$

where $[u^\alpha_i]$ is the inverse of $[u^i_\alpha]$ and $W_{ij} = \delta_{\alpha\beta} u^\alpha_i u^\beta_j$. Define now $W^{kl} = \delta^{\alpha\beta} u^k_\alpha u^l_\beta$, so that $W^{ir} W_{rj} = \delta^i_j$ and $W_{ir} W^{rj} = \delta^j_i$. We can then write the metric g_{FM} directly as

$$g_{FM}(\xi_h D^h + \xi_{h_\gamma} D^{h_\gamma}) = W^{ih} \xi_h D_i, \tag{11}$$

because $\langle w, g_{FM}(\xi) \rangle = \left\langle w, W^{jh} \xi_h D_j \right\rangle = W_{ij} w^i W^{jh} \xi_h = w^i \xi_i = \xi_h D^h (w^j D_j) = \xi(w)$. One clearly recognizes the dependence on the horizontal H^*FM part of T^*FM only, and the fact that g_{FM} has image in HFM. The sub-Riemannian energy of an almost everywhere horizontal path u_t is

$$l_{FM}(u_t) = \int g_{FM}(\dot{u}_t, \dot{u}_t) dt;$$

i.e., the line element is $ds^2 = W_{ij} D^i D^j$ in adapted coordinates. The corresponding distance is given by

$$d_{FM}(u_1, u_2) = \inf\{ l_{FM}(\gamma) \mid \gamma(0) = u_1, \gamma(1) = u_2 \}.$$

If we wish to express g_{FM} in canonical coordinates on T^*FM, we can switch between the adapted frame and the coordinates $(x^i, u^i_\alpha, \xi^i, \xi^i_\alpha)$. From (11), g_{FM} has D, D^* components

$$_D g_{FM,D^*} = \begin{bmatrix} W^{-1} & 0 \\ 0 & 0 \end{bmatrix}.$$

Therefore, g_{FM} has the following components in the coordinates $(x^i, u^i_\alpha, \xi_h, \xi_{h_\gamma})$

$$_{(\partial_{x^i}, \partial_{u^i_\alpha})} g_{FM, (\partial_x, \partial_{u^i_\alpha})^*} = {}_{(\partial_{x^i}, \partial_{u^i_\alpha})} A_D \, {}_D g_{FM,D^*} \, {}_{D^*} A_{(\partial_{x^i}, \partial_{u^i_\alpha})^*} = \begin{bmatrix} W^{-1} & -W^{-1}\Gamma^T \\ -\Gamma W^{-1} & \Gamma W^{-1} \Gamma^T \end{bmatrix}$$

or $g^{ij}_{FM} = W^{ij}$, $g^{ij_\beta}_{FM} = -W^{ih} \Gamma^{j_\beta}_h$, $g^{i_\alpha j}_{FM} = -\Gamma^{i_\alpha}_h W^{hj}$, and $g^{i_\alpha j_\beta}_{FM} = \Gamma^{i_\alpha}_k W^{kh} \Gamma^{j_\beta}_h$.

3.2. Covariance and Nonholonomicity

The metric g_{FM} encodes the anisotropic weighting given the frame u, thus up to an affine transformation measuring the energy of horizontal paths equivalently to the negative log-probability of sample paths of Euclidean anisotropic diffusions as formally given in (4). In addition, the requirement that paths must stay horizontal almost everywhere enforces that $C(\dot{u}_t) = 0$ a.e., i.e., that *no change of the covariance is measured by the connection*. The intuitive effect is that covariance is covariantly constant as seen by the connection. Globally, *curvature* of C will imply that the covariance changes when transported along closed loops, and *torsion* will imply that the base point "slips" when travelling along covariantly closed loops on M. However, the zero acceleration requirement implies that the covariance is as close to spatially constant as possible with the given connection. This is enabled by the parallel transport of the frame, and it ensures that the model closely resembles the Euclidean case with spatially constant stochastic generator.

With non-zero curvature of C, the horizontal distribution is non-integrable (i.e., the brackets $[H_i, H_j]$ are non-zero for some i, j). This prevents integrability of the horizontal distribution HFM in the sense of the Frobenius theorem. In this case, the horizontal constraint is *nonholonomic* similarly to nonholonomic constraints appearing in geometric mechanics (e.g., [24]). The requirement of covariantly constant covariance thus results in a nonholonomic system.

3.3. Riemannian Metrics on FM

If the horizontality constraint is relaxed, a related Riemannian metric on FM can be defined by pulling back a metric on $\mathfrak{gl}(n)$ to each fiber using the isomorphism $\psi(u, \cdot)^{-1} : V_u FM \to \mathfrak{gl}(n)$.

Therefore, the metric on HFM can be extended to a Riemannian metric on FM. Such metrics incorporate the anisotropically weighted metric on HFM, however, allowing vertical variations and thus that covariances can change unrestricted.

When M is Riemannian, the metric g_{FM} is in addition related to the Sasaki–Mok metric on FM [18] that extends the Sasaki metric on TM. As for the above Riemannian metric on FM, the Sasaki–Mok metric allows paths in FM to have derivatives in the vertical space VFM. On HFM, the Riemannian metric g_R is here lifted to the metric $g_{SM} = (v_u, w_u) = g_R(\pi_*(v_u), \pi_*(w_u))$ (i.e., the metric is not anisotropically weighted). The line element is in this case $ds^2 = g_{ij}dx^i dx^j + X_{\beta\alpha} g_{ij} D^{\alpha_i} D^{\beta_j}$.

Geodesics for g_{SM} are lifts of Riemannian geodesics for g_R on M, in contrast to the sub-Riemannian normal geodesics for g_{FM} which we will characterize below. The family of curves arising as projections to M of normal geodesics for g_{FM} includes Riemannian geodesics for g_R (and thus projections of geodesics for g_{SM}), but the family is in general larger than geodesics for g_R.

4. Constrained Evolutions

Extremal paths for (5) can be interpreted as most probable paths for the driving process W_t when u_0 defines an anisotropic diffusion. This is captured in the following definition [3]:

Definition 1. *A most probable path for the driving process (MPP) from $x = \pi(u_0) \in M$ to $y \in M$ is a smooth path $x_t : [0,1] \to M$ with $x_0 = x$ and $x_1 = y$ such that its anti-development $\varphi_{u_0}^{-1}(x_t)$ is a most probable path for W_t; i.e.,*

$$x_t \in argmin_{\sigma, \sigma_0 = x, \sigma_1 = y} \int_0^1 -L_{\mathbb{R}^d}(\varphi_{u_0}^{-1}(\sigma_t), \tfrac{d}{dt}\varphi_{u_0}^{-1}(\sigma_t))\, dt$$

with $L_{\mathbb{R}^d}$ being the Onsager–Machlup function for the process W_t on $\mathbb{R}^d$ [22].

The definition uses the one-to-one relation between $W(\mathbb{R}^d)$ and $W(M)$ provided by φ_{u_0} to characterize the paths using the $\mathbb{R}^d$ Onsager–Machlup function $L_{\mathbb{R}^d}$. When M is Riemannian with metric g_R, the Onsager–Machlup function for a g-Brownian motion on M is $L(x_t, \dot{x}_t) = -\frac{1}{2}\|\dot{x}_t\|^2_{g_R} + \frac{1}{12} S_{g_R}(x_t)$ with S_{g_R} denoting the scalar curvature. This curvature term vanishes on $\mathbb{R}^d$, and therefore $L_{\mathbb{R}^d}(\gamma_t, \dot{\gamma}_t) = -\frac{1}{2}\|\dot{\gamma}_t\|^2$ for a curve $\gamma_t \in \mathbb{R}^d$.

By pulling $x_t \in M$ back to $\mathbb{R}^d$ using $\varphi_{u_0}^{-1}$, the construction removes the $\frac{1}{12} S_{g_R}(x_t)$ scalar curvature correction term present in the non-Euclidean Onsager–Machlup function. It thereby provides a relation between geodesic energy and most probable paths for the driving process. This is contained in the following characterization of most probable paths for the driving process as extremal paths of the sub-Riemannian distance [3] that follows from the Euclidean space Onsager–Machlup theorem [22].

Theorem 1 ([3])**.** *Let $Q(u_0)$ denote the principal sub-bundle of FM of points $z \in FM$ reachable from $u_0 \in FM$ by horizontal paths. Suppose the Hörmander condition is satisfied on $Q(u_0)$, and that $Q(u_0)$ has compact fibers. Then, most probable paths from x_0 to $y \in M$ for the driving process of X_t exist, and they are projections of sub-Riemannian geodesics in FM minimizing the sub-Riemannian distance from u_0 to $\pi^{-1}(y)$.*

Below, we will derive evolution equations for the set of such extremal paths that correspond to normal sub-Riemannian geodesics.

4.1. Normal Geodesics for g_{FM}

Connected to the metric g_{FM} is the Hamiltonian

$$H(z) = \frac{1}{2}(z | g_{FM}(z)) \tag{12}$$

on the symplectic space T^*FM. Letting $\hat{\pi}$ denote the projection on the bundle $T^*FM \to FM$, (8) gives

$$H(z) = \frac{1}{2}\langle g_{FM}(z)|g_{FM}(z)\rangle = \frac{1}{2}\|z \circ h_{\hat{\pi}(z)} \circ \hat{\pi}(z)\|^2_{(\mathbb{R}^d)^*} = \frac{1}{2}\sum_{i=1}^{d} \xi(H_i(u))^2.$$

Normal geodesics in sub-Riemannian manifolds satisfy the Hamilton–Jacobi equations [23] with Hamiltonian flow

$$\dot{z}_t = X_H = \Omega^\# dH(z) \tag{13}$$

where Ω here is the canonical symplectic form on T^*FM (e.g., [25]). We denote (13) the MPP equations, and we let projections $x_t = \pi_{T^*FM}(z_t)$ of minimizing curves satisfying (13) be denoted normal MPPs. The system (13) has $2(d+d^2)$ degrees of freedom, in contrast to the usual $2d$ degrees of freedom for the classical geodesic equation. Of these, d^2 describes the current frame at time t, while the remaining d^2 allows the curve to "twist" while still being horizontal. We will see this effect visualized in Section 6.

In a local canonical trivialization $z = (u, \xi)$, (13) gives the Hamilton–Jacobi equations

$$\begin{aligned}
\dot{u} &= \partial_\xi H(u,\xi) = g_{FM}(u,\xi) = h_u(u(\,\xi(H_1(u)),\ldots,\xi(H_d(u))\,)^T) \\
\dot{\xi} &= -\partial_u H(u,\xi) = -\partial_u \frac{1}{2}\|\xi \circ h_u \circ u\|^2_{(\mathbb{R}^d)^*} = -\partial_u \frac{1}{2}\sum_{i=1}^{d} \xi(H_i(u))^2.
\end{aligned} \tag{14}$$

Using (3), we have for the second equation

$$\begin{aligned}
\dot{\xi} &= -\sum_{i=1}^{d} \xi(H_i(u))\xi(\partial_u h_u(ue_i)) \\
&= -\sum_{i=1}^{d} \xi(H_i(u))\xi(\psi(u, R_u(ue_i, \pi_*(\partial_u))) + \partial_{h_u(ue_i)}\psi(u, \psi^{-1}(C(\partial_u))) + \partial_{h_u(ue_i)} h_u(\pi_*(\partial_u))) \\
&= -\xi(\psi(u, R_u(\pi_*(\dot{u}), \pi_*(\partial_u))) + \partial_{\dot{u}}\psi(u, \psi^{-1}(C(\partial_u))) + \partial_{\dot{u}} h_u(\pi_*(\partial_u))).
\end{aligned} \tag{15}$$

Here $\partial_{\dot{u}}$ denotes u-derivative in the direction $\dot{u}$, equivalently $\partial_{\dot{u}} h_u(v) = \partial_t(h_u)(v)$. While the first equation of (14) involves only the horizontal part of ξ, the second equation couples the vertical part of ξ through the evaluation of ξ on the term $\psi(u, R_u(\pi_*(\dot{u}), \pi_*(\partial_u)))$. If the connection is curvature-free, which in non-flat cases implies that it carries torsion, this vertical term vanishes. Conversely, when M is Riemannian, C the g_R Levi–Civita connection, and u_0 is g_R orthonormal, $g_{FM}(h_u(v), h_u(w)) = g_R(v,w)$ for all $v, w \in T_{\pi(u_t)}M$. In this case, a normal MPP $\pi(u_t)$ will be a Riemannian g_R geodesic.

4.2. Evolution in Coordinates

In coordinates $u = (x^i, u^i_\alpha, \xi_i, \xi_{i_\alpha})$ for T^*FM, we can equivalently write

$$\begin{aligned}
\dot{x}^i &= g^{ij}\xi_j + g^{ij_\beta}\xi_{j_\beta} = W^{ij}\xi_j - W^{ih}\Gamma^{j_\beta}_h \xi_{j_\beta} \\
\dot{X}^i_\alpha &= g^{i_\alpha j}\xi_j + g^{i_\alpha j_\beta}\xi_{j_\beta} = -\Gamma^{i_\alpha}_h W^{hj}\xi_j + \Gamma^{i_\alpha}_k W^{kh}\Gamma^{j_\beta}_h \xi_{j_\beta} \\
\dot{\xi}_i &= -\frac{1}{2}\left(\partial_{y^i} g^{hk}_y \xi_h \xi_k + \partial_{y^i} g^{hk_\delta}_y \xi_h \xi_{k_\delta} + \partial_{y^i} g^{h_\gamma k}_y \xi_{h_\gamma}\xi_k + \partial_{y^i} g^{h_\gamma k_\delta}_y \xi_{h_\gamma}\xi_{k_\delta}\right) \\
\dot{\xi}_{i_\alpha} &= -\frac{1}{2}\left(\partial_{y^{i_\alpha}} g^{hk}_y \xi_h \xi_k + \partial_{y^{i_\alpha}} g^{hk_\delta}_y \xi_h \xi_{k_\delta} + \partial_{y^{i_\alpha}} g^{h_\gamma k}_y \xi_{h_\gamma}\xi_k + \partial_{y^{i_\alpha}} g^{h_\gamma k_\delta}_y \xi_{h_\gamma}\xi_{k_\delta}\right)
\end{aligned}$$

with $\Gamma^{h_\gamma}_{k,i}$ for $\partial_{y^i}\Gamma^{h_\gamma}_k$, and where

$$\partial_{y^l} g^{ij} = 0\,, \quad \partial_{y^l} g^{ij_\beta} = -W^{ih}\Gamma^{j_\beta}_{h,l}\,, \quad \partial_{y^l} g^{i_\alpha j} = -\Gamma^{i_\alpha}_{h,l} W^{hj}\,, \quad \partial_{y^l} g^{i_\alpha j_\beta} = \Gamma^{i_\alpha}_{k,l} W^{kh}\Gamma^{j_\beta}_h + \Gamma^{i_\alpha}_k W^{kh}\Gamma^{j_\beta}_{h,l}\,,$$

$$\partial_{y^{l_\zeta}} g^{ij} = W^{ij}{}_{,l_\zeta}\,, \quad \partial_{y^{l_\zeta}} g^{ij_\beta} = -W^{ih}{}_{,l_\zeta}\Gamma^{j_\beta}_h - W^{ih}\Gamma^{j_\beta}_{h,l_\zeta}\,, \quad \partial_{y^{l_\zeta}} g^{i_\alpha j} = -\Gamma^{i_\alpha}_{h,l_\zeta} W^{hj} - \Gamma^{i_\alpha}_h W^{hj}{}_{,l_\zeta}\,,$$
$$\partial_{y^{l_\zeta}} g^{i_\alpha j_\beta} = \Gamma^{i_\alpha}_{k,l_\zeta} W^{kh}\Gamma^{j_\beta}_h + \Gamma^{i_\alpha}_k W^{kh}{}_{,l_\zeta}\Gamma^{j_\beta}_h + \Gamma^{i_\alpha}_k W^{kh}\Gamma^{j_\beta}_{h,l_\zeta}\,,$$
$$\Gamma^{i_\alpha}_{h,l_\zeta} = \partial_{y^{l_\zeta}}\left(\Gamma^i{}_{hk}u^k_\alpha\right) = \delta^{\zeta\alpha}\Gamma^i{}_{hl}\,, \quad W^{ij}{}_{,l_\zeta} = \delta^{il}u^j_\zeta + \delta^{jl}u^i_\zeta\,.$$

Combining these expressions, we obtain

$$\dot{x}^i = W^{ij}\xi_j - W^{ih}\Gamma^{j_\beta}_h\xi_{j_\beta}\quad,\quad \dot{X}^i_\alpha = -\Gamma^{i_\alpha}_h W^{hj}\xi_j + \Gamma^{i_\alpha}_k W^{kh}\Gamma^{j_\beta}_h\xi_{j_\beta}$$
$$\dot{\xi}_i = W^{hl}\Gamma^{k_\delta}_{l,i}\xi_h\xi_{k_\delta} - \frac{1}{2}\left(\Gamma^{h_\gamma}_{k,i}W^{kh}\Gamma^{k_\delta}_h + \Gamma^{h_\gamma}_k W^{kh}\Gamma^{k_\delta}_{h,i}\right)\xi_{h_\gamma}\xi_{k_\delta}$$
$$\dot{\xi}_{i_\alpha} = \Gamma^{h_\delta}_{k,i_\alpha}W^{kh}\Gamma^{k_\delta}_h\xi_{h_\gamma}\xi_{k_\delta} - \left(W^{hl}{}_{,i_\alpha}\Gamma^{k_\delta}_l + W^{hl}\Gamma^{k_\delta}_{l,i_\alpha}\right)\xi_h\xi_{k_\delta} - \frac{1}{2}\left(W^{hk}{}_{,i_\alpha}\xi_h\xi_k + \Gamma^{h_\delta}_k W^{kh}{}_{,i_\alpha}\Gamma^{k_\delta}_h\xi_{h_\gamma}\xi_{k_\delta}\right).$$

4.3. Acceleration and Polynomials for C

We can identify the covariant acceleration $\nabla_{\dot{x}_t}\dot{x}_t$ of curves satisfying the MPP equations, and hence normal MPPs through their frame coordinates. Let (u_t, ξ_t) satisfy (13). Then, u_t is a horizontal lift of $x_t = \pi(u_t)$ and hence by (1), (3), (10), and (15),

$$\begin{aligned}
u_t^{-1}\nabla_{\dot{x}_t}\dot{x}_t &= \frac{d}{dt}\begin{pmatrix}\xi(h_{u_t}(u_te_1))\\ \vdots \\ \xi(h_{u_t}(u_te_d))\end{pmatrix} = \begin{pmatrix}\dot{\xi}(h_{u_t}(u_te_1))\\ \vdots \\ \dot{\xi}(h_{u_t}(u_te_d))\end{pmatrix} + \begin{pmatrix}\xi(\partial_t h_{u_t}(u_te_1))\\ \vdots \\ \xi(\partial_t h_{u_t}(u_te_d))\end{pmatrix}\\
&= -\begin{pmatrix}\xi(\partial_{h_{u_t}(u_te_1)}h_{u_t}(\pi_*(\dot{u}_t)))\\ \vdots \\ \xi(\partial_{h_{u_t}(u_te_d)}h_{u_t}(\pi_*(\dot{u}_t)))\end{pmatrix} + \begin{pmatrix}\xi(\partial_{h_{u_t}(\pi_*(\dot{u}_t))}h_{u_t}(u_te_1))\\ \vdots \\ \xi(\partial_{h_{u_t}(\pi_*(\dot{u}_t))}h_{u_t}(u_te_d))\end{pmatrix}\\
&= \begin{pmatrix}\xi(\psi(u_t, R_{u_t}(u_te_1, \pi_*(\dot{u}_t))))\\ \vdots \\ \xi(\psi(u_t, R_{u_t}(u_te_d, \pi_*(\dot{u}_t))))\end{pmatrix}.
\end{aligned} \tag{16}$$

The fact that the covariant derivative vanishes for classical geodesic leads to a definition of higher-order polynomials through the covariant derivative by requiring $(\nabla_{\dot{x}_t})^k\dot{x}_t = 0$ for a kth order polynomial (e.g., [26,27]). As discussed above, compared to classical geodesics, curves satisfying the MPP equations have extra d^2 degrees of freedom, allowing the curves to twist and deviate from being geodesic with respect to C while still satisfying the horizontality constraint on FM. This makes it natural to ask if normal MPPs relate to polynomials defined using C. For curves satisfying the MPP equations, using (16) and (15), we have

$$u_t^{-1}(\nabla_{\dot{x}_t})^2\dot{x}_t = \frac{d}{dt}\begin{pmatrix}\xi(\psi(u_t, R_{u_t}(u_te_1, \pi_*(\dot{u}_t))))\\ \vdots \\ \xi(\psi(u_t, R_{u_t}(u_te_d, \pi_*(\dot{u}_t))))\end{pmatrix} = \begin{pmatrix}\xi(\psi(u_t, \frac{d}{dt}R_{u_t}(u_te_1, \pi_*(\dot{u}_t))))\\ \vdots \\ \xi(\psi(u_t, \frac{d}{dt}R_{u_t}(u_te_d, \pi_*(\dot{u}_t))))\end{pmatrix}.$$

Thus, in general, normal MPPs are not second order polynomials in the sense $(\nabla_{\dot{x}_t})^2\dot{x}_t = 0$ unless the curvature $R_{u_t}(u_te_i, \pi_*(\dot{u}_t))$ is constant in t.

For comparison, in the Riemannian case, a variational formulation placing a cost on covariant acceleration [28,29] leads to cubic splines

$$(\nabla_{\dot{x}_t})^2\dot{x}_t = -R(\nabla_{\dot{x}_t}\dot{x}_t, x_t,)\dot{x}_t\,.$$

In (16), the curvature terms appear in the covariant acceleration for normal MPPs, while cubic splines leads to the curvature term appearing in the third order derivative.

5. Cometric Formulation and Low-Rank Generator

We now investigate a cometric $g_{F^kM} + \lambda g_R$, where g_R is Riemannian, g_{F^kM} is a rank k positive semi-definite inner product arising from k linearly independent tangent vectors, and $\lambda > 0$ a weight. We assume that g_{F^kM} is chosen so that $g_{F^kM} + \lambda g_R$ is invertible, even though g_{F^kM} is rank-deficient. The situation corresponds to extracting the first k eigenvectors in Euclidean space PCA. If the eigenvectors are estimated statistically from observed data, this allows the estimation to be restricted to only the first k eigenvectors. In addition, an important practical implication of the construction is that a numerical implementation need not transport a full $d \times d$ matrix for the frame, but a potentially much lower dimensional $d \times k$ matrix. This point is essential when dealing with high-dimensional data, examples of which are landmark manifolds as discussed in Section 6.

When using the frame bundle to model covariances, the sum formulation is natural to express as a cometric compared to a metric because, with the cometric formulation, $g_{F^kM} + \lambda g_R$ represents a sum of covariance matrices instead of a sum of precision matrices. Thus, $g_{F^kM} + \lambda g_R$ can be intuitively thought of as adding isotropic noise of variance λ to the covariance represented by g_{F^kM}.

To pursue this, let F^kM denote the bundle of rank k linear maps $\mathbb{R}^k \to T_xM$. We define a cometric by

$$\langle \xi, \tilde{\xi} \rangle = \delta^{\alpha\beta}(\xi | h_u(u_\alpha))(\tilde{\xi} | h_u(u_\beta)) + \lambda \langle \xi, \tilde{\xi} \rangle_{g_R}$$

for $\xi, \tilde{\xi} \in T_u^*F^kM$. The sum over α, β is for $\alpha, \beta = 1, \ldots, k$. The first term is equivalent to the lift (9) of the cometric $\langle \xi, \tilde{\xi} \rangle = (\xi | g_u(\tilde{\xi}))$ given $u : \mathbb{R}^k \to T_xM$. Note that in the definition (6) of g_u, the map u is not inverted; thus, the definition of the metric immediately carries over to the rank-deficient case.

Let (x^i, u^i_α), $\alpha = 1, \ldots, k$ be a coordinate system on F^kM. The vertical distribution is in this case spanned by the dk vector fields $D_{j_\beta} = \partial_{u^j_\beta}$. Except for index sums being over k instead of d terms, the situation is thus similar to the full-rank case. Note that $(\xi | \pi_*^{-1} w) = (\xi | w^j D_j) = w^i \xi_i$. The cometric in coordinates is

$$\langle \xi, \tilde{\xi} \rangle = \delta^{\alpha\beta} u^i_\alpha \xi_i u^j_\beta \tilde{\xi}_j + \lambda g_R^{ij} \xi_i \tilde{\xi}_j = \xi_i \left(\delta^{\alpha\beta} u^i_\alpha u^j_\beta + \lambda g_R^{ij} \right) \tilde{\xi}_j = \xi_i W^{ij} \tilde{\xi}_j$$

with $W^{ij} = \delta^{\alpha\beta} u^i_\alpha u^j_\beta + \lambda g_R^{ij}$. We can then write the corresponding sub-Riemannian metric g_{F^kM} in terms of the adapted frame D

$$g_{F^kM}(\xi_h D^h + \xi_{h_\gamma} D^{h_\gamma}) = W^{ih} \xi_h D_i \tag{17}$$

because $(\xi | g_{F^kM}(\tilde{\xi})) = \langle \xi, \tilde{\xi} \rangle = \xi_i W^{ij} \tilde{\xi}_j$. That is, the situation is analogous to (11), except the term λg_R^{ij} is added to W^{ij}.

The geodesic system is again given by the Hamilton–Jacobi equations. As in the full-rank case, the system is specified by the derivatives of g_{F^kM}:

$$\begin{aligned}
&\partial_{y^l} g^{ij}_{F^kM} = W^{ij}{}_{,l}\,, \quad \partial_{y^l} g^{ij_\beta}_{F^kM} = -W^{ih}{}_{,l} \Gamma^{j_\beta}_h - W^{ih} \Gamma^{j_\beta}_{h,l}\,, \quad \partial_{y^l} g^{i_\alpha j}_{F^kM} = -\Gamma^{i_\alpha}_{h,l} W^{hj} - \Gamma^{i_\alpha}_h W^{hj}{}_{,l}\,, \\
&\partial_{y^l} g^{i_\alpha j_\beta}_{F^kM} = \Gamma^{i_\alpha}_{k,l} W^{kh} \Gamma^{j_\beta}_h + \Gamma^{i_\alpha}_k W^{kh}{}_{,l} \Gamma^{j_\beta}_h + \Gamma^{i_\alpha}_k W^{kh} \Gamma^{j_\beta}_{h,l}\,, \\
&\partial_{y^{l_\zeta}} g^{ij}_{F^kM} = W^{ij}{}_{,l_\zeta}\,, \quad \partial_{y^{l_\zeta}} g^{ij_\beta}_{F^kM} = -W^{ih}{}_{,l_\zeta} \Gamma^{j_\beta}_h - W^{ih} \Gamma^{j_\beta}_{h,l_\zeta}\,, \quad \partial_{y^{l_\zeta}} g^{i_\alpha j}_{F^kM} = -\Gamma^{i_\alpha}_h W^{hj}{}_{,l_\zeta} - \Gamma^{i_\alpha}_{h,l_\zeta} W^{hj}\,, \\
&\partial_{y^{l_\zeta}} g^{i_\alpha j_\beta}_{F^kM} = \Gamma^{i_\alpha}_{k,l_\zeta} W^{kh} \Gamma^{j_\beta}_h + \Gamma^{i_\alpha}_k W^{kh}{}_{,l_\zeta} \Gamma^{j_\beta}_h + \Gamma^{i_\alpha}_k W^{kh} \Gamma^{j_\beta}_{h,l_\zeta}\,, \\
&\Gamma^{i_\alpha}_{h,l_\zeta} = \partial_{y^{l_\zeta}} \left(\Gamma^i{}_{hk} u^k_\alpha \right) = \delta^{\zeta\alpha} \Gamma^i{}_{hl}\,, \quad W^{ij}{}_{,l} = \lambda g_R{}^{ij}{}_{,l}\,, \quad W^{ij}{}_{,l_\zeta} = \delta^{il} u^j_\zeta + \delta^{jl} u^i_\zeta\,.
\end{aligned}$$

Note that the introduction of the Riemannian metric g_R implies that W^{ij} are now dependent on the manifold coordinates x^i.

6. Numerical Experiments

We aim at visualizing most probable paths for the driving process and projections of curves satisfying the MPP Equation (13) in two cases: On 2D surfaces embedded in $\mathbb{R}^3$ and on finite dimensional landmark manifolds that arise from equipping a subset of the diffeomorphism group with a right-invariant metric and letting the action descend to the landmarks by a left action. The surface examples are implemented in Python using the Theano [30] framework for symbolic operations, automatic differentiation, and numerical evaluation. The landmark equations are detailed below and implemented in Numpy using Numpy's standard ODE integrators. The code for running the experiments is available at http://bitbucket.com/stefansommer/mpps/.

6.1. Embedded Surfaces

We visualize normal MPPs and projections of curves satisfying the MPP Equation (13) on surfaces embedded in $\mathbb{R}^3$ in three cases: The sphere $\mathbb{S}^2$, on an ellipsoid, and on a hyperbolic surface. The surfaces are chosen in order to have both positive and negative curvature, and to have varying degree of symmetry. In all cases, an open subset of the surfaces are represented in a single chart by a map $F : \mathbb{R}^2 \to \mathbb{R}^3$. For the sphere and ellipsoid, this gives a representation of the surface, except for the south pole. The metric and Christoffel symbols are calculated using the symbolic differentiation features of Theano. The integration are performed by a simple Euler integrator.

Figures 4–6 show families of curves satisfying the MPP equations in three cases: (1) With fixed starting point $x_0 \in M$ and initial velocity $\dot{x}_0 \in TM$ but varying anisotropy represented by changing frame u in the fiber above x_0; (2) minimizing normal MPPs with fixed starting point and endpoint $x_0, x_1 \in M$ but changing frame u above x_0; (3) fixed starting point $x_0 \in M$ and frame u but varying V^*FM vertical part of the initial momentum $\xi_0 \in T^*FM$. The first and second cases thus show the effect of varying anisotropy, while the third case illustrates the effect of the "twist" that the d^2 degrees in the vertical momentum allows. Note the displayed anti-developed curves in $\mathbb{R}^2$ that for classical C geodesics would always be straight lines.

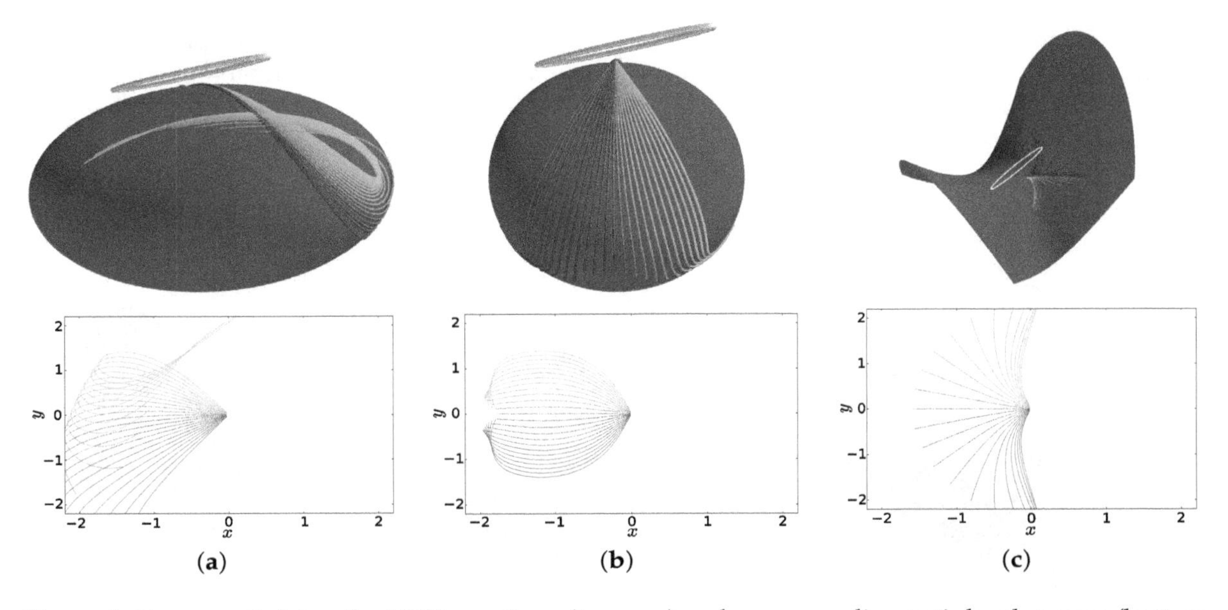

Figure 4. Curves satisfying the MPP equations (top row) and corresponding anti-development (bottom row) on three surfaces embedded in $\mathbb{R}^3$: (**a**) An ellipsoid; (**b**) a sphere; (**c**) a hyperbolic surface. The family of curves is generated by rotating by $\pi/2$ radians the anisotropic covariance represented in the initial frame u_0 and displayed in the gray ellipse.

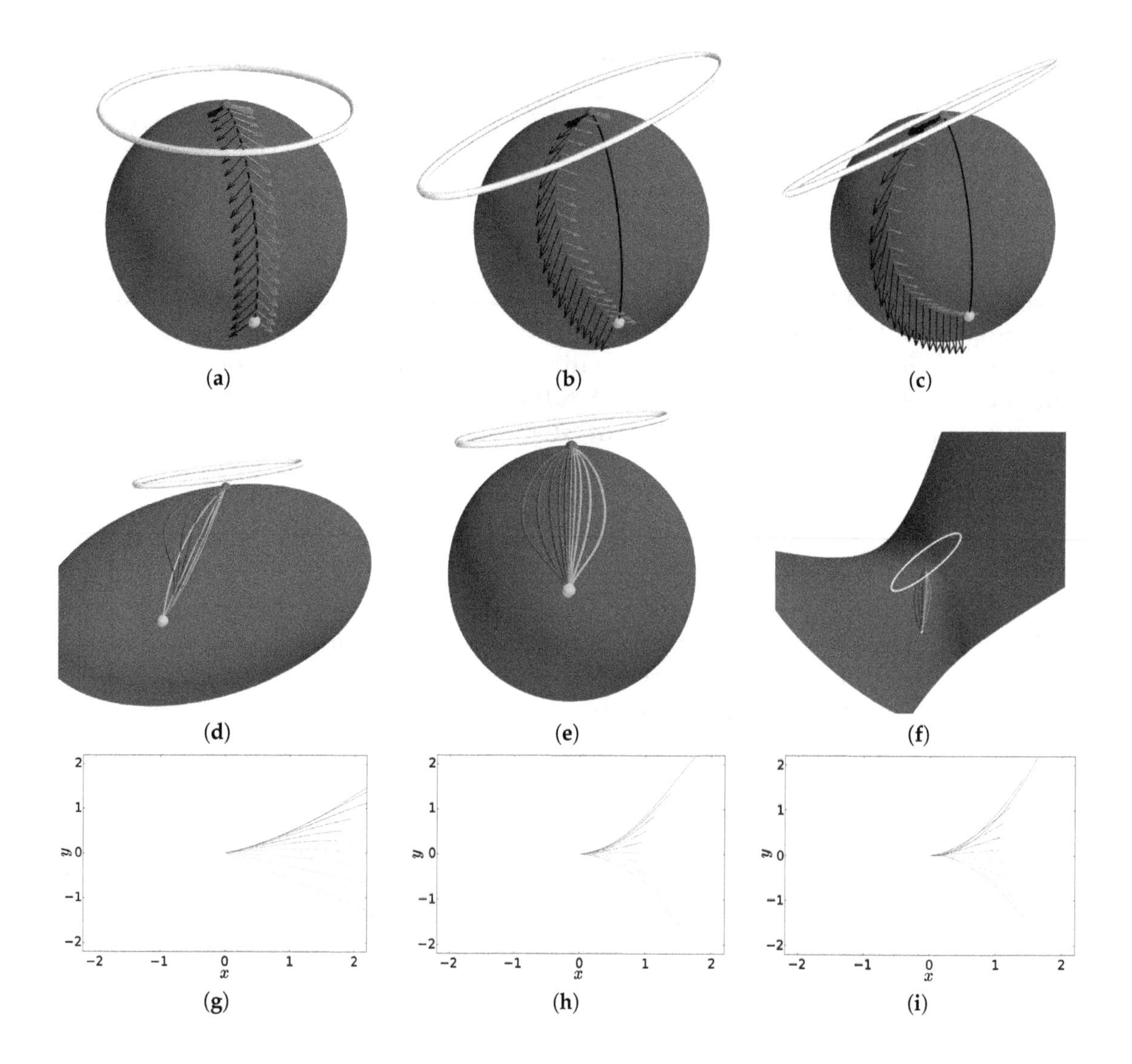

Figure 5. Minimizing normal MPPs between two fixed points (red/cyan). From isotropic covariance (top row, (**a**)) to anisotropic (top row, (**c**)) on $\mathbb{S}^2$. Compare with minimizing Riemannian geodesic (black curve). The MPP travels longer in the directions of high variance. Families of curves (middle row, (**d–f**)) and corresponding anti-development (bottom row, (**g–i**)) on the three surfaces in Figure 4. The family of curves is generated by rotating the covariance matrix as in Figure 4. Notice how the varying anisotropy affects the resulting minimizing curves, and how the anti-developed curves end at different points in $\mathbb{R}^2$.

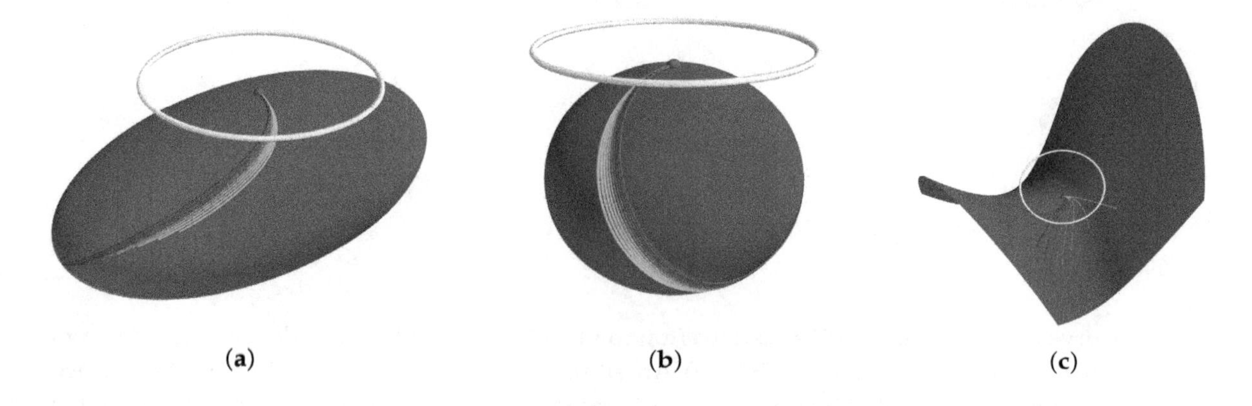

Figure 6. *Cont.*

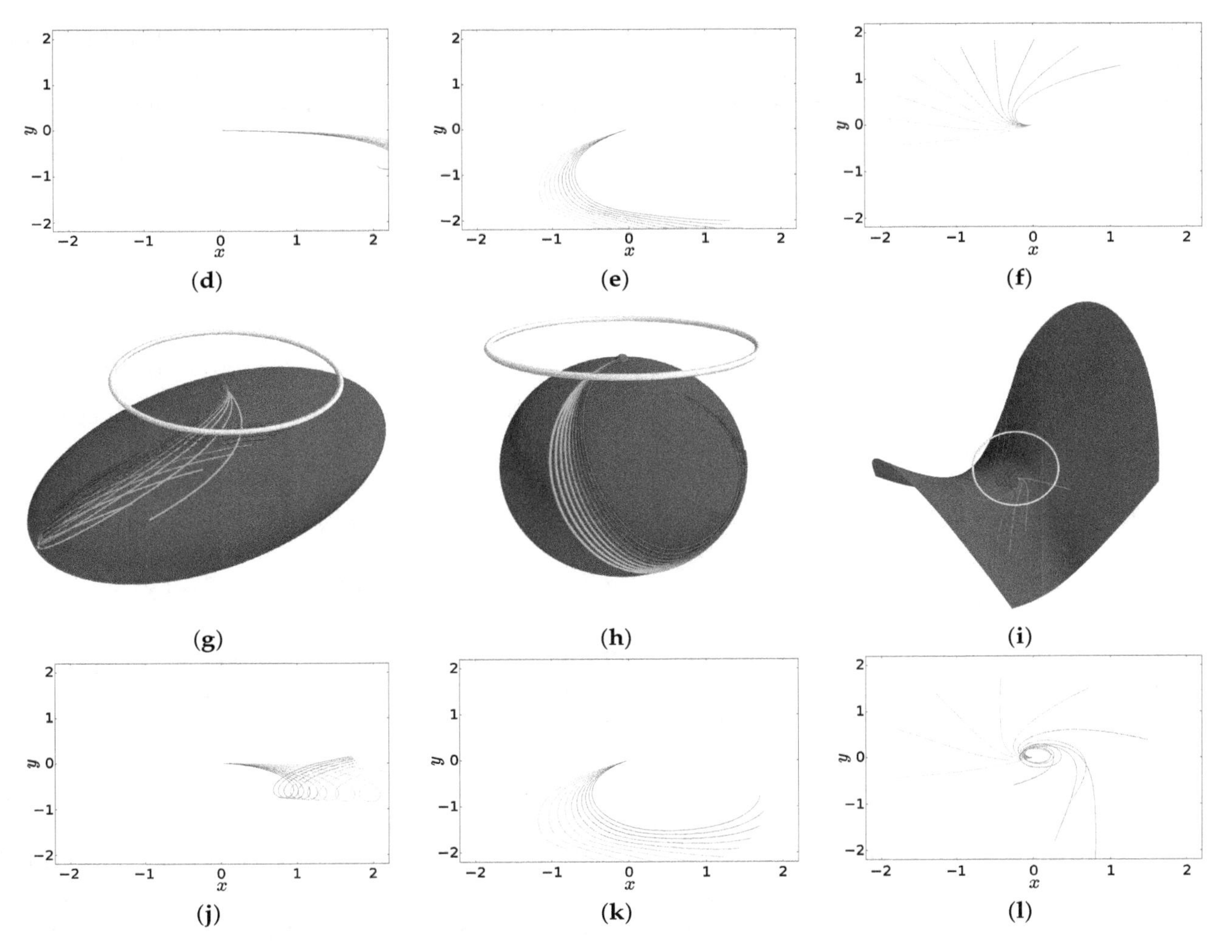

Figure 6. (**a–l**) With the setup of Figures 4 and 5, generated families of curves by varying the vertical V^*FM part of the initial momentum $\xi_0 \in T^*FM$ but keeping the base point and frame u_0 fixed. The vertical part allows varying degree of "twisting" of the curve.

6.2. LDDMM Landmark Equations

We here give a example of the MPP equations using the finite dimensional landmark manifolds that arise from right invariant metrics on subsets of the diffeomorphism group in the Large Deformation Diffeomorphic Metric Mapping (LDDMM) framework [8]. The LDDMM metric can be conveniently expressed as a cometric, and, using a rank-deficient inner product g_{F^kM} as discussed in Section 5, we can obtain a reduction of the system of equations to $2(2N+2Nk)$ compared to $2(2N+(2N)^2)$ with N landmarks in $\mathbb{R}^2$.

Let $\{p_1,\ldots,p_N\}$ be landmarks in a subset $\Omega \subset \mathbb{R}^d$. The diffeomorphism group $\mathrm{Diff}(\Omega)$ acts on the left on landmarks with the action $\varphi.\{p_1,\ldots,p_N\} = \{\varphi(p_1),\ldots,\varphi(p_N)\}$. In LDDMM, a Hilbert space structure is imposed on a linear subspace V of $L^2(\Omega,\mathbb{R}^d)$ using a self-adjoint operator $L: V \to V^* \subset L^2(\Omega,\mathbb{R}^d)$ and defining the inner product $\langle\cdot,\cdot\rangle_V$ by

$$\langle v,w\rangle_V = \langle Lv,w\rangle_{L^2}\,.$$

Under sufficient conditions on L, V is reproducing and admits a kernel K inverse to L. K is a Green's kernel when L is a differential operator, or K can be a Gaussian kernel. The Hilbert structure on V gives a Riemannian metric on a subset $G_V \subset \mathrm{Diff}(\Omega)$ by setting $\|v\|_\varphi^2 = \|v\circ\varphi^{-1}\|_V^2$; i.e., regarding $\langle\cdot,\cdot\rangle_V$ an inner product on $T_{\mathrm{Id}}G_V$ and extending the metric to G_V by right-invariance. This Riemannian metric descends to a Riemannian metric on the landmark space.

Let M be the manifold $M = \{(p_1^1, \ldots, p_1^d, \ldots, p_N^1, \ldots, p_N^d) | (p_i^1, \ldots, p_i^d) \in \mathbb{R}^d\}$. The LDDMM metric on the landmark manifold M is directly related to the kernel K when written as a cometric $g_p(\xi, \eta) = \sum_{i,j=1}^N \xi^i K(p_i, p_j) \eta^j$. Letting i^k denote the index of the kth component of the ith landmark, the cometric is in coordinates $g_p^{i^k j^l} = K(p_i, p_j)_k^l$. The Christoffel symbols can be written in terms of derivatives of the cometric g^{ij} [31] (recall that $\delta_j^i = g^{ik} g_{kj} = g_{jk} g^{ki}$)

$$\Gamma^k{}_{ij} = \frac{1}{2} g_{ir} \left(g^{kl} g^{rs}{}_{,l} - g^{sl} g^{rk}{}_{,l} - g^{rl} g^{ks}{}_{,l} \right) g_{sj} \,. \tag{18}$$

This relation comes from the fact that $g_{jm,k} = -g_{jr} g^{rs}{}_{,k} g_{sm}$ gives the derivative of the metric. The derivatives of the cometric is simply $g^{i^k j^l}{}_{,r^q} = (\delta_r^i + \delta_r^j) \partial_{p_r^q} K(p_i, p_j)_k^l$. Using (18), derivatives of the Christoffel symbols can be computed

$$\begin{aligned} \Gamma^k{}_{ij,\xi} = {} & \frac{1}{2} g_{ir,\xi} \left(g^{kl} g^{rs}{}_{,l} - g^{sl} g^{rk}{}_{,l} - g^{rl} g^{ks}{}_{,l} \right) g_{sj} + \frac{1}{2} g_{ir} \left(g^{kl} g^{rs}{}_{,l} - g^{sl} g^{rk}{}_{,l} - g^{rl} g^{ks}{}_{,l} \right) g_{sj,\xi} \\ & + \frac{1}{2} g_{ir} \left(g^{kl}{}_{,\xi} g^{rs}{}_{,l} + g^{kl} g^{rs}{}_{,l\xi} - g^{sl}{}_{,\xi} g^{rk}{}_{,l} - g^{sl} g^{rk}{}_{,l\xi} - g^{rl}{}_{,\xi} g^{ks}{}_{,l} - g^{rl} g^{ks}{}_{,l\xi} \right) g_{sj} \,. \end{aligned}$$

This provides the full data for numerical integration of the evolution equations on $F^k M$.

In Figure 7 (top row), we plot minimizing normal MPPs on the landmark manifold with two landmarks and varying covariance in the $\mathbb{R}^2$ horizontal and vertical direction. The plot shows the landmark equivalent of the experiment in Figure 5. Note how adding covariance in the horizontal and vertical direction, respectively, allows the minimizing normal MPP to vary more in these directions because the anisotropically-weighted metric penalizes high-covariance directions less.

Figure 7 (bottom row) shows five curves satisfying the MPP equations with varying vertical V^*FM initial momentum similarly to the plots in Figure 6. Again, we see how the extra degrees of freedom allows the paths to twist, generating a higher-dimensional family than classical geodesics with respect to $\mathcal{C}$.

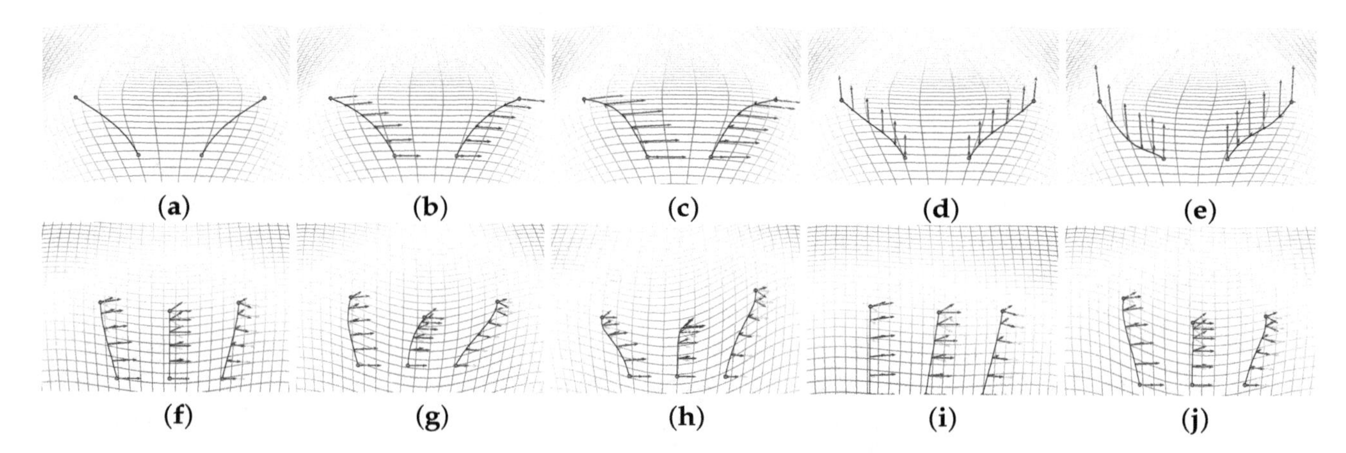

Figure 7. (Top row) Matching of two landmarks (green) to two landmarks (red) by (**a**) computing a minimizing Riemannian geodesic on the landmark manifold, and (**b–e**) minimizing MPPs with added covariance (arrows) in $\mathbb{R}^2$ horizontal direction (**b**,**c**) and vertical (**d**,**e**). The action of the corresponding diffeomorphisms on a regular grid is visualized by the deformed grid which is colored by the warp strain. The added covariance allows the paths to have more movement in the horizontal and vertical direction, respectively, because the anisotropically weighted metric penalizes high-covariance directions less. (bottom row, (**f–j**)) Five landmark trajectories with fixed initial velocity and anisotropic covariance but varying V^*FM vertical initial momentum ξ_0. Changing the vertical momentum "twists" the paths.

7. Discussion and Concluding Remarks

Incorporating anisotropy in models for data in non-linear spaces via the frame bundle as pursued in this paper leads to a sub-Riemannian structure and metric. A direct implication is that most probable paths to observed data in the sense of sequences of stochastic steps of a driving semi-martingale are not related to geodesics in the classical sense. Instead, a best estimate of the sequence of steps $w_t \in \mathbb{R}^d$ that leads to an observation $x = \varphi_u(w_t)|_{t=1}$ is an MPP in the sense of Definition 1. As shown in the paper, these paths are generally not geodesics or polynomials with respect to the connection on the manifold. In particular, if M has a Riemannian structure, the MPPs are generally neither Riemannian geodesics nor Riemannian polynomials. Below, we discuss the statistical implications of this result.

7.1. Statistical Estimators

Metric distances and Riemannian geodesics have been the traditional vehicle for representing observed data in non-linear spaces. Most fundamentally, the sample Frechét mean

$$\hat{x} = \text{argmin}_{x \in M} \sum_{i=1}^{N} d_{g_R}(x, x_i)^2 \tag{19}$$

of observed data $x_1, \ldots, x_N \in M$ relies crucially on the Riemannian distance d_{g_R} connected to the metric g_R. Many PCA constructs (e.g., Principal Geodesics Analysis [6]) use the Riemannian Exp. and Log maps to map between linear tangent spaces and the manifold. These maps are defined from the Riemannian metric and Riemannian geodesics. Distributions modelled as in the random orbit model [32] or Bayesian models [15,33] again rely on geodesics with random initial conditions.

Using the frame bundle sub-Riemannian metric g_{FM}, we can define an estimator analogous to the Riemannian Frechét mean estimator. Assuming the covariance is a priori known, the estimator

$$\hat{x} = \text{argmin}_{u \in s(M)} \sum_{i=1}^{N} d_{FM}\left(u, \pi^{-1}(x_i)\right)^2 \tag{20}$$

acts correspondingly to the Frechét mean estimator (19). Here $s \in \Gamma(FM)$ is a (local) section of FM that to $x \in M$ connects the known covariance represented by $s(x) \in FM$. The distances $d_{FM}\left(u, \pi^{-1}(x_i)\right)$, $u = s(x)$ are realized by MPPs from the mean candidate x to the fibers $\pi^{-1}(x_i)$. The Frechét mean problem is thus lifted to the frame bundle with the anisotropic weighting incorporated in the metric g_{FM}. This metric is not related to g_R, except for its dependence on the connection C that can be defined as the Levi–Civita connection of g_R. The fundamental role of the distance d_{g_R} and g_R geodesics in (19) is thus removed.

Because covariance is an integral part of the model, sample covariance can also be estimated directly along with the sample mean. In [3], the estimator

$$\hat{u} = \text{argmin}_{u \in FM} \sum_{i=1}^{N} d_{FM}\left(u, \pi^{-1}(x_i)\right)^2 - N \log(\det_{g_R} u) \tag{21}$$

is suggested. The normalizing term $-N \log(\det_{g_R} u)$ is derived such that the estimator exactly corresponds to the maximum likelihood estimator of mean and covariance for Euclidean Gaussian distributions. The determinant is defined via g_R, and the term acts to prevent the covariance from approaching infinity. Maximum likelihood estimators of mean and covariance for normally distributed Euclidean data have unique solutions in the sample mean and sample covariance matrix, respectively. Uniqueness of the Frechét mean (19) is only ensured for sufficiently concentrated data. For the estimator (21), existence and uniqueness properties are not immediate, and more work is needed in order to find necessary and sufficient conditions.

7.2. Priors and Low-Rank Estimation

The low-rank cometric formulation pursued in Section 5 gives a natural restriction of (21) to $u \in F^kM, 1 \leq k \leq d$. As for Euclidean PCA, most variance is often captured in the span of the first k eigenvectors with $k \ll d$. Estimates of the remaining eigenvectors are generally ignored, as the variance of the eigenvector estimates increases as the noise captured in the span of the last eigenvectors becomes increasingly uniform. The low-rank cometric restricts the estimation to only the first k eigenvectors, and thus builds the construction directly into the model. In addition, it makes numerical implementation feasible, because a numerical representation need only store and evolve $d \times k$ matrices. As a different approach for regularizing the estimator (21), the normalizing term $-N \log(\det_{g_R} u)$ can be extended with other priors (e.g., an L^1-type penalizing term). Such priors can potentially partly remove existence and uniqueness issues, and result in additional sparsity properties that can benefit numerical implementations. The effects of such priors have yet to be investigated.

In the $k = d$ case, the number of degrees of freedom for the MPPs grows quadratically in the dimension d. This naturally increases the variance of any MPP estimate given only one sample from its trajectory. The low-rank cometric formulation reduces the growth to linear in d. The number of degrees of freedom is however still k times larger than for Riemannian geodesics. With longitudinal data, more samples per trajectory can be obtained, reducing the variance and allowing a better estimate of the MPP. However, for the estimators (20) and (21) above, estimates of the actual optimal MPPs are not needed—only their squared length. It can be hypothesized that the variance of the length estimates is lower than the variance of the estimates of the corresponding MPPs. Further investigation regarding this will be the subject of future work.

7.3. Conclusions

The underlying model of anisotropy used in this paper originates from the anisotropic normal distributions formulated in [2] and the diffusion PCA framework [1]. Because many statistical models are defined using normal distributions, this approach to incorporating anisotropy extends to models such as linear regression. We expect that finding most probable paths in other statistical models such as regressions models can be carried out with a program similar to the program presented in this paper.

The difference between MPPs and geodesics shows that the geometric and metric properties of geodesics, zero acceleration, and local distance minimization are not directly related to statistical properties such as maximizing path probability. Whereas the concrete application and model determines if metric or statistical properties are fundamental, most statistical models are formulated without referring to metric properties of the underlying space. It can therefore be argued that the direct incorporation of anisotropy and the resulting MPPs are natural in the context of many models of data variation in non-liner spaces.

Acknowledgments: The author wishes to thank Peter W. Michor and Sarang Joshi for suggestions for the geometric interpretation of the sub-Riemannian metric on FM and discussions on diffusion processes on manifolds. The work was supported by the Danish Council for Independent Research, the CSGB Centre for Stochastic Geometry and Advanced Bioimaging funded by a grant from the Villum foundation, and the Erwin Schrödinger Institute in Vienna.

References

1. Sommer, S. Diffusion Processes and PCA on Manifolds. Available online: https://www.mfo.de/document/1440a/OWR_2014_44.pdf (accessed on 24 November 2016).
2. Sommer, S. Anisotropic distributions on manifolds: Template estimation and most probable paths. In *Information Processing in Medical Imaging*; Lecture Notes in Computer Science; Springer: Berlin/Heidelberg, Germany, 2015; Volume 9123, pp. 193–204.
3. Sommer, S.; Svane, A.M. Modelling anisotropic covariance using stochastic development and sub-riemannian frame bundle geometry. *J. Geom. Mech.* **2016**, in press.

4. Hsu, E.P. *Stochastic Analysis on Manifolds*; American Mathematical Society: Providence, RI, USA, 2002.
5. Fréchet, M. Les éléments aléatoires de nature quelconque dans un espace distancie. *Annales de l'Institut Henri Poincaré* **1948**, *10*, 215–310.
6. Fletcher, P.; Lu, C.; Pizer, S.; Joshi, S. Principal geodesic analysis for the study of nonlinear statistics of shape. *IEEE Trans. Med. Imaging* **2004**, *23*, 995–1005.
7. Vaillant, M.; Miller, M.; Younes, L.; Trouvé, A. Statistics on diffeomorphisms via tangent space representations. *NeuroImage* **2004**, *23*, S161–S169.
8. Younes, L. *Shapes and Diffeomorphisms*; Springer: Berlin/Heidelberg, Germany, 2010.
9. Pennec, X. Intrinsic statistics on riemannian manifolds: Basic tools for geometric measurements. *J. Math. Imaging Vis.* **2006**, *25*, 127–154.
10. Karcher, H. Riemannian center of mass and mollifier smoothing. *Commun. Pure Appl. Math.* **1977**, *30*, 509–541.
11. Huckemann, S.; Hotz, T.; Munk, A. Intrinsic shape analysis: Geodesic PCA for Riemannian manifolds modulo isometric lie group actions. *Stat. Sin.* **2010**, *20*, 1–100.
12. Jung, S.; Dryden, I.L.; Marron, J.S. Analysis of principal nested spheres. *Biometrika* **2012**, *99*, 551–568.
13. Pennec, X. Barycentric subspaces and affine spans in manifolds. In Proceedings of the Second International Conference on Geometric Science of Information, Paris, France, 28–30 October 2015; Nielsen, F., Barbaresco, F., Eds.; Lecture Notes in Computer Science; pp. 12–21.
14. Sommer, S. Horizontal dimensionality reduction and iterated frame bundle development. In *Geometric Science of Information*; Springer: Berlin/Heidelberg, Germany, 2013; pp. 76–83.
15. Zhang, M.; Fletcher, P. Probabilistic principal geodesic analysis. In Proceedings of the 26th International Conference on Neural Information Processing Systems, Lake Tahoe, Nevada, 5–10 December 2013; pp. 1178–1186.
16. Tipping, M.E.; Bishop, C.M. Probabilistic principal component analysis. *J. R. Stat. Soc. Ser. B* **1999**, *61*, 611–622.
17. Elworthy, D. Geometric aspects of diffusions on manifolds. In *École d'Été de Probabilités de Saint-Flour XV–XVII, 1985–1987*; Hennequin, P.L., Ed.; Number 1362 in Lecture Notes in Mathematics; Springer: Berlin/Heidelberg, Germany, 1988; pp. 277–425.
18. Mok, K.P. On the differential geometry of frame bundles of Riemannian manifolds. *J. Reine Angew. Math.* **1978**, *1978*, 16–31.
19. Taubes, C.H. *Differential Geometry: Bundles, Connections, Metrics and Curvature*, 1st ed.; Oxford University Press: Oxford, UK; New York, NY, USA, 2011.
20. Kolář, I.; Slovák, J.; Michor, P.W. *Natural Operations in Differential Geometry*; Springer: Berlin/Heidelberg, Germany, 1993.
21. Andersson, L.; Driver, B.K. Finite dimensional approximations to wiener measure and path integral formulas on manifolds. *J. Funct. Anal.* **1999**, *165*, 430–498.
22. Fujita, T.; Kotani, S.i. The Onsager-Machlup function for diffusion processes. *J. Math. Kyoto Univ.* **1982**, *22*, 115–130.
23. Strichartz, R.S. Sub-Riemannian geometry. *J. Differ. Geom.* **1986**, *24*, 221–263.
24. Bloch, A.M. Nonholonomic mechanics and control. In *Interdisciplinary Applied Mathematics*; Springer: New York, NY, USA, 2003; Volume 24,
25. Marsden, J.E.; Ratiu, T.S. Introduction to mechanics and symmetry. In *Texts in Applied Mathematics*; Springer: New York, NY, USA, 1999; Volume 17,
26. Leite, F.S.; Krakowski, K.A. *Covariant Differentiation under Rolling Maps*; Centro de Matemática da Universidade de Coimbra: Coimbra, Portugal, 2008.
27. Hinkle, J.; Fletcher, P.T.; Joshi, S. Intrinsic polynomials for regression on riemannian manifolds. *J. Math. Imaging Vis.* **2014**, *50*, 32–52.
28. Noakes, L.; Heinzinger, G.; Paden, B. Cubic splines on curved spaces. *IMA J. Math. Control Inf.* **1989**, *6*, 465–473.
29. Camarinha, M.; Silva Leite, F.; Crouch, P. On the geometry of Riemannian cubic polynomials. *Differ. Geom. Appl.* **2001**, *15*, 107–135.
30. Team, T.T.D.; Al-Rfou, R.; Alain, G.; Almahairi, A.; Angermueller, C.; Bahdanau, D.; Ballas, N.; Bastien, F.; Bayer, J.; Belikov, A.; et al. Theano: A Python framework for fast computation of mathematical expressions. *arXiv* **2016**, arXiv:1605.02688.

31. Micheli, M. The Differential Geometry of Landmark Shape Manifolds: Metrics, Geodesics, and Curvature. Ph.D. Thesis, Brown University, Providence, RI, USA, 2008.
32. Miller, M.; Banerjee, A.; Christensen, G.; Joshi, S.; Khaneja, N.; Grenander, U.; Matejic, L. Statistical methods in computational anatomy. *Stat. Methods Med. Res.* **1997**, *6*, 267–299.
33. Zhang, M.; Singh, N.; Fletcher, P.T. Bayesian estimation of regularization and atlas building in diffeomorphic image registration. In *Information Processing for Medical* Imaging (IPMI); Lecture Notes in Computer Science; Springer: Berlin/Heidelberg, Germany, 2013; pp. 37–48.

6

Geometric Theory of Heat from Souriau Lie Groups Thermodynamics and Koszul Hessian Geometry: Applications in Information Geometry for Exponential Families

Frédéric Barbaresco
Advanced Radar Concepts Business Unit, Thales Air Systems, Limours 91470, France;
frederic.barbaresco@thalesgroup.com

Academic Editor: Adom Giffin

Abstract: We introduce the symplectic structure of information geometry based on Souriau's Lie group thermodynamics model, with a covariant definition of Gibbs equilibrium via invariances through co-adjoint action of a group on its moment space, defining physical observables like energy, heat, and moment as pure geometrical objects. Using geometric Planck temperature of Souriau model and symplectic cocycle notion, the Fisher metric is identified as a Souriau geometric heat capacity. The Souriau model is based on affine representation of Lie group and Lie algebra that we compare with Koszul works on G/K homogeneous space and bijective correspondence between the set of G-invariant flat connections on G/K and the set of affine representations of the Lie algebra of G. In the framework of Lie group thermodynamics, an Euler-Poincaré equation is elaborated with respect to thermodynamic variables, and a new variational principal for thermodynamics is built through an invariant Poincaré-Cartan-Souriau integral. The Souriau-Fisher metric is linked to KKS (Kostant–Kirillov–Souriau) 2-form that associates a canonical homogeneous symplectic manifold to the co-adjoint orbits. We apply this model in the framework of information geometry for the action of an affine group for exponential families, and provide some illustrations of use cases for multivariate gaussian densities. Information geometry is presented in the context of the seminal work of Fréchet and his Clairaut-Legendre equation. The Souriau model of statistical physics is validated as compatible with the Balian gauge model of thermodynamics. We recall the precursor work of Casalis on affine group invariance for natural exponential families.

Keywords: Lie group thermodynamics; moment map; Gibbs density; Gibbs equilibrium; maximum entropy; information geometry; symplectic geometry; Cartan-Poincaré integral invariant; geometric mechanics; Euler-Poincaré equation; Fisher metric; gauge theory; affine group

---

> Lorsque le fait qu'on rencontre est en opposition avec une théorie régnante, il faut accepter le fait et abandonner la théorie, alors même que celle-ci, soutenue par de grands noms, est généralement adoptée
>
> —Claude Bernard in "Introduction à l'Étude de la Médecine Expérimentale" [1]

> Au départ, la théorie de la stabilité structurelle m'avait paru d'une telle ampleur et d'une telle généralité, qu'avec elle je pouvais espérer en quelque sorte remplacer la thermodynamique par la géométrie, géométriser en un certain sens la thermodynamique, éliminer des considérations thermodynamiques tous les aspects à caractère mesurable et stochastiques pour ne conserver que la caractérisation géométrique correspondante des attracteurs.
>
> —René Thom in "Logos et théorie des Catastrophes" [2]

1. Introduction

This MDPI Entropy Special Issue on "Differential Geometrical Theory of Statistics" collects a limited number of selected invited and contributed talks presented during the GSI'15 conference on "*Geometric Science of Information*" in October 2015. This paper is an extended version of the paper [3] "*Symplectic Structure of Information Geometry: Fisher Metric and Euler-Poincaré Equation of Souriau Lie Group Thermodynamics*" published in GSI'15 Proceedings. At GSI'15 conference, a special session was organized on "*lie groups and geometric mechanics/thermodynamics*", dedicated to Jean-Marie Souriau's works in statistical physics, organized by Gery de Saxcé and Frédéric Barbaresco, and an invited talk on "*Actions of Lie groups and Lie algebras on symplectic and Poisson manifolds. Application to Lagrangian and Hamiltonian systems*" by Charles-Michel Marle, addressing "*Souriau's thermodynamics of Lie groups*". In honor of Jean-Marie Souriau, who died in 2012 and Claude Vallée [4–6], who passed away in 2015, this Special Issue will publish three papers on Souriau's thermodynamics: Marle's paper on "*From Tools in Symplectic and Poisson Geometry to Souriau's Theories of Statistical Mechanics and Thermodynamics*" [7], de Saxcé's paper on "*Link between Lie Group Statistical Mechanics and Thermodynamics of Continua*" [8] and this publication by Barbaresco. This paper also proposes new developments, compared to paper [9] where relations between Souriau and Koszul models have been initiated.

This paper, similar to the goal of the papers of Marle and de Saxcé in this Special Issue, is intended to honor the memory of the French Physicist Jean-Marie Souriau and to popularize his works, currently little known, on statistical physics and thermodynamics. Souriau is well known for his seminal and major contributions in geometric mechanics, the discipline he created in the 1960s, from previous Lagrange's works that he conceptualized in the framework of symplectic geometry, but very few people know or have exploited Souriau's works contained in Chapter IV of his book "*Structure des systèmes dynamiques*" published in 1970 [10] and only translated into English in 1995 in the book "*Structure of Dynamical Systems: A Symplectic View of Physics*" [11], in which he applied the formalism of geometric mechanics to statistical physics. The personal author's contribution is to place the work of Souriau in the broader context of the emerging "*Geometric Science of Information*" [12] (addressed in GSI'15 conference), for which the author will show that the Souriau model of statistical physics is particularly well adapted to generalize "*information geometry*", that the author illustrates for exponential densities family and multivariate gaussian densities. The author will observe that the Riemannian metric introduced by Souriau is a generalization of Fisher metric, used in "*information geometry*", as being identified to the hessian of the logarithm of the generalized partition function (Massieu characteristic function), for the case of densities on homogeneous manifolds where a non-abelian group acts transively. For a group of time translation, we recover the classical thermodynamics and for the Euclidean space, we recover the classical Fisher metric used in Statistics. The author elaborates a new Euler-Poincaré equation for Souriau's thermodynamics, action on "geometric heat" variable Q (element of dual Lie algebra), and parameterized by "geometric temperature" (element of Lie algebra). The author will integrate Souriau thermodynamics in a variational model by defining an extended Cartan-Poincaré integral invariant defined by Souriau "geometric characteristic function" (the logarithm of the generalized Souriau partition function parameterized by geometric temperature). These results are illustrated for multivariate Gaussian densities, where the associated group is identified to compute a Souriau moment map and reduce the Euler-Poincaré equation of geodesics. In addition, the symplectic cocycle and Souriau-Fisher metric are deduced from a Lie group thermodynamics model.

The main contributions of the author in this paper are the following:

- The Souriau model of Lie group thermodynamics is presented with standard notations of Lie group theory, in place of Souriau equations using less classical conventions (that have limited understanding of his work by his contemporaries).
- We prove that Souriau Riemannian metric introduced with symplectic cocycle is a generalization of Fisher metric (called Souriau-Fisher metric in the following) that preserves the property

to be defined as a hessian of partition function logarithm $g_\beta = -\frac{\partial^2 \Phi}{\partial \beta^2} = \frac{\partial^2 \log \psi_\Omega}{\partial \beta^2}$ as in classical information geometry. We then establish the equality of two terms, the first one given by Souriau's definition from Lie group cocycle Θ and parameterized by "geometric heat" Q (element of dual Lie algebra) and "geometric temperature" β (element of Lie algebra) and the second one, the hessian of the characteristic function $\Phi(\beta) = -\log\psi_\Omega(\beta)$ with respect to the variable β:

$$g_\beta\left([\beta, Z_1], [\beta, Z_2]\right) = \langle \Theta(Z_1), [\beta, Z_2] \rangle + \langle Q, [Z_1, [\beta, Z_2]] \rangle = \frac{\partial^2 \log \psi_\Omega}{\partial \beta^2} \tag{1}$$

A dual Souriau-Fisher metric, the inverse of this last one, could be also elaborated with the hessian of "geometric entropy" $s(Q)$ with respect to the variable Q: $\frac{\partial^2 s(Q)}{\partial Q^2}$ For the maximum entropy density (Gibbs density), the following three terms coincide: $\frac{\partial^2 \log \psi_\Omega}{\partial \beta^2}$ that describes the convexity of the log-likelihood function, $I(\beta) = -E\left[\frac{\partial^2 \log p_\beta(\xi)}{\partial \beta^2}\right]$ the Fisher metric that describes the covariance of the log-likelihood gradient, whereas $I(\beta) = E\left[(\xi - Q)(\xi - Q)^T\right] = Var(\xi)$ that describes the covariance of the observables.

- This Souriau-Fisher metric is also identified to be proportional to the first derivative of the heat $g_\beta = -\frac{\partial Q}{\partial \beta}$, and then comparable by analogy to geometric "specific heat" or "calorific capacity".
- We observe that the Souriau metric is invariant with respect to the action of the group $I\left(Ad_g(\beta)\right) = I(\beta)$, due to the fact that the characteristic function $\Phi(\beta)$ after the action of the group is linearly dependent to β. As the Fisher metric is proportional to the hessian of the characteristic function, we have the following invariance:

$$I\left(Ad_g(\beta)\right) = -\frac{\partial^2\left(\Phi - \langle \theta\left(g^{-1}\right), \beta \rangle\right)}{\partial \beta^2} = -\frac{\partial^2 \Phi}{\partial \beta^2} = I(\beta) \tag{2}$$

- We have proposed, based on Souriau's Lie group model and on analogy with mechanical variables, a variational principle of thermodynamics deduced from Poincaré-Cartan integral invariant. The variational principle holds on $\mathfrak{g}$ the Lie algebra, for variations $\delta\beta = \dot{\eta} + [\beta, \eta]$, where $\eta(t)$ is an arbitrary path that vanishes at the endpoints, $\eta(a) = \eta(b) = 0$:

$$\delta \int_{t_0}^{t_1} \Phi\left(\beta(t)\right) \cdot dt = 0 \tag{3}$$

where the Poincaré-Cartan integral invariant $\int_{C_a} \Phi(\beta) \cdot dt = \int_{C_b} \Phi(\beta) \cdot dt$ is defined with $\Phi(\beta)$, the Massieu characteristic function, with the 1-form $\omega = \Phi(\beta) \cdot dt = (\langle Q, \beta \rangle - s) \cdot dt = \langle Q, (\beta \cdot dt) \rangle - s \cdot dt$

- We have deduced Euler-Poincaré equations for the Souriau model:

$$\frac{dQ}{dt} = ad^*_\beta Q \text{ and } \begin{cases} s(Q) = \langle \beta, Q \rangle - \Phi(\beta) \\ \beta = \frac{\partial s(Q)}{\partial Q} \in \mathfrak{g},\ Q = \frac{\partial \Phi(\beta)}{\partial \beta} \in \mathfrak{g}^* \end{cases} \text{ and } \frac{d}{dt}\left(Ad^*_g Q\right) = 0$$
$$\text{with } \begin{cases} \mathfrak{g}^* : \text{ dual Lie algebra} \\ ad^*_X Y : \text{ Coadjoint operator} \end{cases} \tag{4}$$

where Q is the Souriau geometric heat (element of dual Lie algebra) and β is the Souriau geometric temperature (element of the Lie algebra). The second equation is linked to the result of Souriau based on the moment map that a symplectic manifold is always a coadjoint orbit, affine of its group of Hamiltonian transformations (a symplectic manifold homogeneous under the action of a Lie group, is isomorphic, up to a covering, to a coadjoint orbit; symplectic leaves are the orbits of the affine action that makes the moment map equivariant).

- We have established that the affine representation of Lie group and Lie algebra by Jean-Marie Souriau is equivalent to Jean-Louis Koszul's affine representation developed in the framework of hessian geometry of convex sharp cones. Both Souriau and Koszul have elaborated equations requested for Lie group and Lie algebra to ensure the existence of an affine representation. We have compared both approaches of Souriau and Koszul in a table.
- We have applied the Souriau model for exponential families and especially for multivariate Gaussian densities.
- We have applied the Souriau-Koszul model Gibbs density to compute the maximum entropy density for symmetric positive definite matrices, using the inner product $\langle \eta, \xi \rangle = Tr\left(\eta^T \xi\right), \forall \eta, \xi \in Sym(n)$ given by Cartan-Killing form. The Gibbs density (generalization of Gaussian law for theses matrices and defined as maximum entropy density):

$$p_{\hat{\xi}}(\xi) = e^{-\langle \Theta^{-1}(\hat{\xi}), \xi \rangle + \Phi(\Theta^{-1}(\hat{\xi}))} = \psi_\Omega\left(I_d\right) \cdot \left[\det\left(\alpha \hat{\xi}^{-1}\right)\right] \cdot e^{-Tr\left(\alpha \hat{\xi}^{-1} \xi\right)}$$
$$\text{with } \alpha = \frac{n+1}{2} \quad (5)$$

- For the case of multivariate Gaussian densities, we have considered $GA(n)$ a sub-group of affine group, that we defined by a $(n+1) \times (n+1)$ embedding in matrix Lie group G_{aff}, and that acts for multivariate Gaussian laws by:

$$\begin{bmatrix} Y \\ 1 \end{bmatrix} = \begin{bmatrix} R^{1/2} & m \\ 0 & 1 \end{bmatrix} \begin{bmatrix} X \\ 1 \end{bmatrix} = \begin{bmatrix} R^{1/2}X + m \\ 1 \end{bmatrix}, \begin{cases} (m, R) \in R^n \times Sym^+(n) \\ M = \begin{bmatrix} R^{1/2} & m \\ 0 & 1 \end{bmatrix} \in G_{aff} \end{cases} \quad (6)$$
$$X \approx \aleph(0, I) \rightarrow Y \approx \aleph(m, R)$$

- For multivariate Gaussian densities, as we have identified the acting sub-group of affine group M, we have also developed the computation of the associated Lie algebras η_L and η_R, adjoint and coadjoint operators, and especially the Souriau "moment map" Π_R:

$$\langle n_L, M^{-1} n_R M \rangle = \langle \Pi_R, n_R \rangle$$
$$\text{with } M = \begin{bmatrix} R^{1/2} & m \\ 0 & 1 \end{bmatrix}, n_L = \begin{bmatrix} R^{-1/2}\dot{R}^{1/2} & R^{-1/2}\dot{m} \\ 0 & 0 \end{bmatrix} \text{ and } \eta_R = \begin{bmatrix} R^{-1/2}\dot{R}^{1/2} & \dot{m} - R^{-1/2}\dot{R}^{1/2} m \\ 0 & 0 \end{bmatrix} \quad (7)$$
$$\Rightarrow \Pi_R = \begin{bmatrix} R^{-1/2}\dot{R}^{1/2} + R^{-1}\dot{m}m^T & R^{-1}\dot{m} \\ 0 & 0 \end{bmatrix}$$

Using Souriau Theorem (geometrization of Noether theorem), we use the property that this moment map Π_R is constant (its components are equal to Noether invariants):

$$\frac{d\Pi_R}{dt} = 0 \Rightarrow \begin{cases} R^{-1}\dot{R} + R^{-1}\dot{m}m^T = B = cste \\ R^{-1}\dot{m} = b = cste \end{cases} \quad (8)$$

to reduce the Euler-Lagrange equation of geodesics between two multivariate Gaussian densities:

$$\begin{cases} \ddot{R} + \dot{m}\dot{m}^T - \dot{R}R^{-1}\dot{R} = 0 \\ \ddot{m} - \dot{R}R^{-1}\dot{m} = 0 \end{cases} \quad (9)$$

to this reduced equation of geodesics:

$$\begin{cases} \dot{m} = Rb \\ \dot{R} = R\left(B - bm^T\right) \end{cases} \quad (10)$$

that we solve by "geodesic shooting" technic based on Eriksen equation of exponential map.

- For the families of multivariate Gaussian densities, that we have identified as homogeneous manifold with the associated sub-group of the affine group $\begin{bmatrix} R^{1/2} & m \\ 0 & 1 \end{bmatrix}$, we have considered the elements of exponential families, that play the role of geometric heat Q in Souriau Lie group thermodynamics, and β the geometric (Planck) temperature:

$$Q = \hat{\xi} = \begin{bmatrix} E[z] \\ E\left[zz^T\right] \end{bmatrix} = \begin{bmatrix} m \\ R + mm^T \end{bmatrix}, \ \beta = \begin{bmatrix} -R^{-1}m \\ \frac{1}{2}R^{-1} \end{bmatrix} \quad (11)$$

We have considered that these elements are homeomorph to the $(n + 1) \times (n + 1)$ matrix elements:

$$Q = \hat{\xi} = \begin{bmatrix} R + mm^T & m \\ 0 & 0 \end{bmatrix} \in \mathfrak{g}^*, \ \beta = \begin{bmatrix} \frac{1}{2}R^{-1} & -R^{-1}m \\ 0 & 0 \end{bmatrix} \in \mathfrak{g} \quad (12)$$

to compute the Souriau symplectic cocycle of the Lie group:

$$\theta(M) = \hat{\xi}\left(Ad_M(\beta)\right) - Ad^*_M\hat{\xi} \quad (13)$$

where the adjoint operator is equal to:

$$Ad_M\beta = \begin{bmatrix} \frac{1}{2}\Omega^{-1} & -\Omega^{-1}n \\ 0 & 0 \end{bmatrix} \text{ with } \Omega = R'^{1/2}RR'^{-1/2} \text{ and } n = \left(\frac{1}{2}m' + R'^{1/2}m\right) \quad (14)$$

with

$$\hat{\xi}\left(Ad_M(\beta)\right) = \begin{bmatrix} \Omega + nn^T & n \\ 0 & 0 \end{bmatrix} \quad (15)$$

and the co-adjoint operator:

$$Ad^*_M\hat{\xi} = \begin{bmatrix} R + mm^T - mm'^T & R'^{1/2}m \\ 0 & 0 \end{bmatrix} \quad (16)$$

- Finally, we have computed the Souriau-Fisher metric $g_\beta\left([\beta, Z_1], [\beta, Z_2]\right) = \widetilde{\Theta}_\beta\left(Z_1, [\beta, Z_2]\right)$ for multivariate Gaussian densities, given by:

$$\begin{aligned} g_\beta\left([\beta, Z_1], [\beta, Z_2]\right) &= \widetilde{\Theta}_\beta\left(Z_1, [\beta, Z_2]\right) = \widetilde{\Theta}\left(Z_1, [\beta, Z_2]\right) + \langle\hat{\xi}, [Z_1, [\beta, Z_2]]\rangle \\ &= \langle\Theta\left(Z_1\right), [\beta, Z_2]\rangle + \langle\hat{\xi}, [Z_1, [\beta, Z_2]]\rangle \end{aligned} \quad (17)$$

with element of Lie algebra given by $Z = \begin{bmatrix} \frac{1}{2}\Omega^{-1} & -\Omega^{-1}n \\ 0 & 0 \end{bmatrix}$.

The plan of the paper is as follows. After this introduction in Section 1, we develop in Section 2 the position of Souriau symplectic model of statistical physics in the historical developments of thermodynamic concepts. In Section 3, we develop and revisit the Lie group thermodynamics model of Jean-Marie Souriau in modern notations. In Section 4, we make the link between Souriau Riemannian metric and Fisher metric defined as a geometric heat capacity of Lie group thermodynamics. In Section 5, we elaborate Euler-Lagrange equations of Lie group thermodynamics and a variational model based on Poincaré-Cartan integral invariant. In Section 6, we explore Souriau affine representation of Lie group and Lie algebra (including the notions of: affine representations and cocycles, Souriau moment map and cocycles, equivariance of Souriau moment map, action of Lie group on a symplectic manifold and dual spaces of finite-dimensional Lie algebras) and we analyze the link and parallelisms with Koszul affine representation, developed in another context (comparison is synthetized in a table). In Section 7, we illustrate Koszul and Souriau Lie group models of information geometry for multivariate Gaussian densities. In Section 8, after identifying the affine group acting for these densities, we compute the Souriau moment map to obtain the Euler-Poincaré equation, solved by geodesic shooting method. In Section 9, Souriau Riemannian metric defined by cocycle for multivariate Gaussian densities is computed. We give a conclusion in Section 10 with research prospects in the framework of affine Poisson geometry [13], Bismut stochastic mechanics [14] and second order extension of the Gibbs state [15,16]. We have three appendices: Appendix A develops the Clairaut(-Legendre) equation of Maurice Fréchet associated to "distinguished functions" as a seminal equation of information geometry; Appendix B is about a Balian Gauge model of thermodynamics and its compliance with the Souriau model; Appendix C is devoted to the link of Casalis-Letac's works on affine group invariance for natural exponential families with Souriau's works.

2. Position of Souriau Symplectic Model of Statistical Physics in Historical Developments of Thermodynamic Concepts

In this Section, we will explain the emergence of thermodynamic concepts that give rise to the generalization of the Souriau model of statistical physics. To understand Souriau's theoretical model of heat, we have to consider first his work in geometric mechanics where he introduced the concept of "moment map" and "symplectic cohomology". We will then introduce the concept of "characteristic function" developed by François Massieu, and generalized by Souriau on homogeneous symplectic manifolds. In his statistical physics model, Souriau has also generalized the notion of "heat capacity" that was initially extended by Pierre Duhem as a key structure to jointly consider mechanics and thermodynamics under the umbrella of the same theory. Pierre Duhem has also integrated, in the corpus, the Massieu's characteristic function as a thermodynamic potential. Souriau's idea to develop a covariant model of Gibbs density on homogeneous manifold was also influenced by the seminal work of Constantin Carathéodory that axiomatized thermodynamics in 1909 based on Carnot's works. Souriau has adapted his geometric mechanical model for the theory of heat, where Henri Poincaré did not succeed in his paper on attempts of mechanical explanation for the principles of thermodynamics.

Lagrange's works on "mécanique analytique (analytic mechanics)" has been interpreted by Jean-Marie Souriau in the framework of differential geometry and has initiated a new discipline called after Souriau, "mécanique géométrique (geometric mechanics)" [17–19]. Souriau has observed that the collection of motions of a dynamical system is a manifold with an antisymmetric flat tensor that is a symplectic form where the structure contains all the pertinent information of the state of the system (positions, velocities, forces, etc.). Souriau said: *"Ce que Lagrange a vu, que n'a pas vu Laplace, c'était la structure symplectique (What Lagrange saw, that Laplace didn't see, was the symplectic structure"* [20]. Using the symmetries of a symplectic manifold, Souriau introduced a mapping which he called the "moment map" [21–23], which takes its values in a space attached to the group of symmetries (in the dual space of its Lie algebra). He [10] called dynamical groups every dimensional group of symplectomorphisms (an isomorphism between symplectic manifolds, a transformation of phase space that is volume-preserving), and introduced Galileo group for classical mechanics and Poincaré group for relativistic mechanics (both are sub-groups of affine group [24,25]). For instance, a Galileo group could be represented in a matrix form by (with A rotation, b the boost, c space translation and e time translation):

$$\begin{bmatrix} x' \\ t \\ 1 \end{bmatrix} = \underbrace{\begin{bmatrix} A & b & c \\ 0 & 1 & e \\ 0 & 0 & 1 \end{bmatrix}}_{\text{GALILEO GROUP}} \begin{bmatrix} x \\ t \\ 1 \end{bmatrix} \text{ with } \begin{cases} A \in SO(3) \\ b, c \in R^3 \\ e \in R \end{cases}, \text{ Lie Algebra } \begin{bmatrix} \omega & \eta & \gamma \\ 0 & 0 & \varepsilon \\ 0 & 0 & 0 \end{bmatrix} \text{ with } \begin{cases} \omega \in so(3) \\ \eta, \gamma \in R^3 \\ \varepsilon \in R^+ \end{cases} \quad (18)$$

Souriau associated to this moment map, the notion of symplectic cohomology, linked to the fact that such a moment is defined up to an additive constant that brings into play an algebraic mechanism (called cohomology). Souriau proved that the moment map is a constant of the motion, and provided geometric generalization of Emmy Noether invariant theorem (invariants of E. Noether theorem are the components of the moment map). For instance, Souriau gave an ontological definition of mass in classical mechanics as the measure of the symplectic cohomology of the action of the Galileo group (the mass is no longer an arbitrary variable but a characteristic of the space). This is no longer true for Poincaré group in relativistic mechanics, where the symplectic cohomology is null, explaining the lack of conservation of mass in relativity. All the details of classical mechanics thus appear as geometric necessities, as ontological elements. Souriau has also observed that the symplectic structure has the property to be able to be reconstructed from its symmetries alone, through a 2-form (called Kirillov–Kostant–Souriau form) defined on coadjoint orbits. Souriau said that the different versions of mechanical science can be classified by the geometry that each implies for space and time; geometry is determined by the covariance of group theory. Thus, Newtonian mechanics is covariant by the group of Galileo, the relativity by the group of Poincaré; General relativity by the "smooth" group (the group of diffeomorphisms of space-time). However, Souriau added "*However, there are some statements of mechanics whose covariance belongs to a fourth group rarely considered: the affine group, a group shown in the following diagram for inclusion. How is it possible that a unitary point of view (which would be necessarily a true thermodynamics), has not yet come to crown the picture? Mystery...*" [26]. See Figure 1.

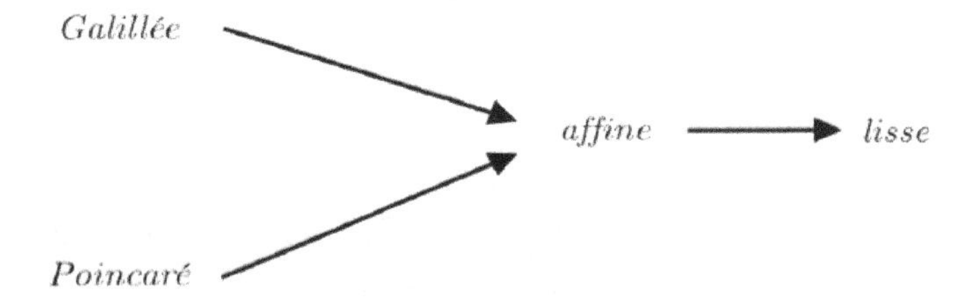

Figure 1. Souriau Scheme about mysterious "affine group" of a true thermodynamics between Galileo group of classical mechanics, Poincaré group of relativistic mechanics and Smooth group of general relativity.

As early as 1966, Souriau applied his theory to statistical mechanics, developed it in the Chapter IV of his book "*Structure of Dynamical Systems*" [11], and elaborated what he called a "Lie group thermodynamics" [10,11,27–37]. Using Lagrange's viewpoint, in Souriau statistical mechanics, a statistical state is a probability measure on the manifold of motions (and no longer in phase space [38]). Souriau observed that Gibbs equilibrium [39] is not covariant with respect to dynamic groups of Physics. To solve this braking of symmetry, Souriau introduced a new "geometric theory of heat" where the equilibrium states are indexed by a parameter β with values in the Lie algebra of the group, generalizing the Gibbs equilibrium states, where β plays the role of a geometric (Planck) temperature. The invariance with respect to the group, and the fact that the entropy s is a convex function of this geometric temperature β, imposes very strict, universal conditions (e.g., there exists necessarily a critical temperature beyond which no equilibrium can exist). Souriau observed that the group of time translations of the classical thermodynamics [40,41] is not a normal subgroup of the Galilei group, proving that if a dynamical system is conservative in an inertial reference frame, it need not be conservative in another. Based on this fact, Souriau generalized the formulation of the Gibbs principle to become compatible with Galileo relativity in classical mechanics and with Poincaré relativity in relativistic mechanics. The maximum entropy principle [42–51] is preserved, and the Gibbs density is given by the density of maximum entropy (among the equilibrium states for which the average value of the energy takes a prescribed value, the Gibbs measures are those which have the largest entropy), but with a new principle "*If a dynamical system is invariant under a Lie subgroup G′ of the Galileo group, then the natural equilibria of the system forms the Gibbs ensemble of the dynamical group G′*" [10]. The classical notion of Gibbs canonical ensemble is extended for a homogneous symplectic manifold on which a Lie group (dynamic group) has a symplectic action. When the group is not abelian (non-commutative group), the symmetry is broken, and new "cohomological" relations should be verified in Lie algebra of the group [52–55]. A natural equilibrium state will thus be characterized by an element of the Lie algebra of the Lie group, determining the equilibrium temperature β. The entropy $s(Q)$, parametrized by Q the geometric heat (mean of energy U, element of the dual Lie algebra) is defined by the Legendre transform [56–59] of the Massieu potential $\Phi(\beta)$ parametrized by β ($\Phi(\beta)$ is the minus logarithm of the partition function $\psi_\Omega(\beta)$):

$$s(Q) = \langle \beta, Q \rangle - \Phi(\beta) \text{ with } \begin{cases} Q = \dfrac{\partial \Phi}{\partial \beta} \in \mathfrak{g}^* \\ \beta = \dfrac{\partial s}{\partial Q} \in \mathfrak{g} \end{cases} \tag{19}$$

$$\begin{aligned} p_{Gibbs}(\xi) &= e^{\Phi(\beta) - \langle \beta, U(\xi) \rangle} = \frac{e^{-\langle \beta, U(\xi) \rangle}}{\int_M e^{-\langle \beta, U(\xi) \rangle} d\omega}, \\ Q &= \frac{\partial \Phi(\beta)}{\partial \beta} = \frac{\int_M U(\xi) e^{-\langle \beta, U(\xi) \rangle} d\omega}{\int_M e^{-\langle \beta, U(\xi) \rangle} d\omega} = \int_M U(\xi) p(\xi) d\omega \quad \text{with } \Phi(\beta) = -\log \int_M e^{-\langle \beta, U(\xi) \rangle} d\omega \end{aligned} \tag{20}$$

Souriau completed his "geometric heat theory" by introducing a 2-form in the Lie algebra, that is a Riemannian metric tensor in the values of adjoint orbit of β, $[\beta, Z]$ with Z an element of the Lie algebra. This metric is given for (β, Q):

$$g_\beta([\beta, Z_1], [\beta, Z_2]) = \langle \Theta(Z_1), [\beta, Z_2] \rangle + \langle Q, [Z_1, [\beta, Z_2]] \rangle \tag{21}$$

where Θ is a cocycle of the Lie algebra, defined by $\Theta = T_e\theta$ with θ a cocycle of the Lie group defined by $\theta(M) = Q(Ad_M(\beta)) - Ad_M^* Q$. We have observed that this metric g_β is also given by the hessian of the Massieu potential $g_\beta = -\dfrac{\partial^2 \Phi}{\partial \beta^2} = \dfrac{\partial \log \psi_\Omega}{\partial \beta^2}$ as Fisher metric in classical information geometry theory [60], and so this is a generalization of the Fisher metric for homogeneous manifold. We call this new metric the Souriau-Fisher metric. As $g_\beta = -\dfrac{\partial Q}{\partial \beta}$, Souriau compared it by analogy with classical thermodynamics to a "geometric specific heat" (geometric calorific capacity).

The potential theory of thermodynamics and the introduction of "characteristic function" (previous function $\Phi(\beta) = -\log \psi_\Omega(\beta)$ in Souriau theory) was initiated by François Jacques Dominique Massieu [61–64]. Massieu was the son of Pierre François Marie Massieu and Thérèse Claire Castel. He married in 1862 with Mlle Morand and had 2 children. He graduated from Ecole Polytechnique in 1851 and Ecole des Mines de Paris in 1956, he has integrated "Corps des Mines". He defended his Ph.D. in 1861 on "*Sur les intégrales algébriques des problèmes de mécanique*" and on "*Sur le mode de propagation des ondes planes et la surface de l'onde élémentaire dans les cristaux biréfringents à deux axes*" [65] with the jury composed of Lamé, Delaunay et Puiseux. In 1870, François Massieu presented his paper to the French Academy of Sciences on "*characteristic functions of the various fluids and the theory of vapors*" [61]. The design of the characteristic function is the finest scientific title of Mr. Massieu. A prominent judge, Joseph Bertrand, do not hesitate to declare, in a statement read to the French Academy of Sciences 25 July 1870, that "*the introduction of this function in formulas that summarize all the possible consequences of the two fundamental theorems seems, for the theory, a similar service almost equivalent to that Clausius has made by linking the Carnot's theorem to entropy"* [66]. The final manuscript was published by Massieu in 1873, *"Exposé des principes fondamentaux de la théorie mécanique de la chaleur (Note destinée à servir d'introduction au Mémoire de l'auteur sur les fonctions caractéristiques des divers fluides et la théorie des vapeurs)"* [63].

Massieu introduced the following potential $\Phi(\beta)$, called "characteristic function", as illustrated in Figure 2, that is the potential used by Souriau to generalize the theory: $s\,(Q) = \langle \beta, Q \rangle - \Phi(\beta) \underset{\beta=\frac{1}{T}}{\Rightarrow} \Phi = \frac{Q}{T} - S$. However, in his third paper, Massieu was influenced by M. Bertrand, as illustrated in Figure 3, to replace the variable $\beta = \frac{1}{T}$ (that he used in his two first papers) by T. We have then to wait 50 years more for the paper of Planck, who introduced again the good variable $\beta = \frac{1}{T}$, and then generalized by Souriau, giving to Planck temperature β an ontological and geometric status as element of the Lie algebra of the dynamic group.

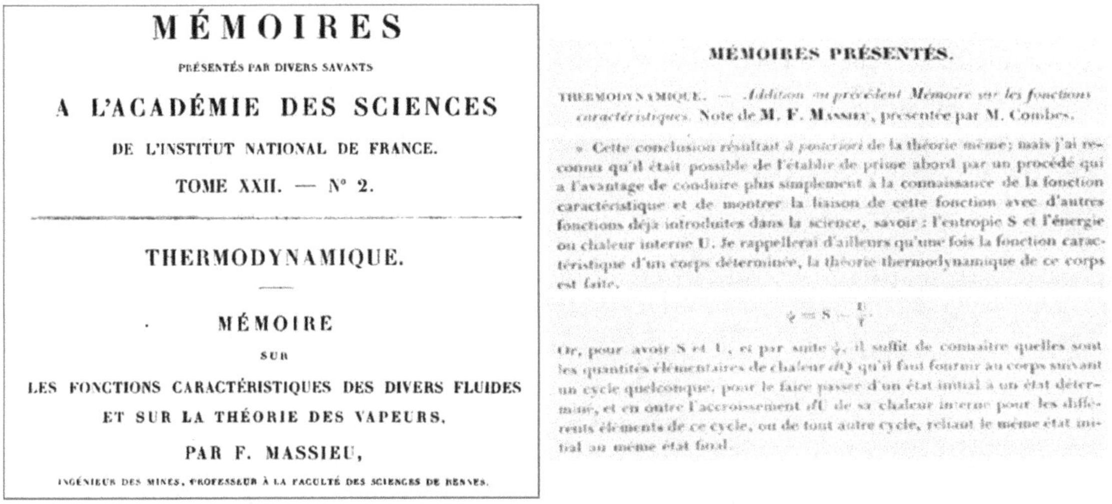

MÉMOIRES

PRÉSENTÉS PAR DIVERS SAVANTS

A L'ACADÉMIE DES SCIENCES

DE L'INSTITUT NATIONAL DE FRANCE.

TOME XXII. — N° 2.

THERMODYNAMIQUE.

MÉMOIRE

SUR

LES FONCTIONS CARACTÉRISTIQUES DES DIVERS FLUIDES ET SUR LA THÉORIE DES VAPEURS.

PAR F. MASSIEU,

INGÉNIEUR DES MINES, PROFESSEUR À LA FACULTÉ DES SCIENCES DE RENNES.

MÉMOIRES PRÉSENTÉS.

THERMODYNAMIQUE. — *Addition au précédent Mémoire sur les fonctions caractéristiques.* Note de M. F. Massieu, présentée par M. Combes.

« Cette conclusion résultait *à posteriori* de la théorie même; mais j'ai reconnu qu'il était possible de l'établir de prime abord par un procédé qui a l'avantage de conduire plus simplement à la connaissance de la fonction caractéristique et de montrer la liaison de cette fonction avec d'autres fonctions déjà introduites dans la science, savoir : l'entropie S et l'énergie ou chaleur interne U. Je rappellerai d'ailleurs qu'une fois la fonction caractéristique d'un corps déterminée, la théorie thermodynamique de ce corps est faite.

$$\psi = S - \frac{U}{T}.$$

Or, pour avoir S et U, et par suite ψ, il suffit de connaître quelles sont les quantités élémentaires de chaleur dQ qu'il faut fournir au corps suivant un cycle quelconque, pour le faire passer d'un état initial à un état déterminé, et en outre l'accroissement dU de sa chaleur interne pour les différents éléments de ce cycle, ou de tout autre cycle, reliant le même état initial au même état final.

Figure 2. Extract from the second paper of François Massieu to the French Academy of Sciences [61,62].

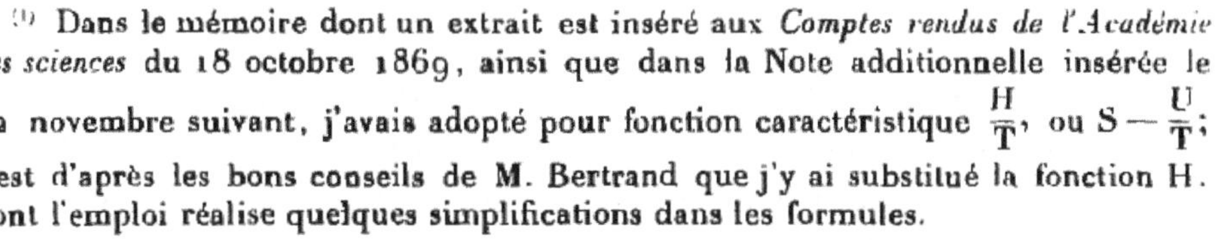

(1) Dans le mémoire dont un extrait est inséré aux *Comptes rendus de l'Académie des sciences* du 18 octobre 1869, ainsi que dans la Note additionnelle insérée le 22 novembre suivant, j'avais adopté pour fonction caractéristique $\frac{H}{T}$, ou $S - \frac{U}{T}$; c'est d'après les bons conseils de M. Bertrand que j'y ai substitué la fonction H. dont l'emploi réalise quelques simplifications dans les formules.

Figure 3. Remark of Massieu in 1876 paper [64], where he explained why he took into account the "good advice" of Bertrand to replace variable $1/T$, used in his initial paper of 1869, by the variable T.

This Lie group thermodynamics of Souriau is able to explain astronomical phenomenon (rotation of celestial bodies: the Earth and the stars rotating about themselves). The geometric temperature β can be also interpreted as a space-time vector (generalization of the temperature vector of Planck), where the temperature vector and entropy flux are in duality unifying heat conduction and viscosity (equations of Fourier and Navier). In case of centrifuge system (e.g., used for enrichment of uranium), the Gibbs Equilibrium state [60,67] are given by Souriau equations as the variation in concentration of the components of an inhomogeneous gas. Classical statistical mechanics corresponds to the dynamical group of time translations, for which we recover from Souriau equations the concepts and principles of classical thermodynamics (temperature, energy, heat, work, entropy, thermodynamic potentials) and of the kinetic theory of gases (pressure, specific heats, Maxwell's velocity distribution, etc.).

Souriau also studied continuous medium thermodynamics, where the "temperature vector" is no longer constrained to be in Lie algebra, but only contrained by phenomenologic equations (e.g., Navier equations, etc.). For thermodynamic equilibrium, the "temperature vector" is then a Killing vector of Space-Time. For each point X, there is a "temperature vector" $\beta(X)$, such it is an infinitesimal conformal transform of the metric of the universe g_{ij}. Conservation equations can then be deduced for components of impulsion-energy tensor T^{ij} and entropy flux S^j with $\hat{\partial}_i T^{ij} = 0$ and $\partial_i S^j = 0$. Temperature and metric are related by the following equations:

$$\begin{cases} \hat{\partial}_i \beta_j + \hat{\partial}_j \beta_i = \lambda g_{ij} \\ \partial_i \beta_j + \partial_j \beta_i - 2\Gamma_{ij}^k \beta_k = \lambda g_{ij} \end{cases} \text{with} \begin{cases} \hat{\partial}_i . : \text{covariant derivative} \\ \beta_j : \text{component of Temperature vector} \end{cases} \quad (22)$$

$$\lambda = 0 \Rightarrow \text{Killing Equation}$$

Leon Brillouin made the link between Boltzmann entropy and Negentropie of information theory [68–71], but before Jean-Marie Souriau, only Constantin Carathéodory and Pierre Duhem [72–75] initiated first theoretical works to generalize thermodynamics.

After three years as lecturer at Lille university, Duhem published a paper in the official revue of the Ecole Normale Supérieure, in 1891, "On general equations of thermodynamics" [72] (Sur les équations générales de la Thermodynamique) in Annales Scientifiques de l'Ecole Normale Supérieure. Duhem generalized the concept of "virtual work" under the action of "external actions" by taking into account both mechanical and thermal actions. In 1894, the design of a generalized mechanics based on thermodynamics was further developed: ordinary mechanics had already become "a particular case of a more general science". Duhem writes "*We made dynamics a special case of thermodynamics, a science that embraces common principles in all changes of state bodies, changes of places as well as changes in physical qualities*" (*Nous avons fait de la dynamique un cas particulier de la thermodynamique, une Science qui embrasse dans des principes communs tous les changements d'état des corps, aussi bien les changements de lieu que les changements de qualités physiques*). In the equations of his generalized mechanics-thermodynamics, some new terms had to be introduced, in order to account for the intrinsic viscosity and friction of the system. As observed by Stefano Bordoni, Duhem aimed at widening the scope of physics: the new physics could not confine itself to "*local motion*" but had to describe what Duhem qualified "*motions of modification*". If Boltzmann had tried to proceed from "local motion" to attain the explanation of more complex transformations, Duhem was trying to proceed from general laws concerning general transformation in order to reach "*local motion*" as a simplified specific case. Four scientists were credited by Duhem with having carried out "the most important researches on that subject": Massieu had managed to derive thermodynamics from a "characteristic function and its partial derivatives"; Gibbs had shown that Massieu's functions "could play the role of potentials in the determination of the states of equilibrium" in a given system; von Helmholtz had put forward "similar ideas"; von Oettingen had given "an exposition of thermodynamics of remarkable generality" based on general duality concept in "Die thermodynamischen Beziehungen antithetisch entwickelt" published at St. Petersburg in 1885. Duhem took into account a system whose elements had the same temperature and where the state of the system could be completely specified by giving its temperature and n other independent quantities. He then introduced some "external forces", and held the system in equilibrium. A virtual work corresponded to such forces, and a set of $n + 1$ equations corresponded to the condition of equilibrium of the physical system. From the thermodynamic point of view, every infinitesimal transformation involving the generalized displacements had to obey to the first law, which could be expressed in terms of the $(n + 1)$ generalized Lagrangian parameters. The amount of heat could be written as a sum of $(n + 1)$ terms. The new alliance between mechanics and thermodynamics led to a sort of symmetry between thermal and mechanical quantities. The $n + 1$ functions played the role of *generalized thermal capacities*, and the last term was nothing other than the ordinary *thermal capacity*. The knowledge of the "*equilibrium equations of a system*" allowed Duhem to compute the partial derivatives of the thermal capacity with regard to all the parameters which described the state of the system, apart from its derivative with regard to temperature. The thermal capacities were therefore known "*except for an unspecified function of temperature*".

The axiomatic approach of thermodynamics was published in 1909 in Mathematische Annalen [76] under the title "*Examination of the Foundations of Thermodynamics*" (*Untersuchungen überdie Grundlagen *der Thermodynamik*) by Constantin Carathéodory based on Carnot's works [77]. Carathéodory introduced entropy through a mathematical approach based on the geometric behavior of a certain class of partial differential equations called Pfaffians. Carathéodory's investigations start by revisiting the first law and reformulating the second law of thermodynamics in the form of two axioms. The first axiom applies to a multiphase system change under adiabatic conditions (axiom of classical thermodynamics due to Clausius [78,79]). The second axiom assumes that in the neighborhood of any equilibrium state of a system (of any number of thermodynamic coordinates), there exist states that are inaccessible by reversible adiabatic processes. In the book of Misha Gromov "*Metric Structures for Riemannian and Non-Riemannian Spaces*", written and edited by Pierre Pansu and Jacques Lafontaine, a new metric is introduced called "*Carnot-Carathéodory metric*". In one of his papers, Misha Gromov [80,81] gives historical remarks "*This result (which seems obvious by the modern standards) appears (in a more general form) in the 1909-paper by Carathéorody on formalization of the classical thermodynamics where horizontal curves roughly correspond to adiabatic processes. In fact, the above proof may be performed in the language of Carnot (cycles) and for this reason the metris distH were christened 'Carnot-Carathéodory' in Gromov-Lafontaine-Pansu book*" [82]. When I ask this question to Pierre Pansu, he gave me the answer that "*The section 4 of [76], entitled Hilfsatz aus der Theorie des Pfaffschen Gleichungen (Lemma from the theory of Pfaffian equations) opens with a statement relating to the differential 1-forms. Carathéodory says, If a Pfaffian equation dx0 + X1 dx1 + X2 dx2 + ... + Xn dxn = 0 is given, in which the Xi are finite, continuous, differentiable functions of the xi, and one knows that in any neighborhood of an arbitrary point P of the space of xi there is a point that one cannot reach along a curve that satisfies this equation then the expression must necessarily possess a multiplier that makes it into a complete differential*". This is confirmed in the introduction of his paper [76], where Carathéodory said "*Finally, in order to be able to treat systems with arbitrarily many degrees of freedom from the outset, instead of the Carnot cycle that is almost always used, but is intuitive and easy to control only for systems with two degrees of freedom, one must employ a theorem from the theory of Pfaffian differential equations, for which a simple proof is given in the fourth section*".

We have also to make reference to Henri Poincaré [83] that published the paper "*On attempts of mechanical explanation for the principles of thermodynamics (Sur les tentatives d'explication mécanique des principes de la thermodynamique)*" at the Comptes rendus de l'Académie des sciences in 1889 [84], in which he tried to consolidate links between mechanics and thermomechanics principles. These elements were also developed in Poincaré's lecture of 1892 [85] on "*thermodynamique*" in Chapter XVII "*Reduction of thermodynamics principles to the general principles of mechanics (Réduction des principes de la Thermodynamique aux principes généraux de la mécanique)*". Poincaré writes in his book [85] "*It is otherwise with the second law of thermodynamics. Clausius was the first to attempt to bring it to the principles of mechanics, but not succeed satisfactorily. Helmholtz in his memoir on the principle of least actions developed a theory much more perfect than that of Clausius. However, it cannot account for irreversible phenomena. (Il en est autrement du second principe de la thermodynamique. Clausius, a le premier, tenté de le ramener aux principes de la Mécanique, mais sans y réussir d'une manière satisfaisante. Helmoltz dans son mémoire sur le principe de la moindre action, a développé une théorie beaucoup plus parfaite que celle de Clausius; cependant elle ne peut rendre compte des phénomènes irréversibles.)*". About Helmoltz work, Poincaré observes [85] "*It follows from these examples that the Helmholtz hypothesis is true in the case of body turning around an axis; So it seems applicable to vortex motions of molecules (Il résulte de ces exemples que l'hypothèse d'Helmoltz est exacte dans le cas de corps tournant autour d'un axe; elle parait donc applicable aux mouvements tourbillonnaires des molecules.)*", but he adds in the following that the Helmoltz model is also true in the case of vibrating motions as molecular motions. However, he finally observes that the Helmoltz model cannot explain the increasing of entropy and concludes [85] "*All attempts of this nature must be abandoned; the only ones that have any chance of success are those based on the intervention of statistical laws, for example, the kinetic theory of gases. This view, which I cannot develop here, can be summed up in a somewhat vulgar way as follows: Suppose we want to place a grain of oats in the middle of a heap of wheat; it will be easy; then suppose we wanted to find it and remove it; we cannot achieve it. All irreversible phenomena, according to some physicists, would be built on this model (Toutes les tentatives de cette nature doivent donc être abandonnées; les seules qui aient quelque chance de succès sont celles qui sont fondées sur l'intervention des lois statistiques comme, par exemple, la théorie cinétique des gaz. Ce point de vue, que je ne puis développer ici, peut se résumer d'une façon un peu vulgaire comme il suit: Supposons que nous voulions placer un grain d'avoine au milieu d'un tas de blé; cela sera facile; supposons que nous voulions ensuite l'y retrouver et l'en retirer; nous ne pourrons y parvenir. Tous les phénomènes irréversibles, d'après certains physiciens, seraient construits sur ce modèle)". In Poincaré's lecture, Massieu has greatly influenced Poincaré to introduce Massieu characteristic function in probability [86]. As we have observed, Poincaré has introduced characteristic function in probability lecture after his lecture on thermodynamics where he discovered in its second edition [85], the Massieu's characteristic function. We can read that* "Since from the functions of Mr. Massieu one can deduce other functions of variables, all equations of thermodynamics can be written so as to only contain these functions and their derivatives; it will thus result in some cases, a great simplification (Puisque des fonctions de M. Massieu on peut déduire les autres fonctions des variables, toutes les équations de la Thermodynamique pourront s'écrire de manière à ne plus renfermer que ces fonctions et leurs dérivées; il en résultera donc, dans certains cas, une grande simplification)." *[85]. He [85] added* "MM. Gibbs von Helmholtz, Duhem have used this function H = U − TS assuming that T and V are constant. Mr. von Helmotz has called it 'free energy' and also proposes to give him the name of "kinetic potential"; Duhem called it 'the thermodynamic potential at constant volume'; this is the most justified naming (MM. Gibbs, von Helmoltz, Duhem ont fait usage de cette function H = TS − U en y supposant T et V constants. M. von Helmotz l'a appellée énergie libre et a propose également de lui donner le nom de potential kinetique; M. Duhem la nomme potentiel thermodynamique à volume constant; c'est la dénomination la plus justifiée)". *In 1906, Henri Poincaré also published a note [87]* "Reflection on The kinetic theory of gases" (Réflexions sur la théorie cinétique des gaz), *where he said that:* "The kinetic theory of gases leaves awkward points for those who are accustomed to mathematical rigor ... One of the points which embarrassed me most was the following one: it is a question of demonstrating that the entropy keeps decreasing, but the reasoning of Gibbs seems to suppose that having made vary the outside conditions we wait that the regime is established before making them vary again. Is this supposition essential, or in other words, we could arrive at opposite results to the principle of Carnot by making vary the outside conditions too fast so that the permanent regime has time to become established?".

Jean-Marie Souriau has elaborated a disruptive and innovative *"théorie géométrique de la chaleur (geometric theory of heat)"* [88] after the works of his predecessors as illustrated in Figure 4: *"théorie analytique de la chaleur (analytic theory of heat)"* by Jean Baptiste Joseph Fourier [88], *"théorie mécanique de la chaleur (mechanic theory of heat)"* by François Clausius [89] and François Massieu and *"théorie mathématique de la chaleur (mathematic theory of heat)"* by Siméon-Denis Poisson [90,91], as illustrated in this figure:

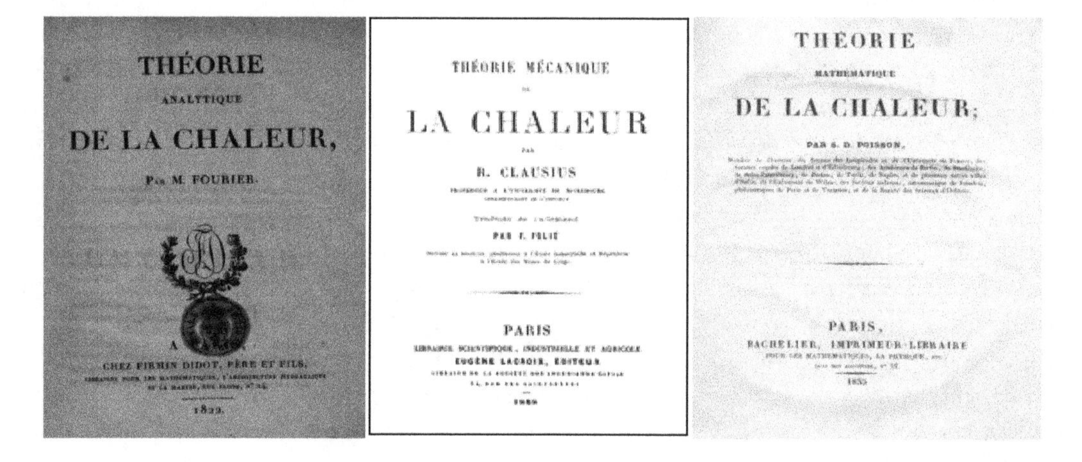

Figure 4. "Théorie analytique de la chaleur (analytic theory of heat)" by Jean Baptiste Joseph Fourier [88], "théorie mécanique de la chaleur (mechanic theory of heat)" by François Clausius [89] and "théorie mathématique de la chaleur (mathematic theory of heat)" by Siméon-Denis Poisson [90].

3. Revisited Souriau Symplectic Model of Statistical Physics

In this Section, we will revisit the Souriau model of thermodynamics but with modern notations, replacing personal Souriau conventions used in his book of 1970 by more classical ones.

In 1970, Souriau introduced the concept of co-adjoint action of a group on its momentum space (or *"moment map"*: mapping induced by symplectic manifold symmetries), based on the orbit method works, that allows to define physical observables like energy, heat and momentum or moment as pure geometrical objects (the moment map takes its values in a space determined by the group of symmetries: the dual space of its Lie algebra). The moment(um) map is a constant of the motion and is associated to symplectic cohomology (assignment of algebraic invariants to a topological space that arises from the algebraic dualization of the homology construction). Souriau introduced the moment map in 1965 in a lecture notes at Marseille University and published it in 1966. Souriau gave the formal definition and its name based on its physical interpretation in 1967. Souriau then studied its properties of equivariance, and formulated the coadjoint orbit theorem in his book in 1970. However, in his book, Souriau also observed in Chapter IV that Gibbs equilibrium states are not covariant by dynamical groups (Galileo or Poincaré groups) and then he developed a covariant model that he called *"Lie group thermodynamics"*, where equilibriums are indexed by a *"geometric (Planck) temperature"*, given by a vector β that lies in the Lie algebra of the dynamical group. For Souriau, all the details of classical mechanics appear as geometric necessities (e.g., mass is the measure of the symplectic cohomology of the action of a Galileo group). Based on this new covariant model of thermodynamic Gibbs equilibrium, Souriau has formulated statistical mechanics and thermodynamics in the framework of symplectic geometry by use of symplectic moments and distribution-tensor concepts, giving a geometric status for temperature, heat and entropy.

There is a controversy about the name "momentum map" or "moment map". Smale [92] referred to this map as the "angular momentum", while Souriau used the French word "moment". Cushman and Duistermaat [93] have suggested that the proper English translation of Souriau's French word was "momentum" which fit better with standard usage in mechanics. On the other hand, Guillemin and Sternberg [94] have validated the name given by Souriau and have used "moment" in English. In this paper, we will see that name "moment" given by Souriau was the most appropriate word. In his Chapter IV of his book [10], studying statistical mechanics, Souriau [10] has ingeniously observed that moments of inertia in mechanics are equivalent to moments in probability in his new geometric model of statistical physics. We will see that in Souriau Lie group thermodynamic model, these statistical moments will be given by the energy and the heat defined geometrically by Souriau, and will be associated with "moment map" in dual Lie algebra.

This work has been extended by Claude Vallée [5,6] and Gery de Saxcé [4,8,95,96]. More recently, Kapranov has also given a thermodynamical interpretation of the moment map for toric varieties [97] and Pavlov, thermodynamics from the differential geometry standpoint [98].

The conservation of the moment of a Hamiltonian action was called by Souriau the *"symplectic or geometric Noether theorem"*. Considering phases space as symplectic manifold, cotangent fiber of configuration space with canonical symplectic form, if Hamiltonian has Lie algebra, then the moment map is constant along the system integral curves. Noether theorem is obtained by considering independently each component of the moment map.

In a first step to establish new foundations of thermodynamics, Souriau [10] has defined a Gibbs canonical ensemble on a symplectic manifold M for a Lie group action on M. In classical statistical mechanics, a state is given by the solution of Liouville equation on the phase space, the partition function. As symplectic manifolds have a completely continuous measure, invariant by diffeomorphisms, the Liouville measure λ, all statistical states will be the product of the Liouville measure by the scalar function given by the generalized partition function $e^{\Phi(\beta)-\langle\beta,U(\xi)\rangle}$ defined by the energy U (defined in the dual of the Lie algebra of this dynamical group) and the geometric temperature β, where Φ is a normalizing constant such the mass of probability is equal to 1,

$\Phi(\beta) = -\log\int\limits_M e^{-\langle\beta,U(\xi)\rangle}d\lambda$ [99]. Jean-Marie Souriau then generalizes the Gibbs equilibrium state to all symplectic manifolds that have a dynamical group. To ensure that all integrals that will be defined could converge, *the canonical Gibbs ensemble is the largest open proper subset (in Lie algebra) where these integrals are convergent. This canonical Gibbs ensemble is convex.* The derivative of Φ, $Q = \frac{\partial\Phi}{\partial\beta}$ (thermodynamic heat) is equal to the mean value of the energy U. The minus derivative of this generalized heat Q, $K = -\frac{\partial Q}{\partial\beta}$ is symmetric and positive (this is a geometric heat capacity). Entropy s is then defined by Legendre transform of Φ, $s = \langle\beta, Q\rangle - \Phi$. If this approach is applied for the group of time translation, this is the classical thermodynamics theory. However, *Souriau* [10] *has observed that if we apply this theory for non-commutative group (Galileo or Poincaré groups), the symmetry has been broken. Classical Gibbs equilibrium states are no longer invariant by this group*. This symmetry breaking provides new equations, discovered by Souriau [10].

We can read in his paper this prophetical sentence "*This Lie group thermodynamics could be also of first interest for mathematics (Peut-être cette Thermodynamique des groups de Lie a-t-elle un intérêt mathématique)*" [30]. He explains that for the dynamic Galileo group with only one axe of rotation, this thermodynamic theory is the theory of centrifuge where the temperature vector dimension is equal to 2 (sub-group of invariance of size 2), used to make "uranium 235" and "ribonucleic acid" [30]. The physical meaning of these two dimensions for vector-valued temperature is "thermic conduction" and "viscosity". Souriau said that the model unifies "heat conduction" and "viscosity" (Fourier and Navier equations) in the same theory of irreversible process. Souriau has applied this theory in detail for relativistic ideal gas with the Poincaré group for the dynamical group.

Before introducing the Souriau Model of Lie group thermodynamics, we will first remind readers of the classical notation of Lie group theory in their application to Lie group thermodynamics:

- The coadjoint representation of G is the contragredient of the adjoint representation. It associates to each $g \in G$ the linear isomorphism $Ad_g^* \in GL(\mathfrak{g}^*)$, which satisfies, for each $\xi \in \mathfrak{g}^*$ and $X \in \mathfrak{g}$:

$$\left\langle Ad_{g^{-1}}^*(\xi), X\right\rangle = \left\langle \xi, Ad_{g^{-1}}(X)\right\rangle \tag{23}$$

- The adjoint representation of the Lie algebra $\mathfrak{g}$ is the linear representation of $\mathfrak{g}$ into itself which associates, to each $X \in \mathfrak{g}$, the linear map $ad_X \in gl(\mathfrak{g})$. ad Tangent application of Ad at neutral element e of G:

$$\begin{gathered} ad = T_e Ad : T_e G \to End(T_e G) \\ X, Y \in T_e G \mapsto ad_X(Y) = [X, Y] \end{gathered} \tag{24}$$

- The coadjoint representation of the Lie algebra $\mathfrak{g}$ is the contragredient of the adjoint representation. It associates, to each $X \in \mathfrak{g}$, the linear map $ad_X^* \in gl(\mathfrak{g}^*)$ which satisfies, for each $\xi \in \mathfrak{g}^*$ and $X \in \mathfrak{g}$:

$$\langle ad_{-X}^*(\xi), Y\rangle = \langle \xi, Ad_{-X}(Y)\rangle \tag{25}$$

We can illustrate for group of matrices for $G = GL_n(K)$ with $K = R$ or C.

$$T_e G = M_n(K),\ X \in M_n(K), g \in G\ \ Ad_g(X) = gXg^{-1} \tag{26}$$

$$X, Y \in M_n(K)\ ad_X(Y) = (T_e Ad)_X(Y) = XY - YX = [X, Y] \tag{27}$$

Then, the curve from $e = I_d = c(0)$ tangent to $X = c(1)$ is given by $c(t) = \exp(tX)$ and transform by Ad: $\gamma(t) = Ad\exp(tX)$

$$ad_X(Y) = (T_e Ad)_X(Y) = \left.\frac{d}{dt}\gamma(t)Y\right|_{t=0} = \left.\frac{d}{dt}\exp(tX)Y\exp(tX)^{-1}\right|_{t=0} = XY - YX \tag{28}$$

For each temperature β, element of the Lie algebra $\mathfrak{g}$, Souriau has introduced a tensor $\widetilde{\Theta}_\beta$, equal to the sum of the cocycle $\widetilde{\Theta}$ and the heat coboundary (with [.,.] Lie bracket):

$$\widetilde{\Theta}_\beta(Z_1, Z_2) = \widetilde{\Theta}(Z_1, Z_2) + \langle Q, ad_{Z_1}(Z_2) \rangle \text{ with } ad_{Z_1}(Z_2) = [Z_1, Z_2] \tag{29}$$

This tensor $\widetilde{\Theta}_\beta$ has the following properties:

- $\widetilde{\Theta}(X,Y) = \langle \Theta(X), Y \rangle$ where the map Θ is the one-cocycle of the Lie algebra $\mathfrak{g}$ with values in $\mathfrak{g}^*$, with $\Theta(X) = T_e\theta(X(e))$ where θ the one-cocycle of the Lie group G. $\widetilde{\Theta}(X,Y)$ is constant on M and the map $\widetilde{\Theta}(X,Y) : \mathfrak{g} \times \mathfrak{g} \to \Re$ is a skew-symmetric bilinear form, and is called the *symplectic cocycle of Lie algebra* $\mathfrak{g}$ associated to the *moment map* J, with the following properties:

$$\widetilde{\Theta}(X,Y) = J_{[X,Y]} - \{J_X, J_Y\} \text{ with } \{.,.\} \text{ Poisson Bracket and } J \text{ the Moment Map} \tag{30}$$

$$\widetilde{\Theta}([X,Y],Z) + \widetilde{\Theta}([Y,Z],X) + \widetilde{\Theta}([Z,X],Y) = 0 \tag{31}$$

where J_X linear application from $\mathfrak{g}$ to differential function on M: $\begin{array}{l}\mathfrak{g} \to C^\infty(M,R) \\ X \to J_X\end{array}$ and the associated differentiable application J, called moment(um) map:

$$\begin{array}{ll} J : M \to \mathfrak{g}^* & \text{such that } J_X(x) = \langle J(x), X \rangle,\ X \in \mathfrak{g} \\ x \mapsto J(x) & \end{array} \tag{32}$$

If instead of J we take the following moment map: $J'(x) = J(x) + Q\ ,\ x \in M$

where $Q \in \mathfrak{g}^*$ is constant, the symplectic cocycle θ is replaced by $\theta'(g) = \theta(g) + Q - Ad_g^* Q$

where $\theta' - \theta = Q - Ad_g^* Q$ is one-coboundary of G with values in $\mathfrak{g}^*$. We also have properties $\theta(g_1 g_2) = Ad_{g_1}^* \theta(g_2) + \theta(g_1)$ and $\theta(e) = 0$.

- The geometric temperature, element of the algebra $\mathfrak{g}$, is in the thekernel of the tensor $\widetilde{\Theta}_\beta$:

$$\beta \in Ker\, \widetilde{\Theta}_\beta, \text{ such that } \widetilde{\Theta}_\beta(\beta, \beta) = 0\ ,\ \forall \beta \in \mathfrak{g} \tag{33}$$

- The following symmetric tensor g_β, defined on all values of $ad_\beta(.) = [\beta, .]$ is positive definite:

$$g_\beta([\beta, Z_1], [\beta, Z_2]) = \widetilde{\Theta}_\beta(Z_1, [\beta, Z_2]) \tag{34}$$

$$g_\beta([\beta, Z_1], Z_2) = \widetilde{\Theta}_\beta(Z_1, Z_2)\ ,\ \forall Z_1 \in \mathfrak{g}, \forall Z_2 \in \mathrm{Im}(ad_\beta(.)) \tag{35}$$

$$g_\beta(Z_1, Z_2) \geq 0\ ,\ \forall Z_1, Z_2 \in \mathrm{Im}(ad_\beta(.)) \tag{36}$$

where the linear map $ad_X \in gl(\mathfrak{g})$ is the adjoint representation of the Lie algebra $\mathfrak{g}$ defined by $X, Y \in \mathfrak{g}(= T_eG) \mapsto ad_X(Y) = [X,Y]$, and the co-adjoint representation of the Lie algebra $\mathfrak{g}$ the linear map $ad_X^* \in gl(\mathfrak{g}^*)$ which satisfies, for each $\xi \in \mathfrak{g}^*$ and $X, Y \in \mathfrak{g}$:$\langle ad_X^*(\xi), Y \rangle = \langle \xi, -ad_X(Y) \rangle$*These equations are universal*, because they are not dependent on the symplectic manifold but only on the dynamical group G, the symplectic cocycle Θ, the temperature β and the heat Q. Souriau called this model "*Lie groups thermodynamics*".

We will give the main theorem of Souriau for this "Lie group thermodynamics":

Theorem 1 (Souriau Theorem of Lie Group Thermodynamics). *Let Ω be the largest open proper subset of g, Lie algebra of G, such that $\int_M e^{-\langle \beta, U(\xi) \rangle} d\lambda$ and $\int_M \xi \cdot e^{-\langle \beta, U(\xi) \rangle} d\lambda$ are convergent integrals, this set Ω is convex and is invariant under every transformation $Ad_g(.)$, where $g \mapsto Ad_g(.)$ is the adjoint representation of G, such that $Ad_g = T_e i_g$ with $i_g : h \mapsto ghg^{-1}$. Let $a : G \times \mathfrak{g}^* \to \mathfrak{g}^*$ a unique affine action a such that linear* *part is a coadjoint representation of G, that is the contragradient of the adjoint representation. It associates to each $g \in G$ the linear isomorphism $Ad_g^* \in GL(\mathfrak{g}^*)$, satisfying, for each:*

$$\xi \in \mathfrak{g}^* \text{ and } X \in \mathfrak{g} : \left\langle Ad_g^*(\xi), X \right\rangle = \left\langle \xi, Ad_{g^{-1}}(X) \right\rangle.$$

Then, the fundamental equations of Lie group thermodynamics are given by the action of the group:

- *Action of Lie group on Lie algebra:*

$$\beta \to Ad_g(\beta) \tag{37}$$

- *Transformation of characteristic function after action of Lie group:*

$$\Phi \to \Phi - \left\langle \theta\left(g^{-1}\right), \beta \right\rangle \tag{38}$$

- *Invariance of entropy with respect to action of Lie group:*

$$s \to s \tag{39}$$

- *Action of Lie group on geometric heat, element of dual Lie algebra:*

$$Q \to a(g, Q) = Ad_g^*(Q) + \theta(g) \tag{40}$$

Souriau equations of Lie group thermodynamics are summarized in the following Figures 5 and 6:

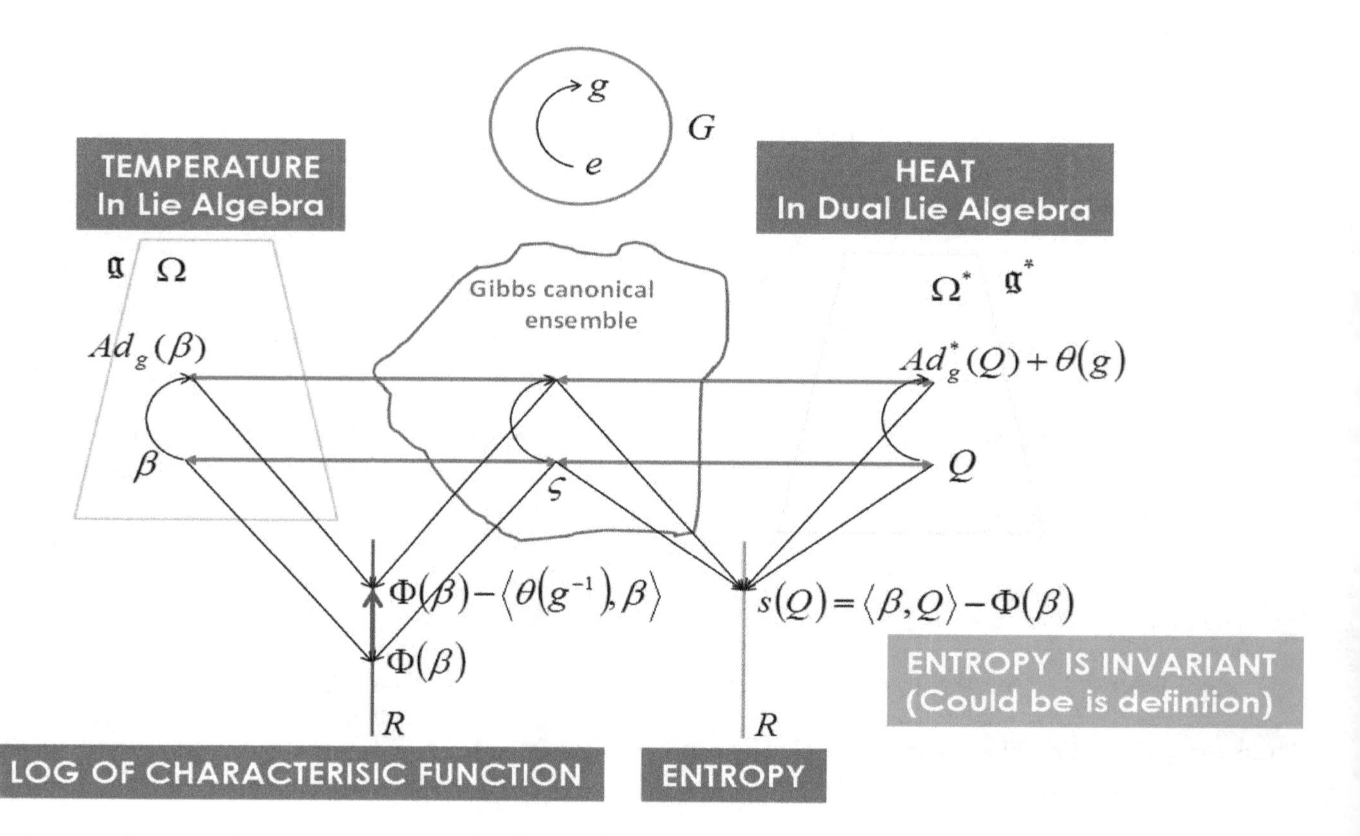

Figure 5. Global Souriau scheme of Lie group thermodynamics.

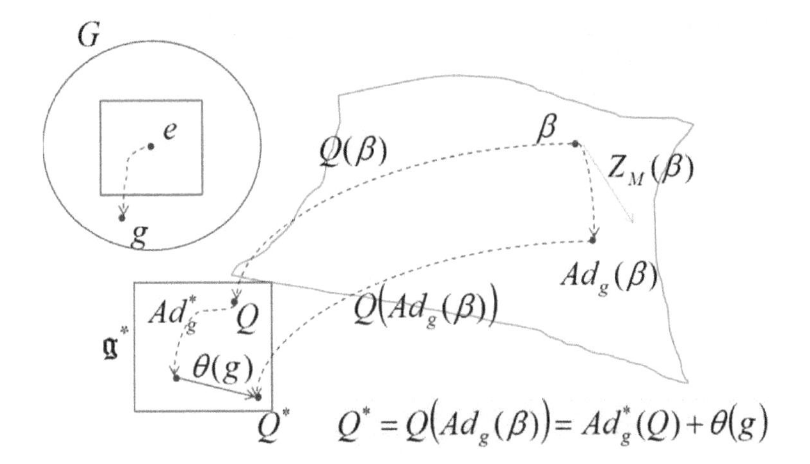

Figure 6. Broken symmetry on geometric heat Q due to adjoint action of the group on temperature β as an element of the Lie algebra.

For Hamiltonian, actions of a Lie group on a connected symplectic manifold, the equivariance of the moment map with respect to an affine action of the group on the dual of its Lie algebra has been studied by Marle and Libermann [100] and Lichnerowics [101,102]:

Theorem 2 (Marle Theorem on Cocycles). *Let G be a connected and simply connected Lie group, $R : G \to GL(E)$ be a linear representation of G in a finite-dimensional vector space E, and $r : \mathfrak{g} \to gl(E)$ be the associated linear representation of its Lie algebra $\mathfrak{g}$. For any one-cocycle $\Theta : \mathfrak{g} \to E$ of the Lie algebra $\mathfrak{g}$ for the linear representation r, there exists a unique one-cocycle $\theta : G \to E$ of the Lie group G for the linear representation R such that $\Theta(X) = T_e\theta\,(X(e))$, which has Θ as associated Lie algebra one-cocycle. The Lie group one-cocycle θ is a Lie group one-coboundary if and only if the Lie algebra one-cocycle Θ is a Lie algebra one-coboundary.*

Let G be a Lie group whose Lie algebra is $\mathfrak{g}$. The skew-symmetric bilinear form $\widetilde{\Theta}$ on $\mathfrak{g} = T_eG$ can be extended into a closed differential two-form on G, since the identity on $\widetilde{\Theta}$ means that its exterior differential $d\widetilde{\Theta}$ vanishes. In other words, $\widetilde{\Theta}$ is a 2-cocycle for the restriction of the de Rham cohomology of G to left invariant differential forms. In the framework of Lie group action on a symplectic manifold, equivariance of moment could be studied to prove that there is a unique action $a(.,.)$ of the Lie group G on the dual $\mathfrak{g}^*$ of its Lie algebra for which the moment map J is equivariant, that means for each $x \in M$:

$$J\left(\Phi_g(x)\right) = a(g, J(x)) = Ad_g^*\left(J(x)\right) + \theta(g) \tag{41}$$

where $\Phi : G \times M \to M$ is an action of Lie group G on differentiable manifold M, the fundamental field associated to an element X of Lie algebra $\mathfrak{g}$ of group G is the vectors field X_M on M:

$$X_M(x) = \frac{d}{dt}\Phi_{\exp(-tX)}\left(x\right)\Big|_{t=0} \tag{42}$$

with $\Phi_{g_1}\left(\Phi_{g_2}(x)\right) = \Phi_{g_1g_2}(x)$ and $\Phi_e(x) = x$. Φ is Hamiltonian on a symplectic manifold M, if Φ is symplectic and if for all $X \in \mathfrak{g}$, the fundamental field X_M is globally Hamiltonian. The cohomology class of the symplectic cocycle θ only depends on the Hamiltonian action Φ, and not on J.

In Appendix B, we observe that Souriau Lie group thermodynamics is compatible with Balian gauge theory of thermodynamics [103], that is obtained by symplectization in dimension $2n + 2$ of contact manifold in dimension $2n + 1$. All elements of the Souriau geometric temperature vector are multiplied by the same gauge parameter.

We conclude this section by this Bourbakiste citation of Jean-Marie Souriau [34]:

It is obvious that one can only define average values on objects belonging to a vector (or affine) space; Therefore—so this assertion may seem Bourbakist—that we will observe and measure average values only as quantity belonging to a set having physically an affine structure. It is clear that this structure is necessarily unique—if not the average values would not be well defined. (Il est évident que l'on ne peut définir de valeurs moyennes que sur des objets appartenant à un espace vectoriel (ou affine); donc—si bourbakiste que puisse sembler cette affirmation—que l'on n'observera et ne mesurera de valeurs moyennes que sur des grandeurs appartenant à un ensemble possédant physiquement une structure affine. Il est clair que cette structure est nécessairement unique—sinon les valeurs moyennes ne seraient pas bien définies.).

4. The Souriau-Fisher Metric as Geometric Heat Capacity of Lie Group Thermodynamics

We observe that Souriau Riemannian metric, introduced with symplectic cocycle, is a generalization of the Fisher metric, that we call the Souriau-Fisher metric, that preserves the property to be defined as a hessian of the partition function logarithm $g_\beta = -\frac{\partial^2 \Phi}{\partial \beta^2} = \frac{\partial^2 \log \psi_\Omega}{\partial \beta^2}$ as in classical information geometry. We will establish the equality of two terms, between Souriau definition based on Lie group cocycle Θ and parameterized by "geometric heat" Q (element of dual Lie algebra) and "geometric temperature" β (element of Lie algebra) and hessian of characteristic function $\Phi(\beta) = -\log \psi_\Omega(\beta)$ with respect to the variable β:

$$g_\beta \left([\beta, Z_1], [\beta, Z_2] \right) = \langle \Theta(Z_1), [\beta, Z_2] \rangle + \langle Q, [Z_1, [\beta, Z_2]] \rangle = \frac{\partial^2 \log \psi_\Omega}{\partial \beta^2} \tag{43}$$

If we differentiate this relation of Souriau theorem $Q\left(Ad_g(\beta)\right) = Ad_g^*(Q) + \theta(g)$, this relation occurs:

$$\frac{\partial Q}{\partial \beta}\left(-[Z_1, \beta], .\right) = \widetilde{\Theta}\left(Z_1, [\beta, .]\right) + \left\langle Q, Ad_{.Z_1}([\beta, .]) \right\rangle = \widetilde{\Theta}_\beta\left(Z_1, [\beta, .]\right) \tag{44}$$

$$-\frac{\partial Q}{\partial \beta}\left([Z_1, \beta], Z_2.\right) = \widetilde{\Theta}\left(Z_1, [\beta, Z_2]\right) + \left\langle Q, Ad_{.Z_1}([\beta, Z_2]) \right\rangle = \widetilde{\Theta}_\beta\left(Z_1, [\beta, Z_2]\right) \tag{45}$$

$$\Rightarrow -\frac{\partial Q}{\partial \beta} = g_\beta\left([\beta, Z_1], [\beta, Z_2]\right) \tag{46}$$

As the entropy is defined by the Legendre transform of the characteristic function, this Souriau-Fisher metric is also equal to the inverse of the hessian of "geometric entropy" $s(Q)$ with respect to the variable Q: $\frac{\partial^2 s(Q)}{\partial Q^2}$

For the maximum entropy density (Gibbs density), the following three terms coincide: $\frac{\partial^2 \log \psi_\Omega}{\partial \beta^2}$ that describes the convexity of the log-likelihood function, $I(\beta) = -E\left[\frac{\partial^2 \log p_\beta(\xi)}{\partial \beta^2}\right]$ the Fisher metric that describes the covariance of the log-likelihood gradient, whereas $I(\beta) = E\left[(\xi - Q)(\xi - Q)^T\right] = Var(\xi)$ that describes the covariance of the observables.

We can also observe that the Fisher metric $I(\beta) = -\frac{\partial Q}{\partial \beta}$ is exactly the Souriau metric defined through symplectic cocycle:

$$I(\beta) = \widetilde{\Theta}_\beta\left(Z_1, [\beta, Z_2]\right) = g_\beta\left([\beta, Z_1], [\beta, Z_2]\right) \tag{47}$$

The Fisher metric $I(\beta) = -\frac{\partial^2 \Phi(\beta)}{\partial \beta^2} = -\frac{\partial Q}{\partial \beta}$ has been considered by Souriau as a *generalization of "heat capacity"*. Souriau called it K the *"geometric capacity"*.

Si l'on divise la quantité que l'on vient de trouver par celle qui est nécessaire pour élever la molécule de la température o à la température 1, on connaîtra l'accroissement de température qui s'opère pendant l'instant dt. Or, cette dernière quantité est C.D $dx\,dy\,dz$: car C désigne la capacité de chaleur de la substance; D sa densité, et $dx\,dy\,dz$ le volume de la molécule. On a donc, pour exprimer le mouvement de la chaleur dans l'intérieur du solide, l'équation

$$\frac{dv}{dt}=\frac{K}{C.D}\left(\frac{d^2v}{dx^2}+\frac{d^2v}{dy^2}+\frac{d^2v}{dz^2}\right) \quad (d)$$

D étant la densité du solide, ou le poids de l'unité de volume, et C la capacité spécifique, ou la quantité de chaleur qui élève l'unité de poids de la température o à la température 1; le produit C.D $dx\,dy\,dz$ exprime combien il faut de chaleur pour élever de o à 1 la molécule dont le volume est $dx\,dy\,dz$. Donc en divisant par ce produit la nouvelle quantité de chaleur que la molécule vient d'acquérir, on aura son accroissement de température. On obtient ainsi

Figure 7. Fourier heat equation in seminal manuscript of Joseph Fourier [88].

For $\beta = \frac{1}{kT}$, $K = -\frac{\partial Q}{\partial \beta} = -\frac{\partial Q}{\partial T}\left(\frac{\partial(1/kT)}{\partial T}\right)^{-1} = kT^2\frac{\partial Q}{\partial T}$ linking the geometric capacity to calorific capacity, then Fisher metric can be introduced in Fourier heat equation (see Figure 7):

$$\frac{\partial T}{\partial t} = \frac{\kappa}{C \cdot D}\Delta T \text{ with } \frac{\partial Q}{\partial T} = C \cdot D \Rightarrow \frac{\partial \beta^{-1}}{\partial t} = \kappa\left[\left(\beta^2/k\right) \cdot I_{Fisher}(\beta)\right]^{-1}\Delta\beta^{-1} \quad (48)$$

We can also observe that Q is related to the mean, and K to the variance of U:

$$K = I(\beta) = -\frac{\partial Q}{\partial \beta} = \mathrm{var}(U) = \int_M U(\xi)^2 \cdot p_\beta(\xi)d\omega - \left(\int_M U(\xi) \cdot p_\beta(\xi)d\omega\right)^2 \quad (49)$$

We observe that the entropy s is unchanged, and Φ is changed but with linear dependence to β, with the consequence that Fisher Souriau metric is invariant:

$$s\left[Q\left(Ad_g(\beta)\right)\right] = s(Q(\beta)) \text{ and } I\left(Ad_g(\beta)\right) = -\frac{\partial^2\left(\Phi - \left\langle\theta\left(g^{-1}\right),\beta\right\rangle\right)}{\partial\beta^2} = -\frac{\partial^2\Phi}{\partial\beta^2} = I(\beta) \quad (50)$$

We have observed that the concept of "heat capacity" is important in the Souriau model because it gives a geometric meaning to its definition. The notion of "heat capacity" has been generalized by Pierre Duhem in his general equations of thermodynamics.

Souriau [34] proposed to define a thermometer (θεϱμόσ) device principle that could measure this geometric temperature using "relative ideal gas thermometer" based on a theory of dynamical group thermometry and has also recovered the (geometric) Laplace barometric law

5. Euler-Poincaré Equations and Variational Principle of Souriau Lie Group Thermodynamics

When a Lie algebra acts locally transitively on the configuration space of a Lagrangian mechanical system, Henri Poincaré proved that the Euler-Lagrange equations are equivalent to a new system of differential equations defined on the product of the configuration space with the Lie algebra. Marle has written about the Euler-Poincaré equations [104], under an intrinsic form, without any reference to a particular system of local coordinates, proving that they can be conveniently expressed in terms of the Legendre and moment maps of the lift to the cotangent bundle of the Lie algebra action on the configuration space. The Lagrangian is a smooth real valued function L defined on the tangent bundle

TM. To each parameterized continuous, piecewise smooth curve $\gamma : [t_0, t_1] \rightarrow M$, defined on a closed interval $[t_0, t_1]$, with values in M, one associates the value at γ of the action integral:

$$I(\gamma) = \int_{t_0}^{t_1} L\left(\frac{d\gamma(t)}{dt}\right) dt \tag{51}$$

The partial differential of the function $L : M \times \mathfrak{g} \rightarrow \Re$ with respect to its second variable $d_2\overline{L}$, which plays an important part in the Euler-Poincaré equation, can be expressed in terms of the moment and Legendre maps: $d_2\overline{L} = p_{g^*} \circ \varphi^t \circ \mathrm{L} \circ \varphi$ with $J = p_{g^*} \circ \varphi^t (\Rightarrow d_2\overline{L} = J \circ \mathrm{L} \circ \varphi)$ the moment map, $p_{\mathfrak{g}^*} : M \times \mathfrak{g}^* \rightarrow \mathfrak{g}^*$ the canonical projection on the second factor, $\mathrm{L} : TM \rightarrow T^*M$ the Legendre transform, with:

$$\varphi : M \times \mathfrak{g} \rightarrow TM / \varphi(x, X) = X_M(x) \text{ and } \varphi^t : T^*M \rightarrow M \times \mathfrak{g}^* / \varphi^t(\xi) = (\pi_M(\xi), J(\xi)) \tag{52}$$

The Euler-Poincaré equation can therefore be written under the form:

$$\left(\frac{d}{dt} - ad^*_{V(t)}\right) (J \circ \mathrm{L} \circ \varphi\left(\gamma(t), V(t)\right)) = J \circ d_1\overline{L}\left(\gamma(t), V(t)\right) \text{ with } \frac{d\gamma(t)}{dt} = \varphi\left(\gamma(t), V(t)\right) \tag{53}$$

with

$$H(\xi) = \left\langle \xi, \mathrm{L}^{-1}(\xi) \right\rangle - L\left(\mathrm{L}^{-1}(\xi)\right), \ \xi \in T^*M, \ \mathrm{L} : TM \rightarrow T^*M, \ H : T^*M \rightarrow R. \tag{54}$$

Following the remark made by Poincaré at the end of his note [105], the most interesting case is when the map $\overline{L} : M \times \mathfrak{g} \rightarrow R$ only depends on its second variable $X \in \mathfrak{g}$. The Euler-Poincaré equation becomes:

$$\left(\frac{d}{dt} - ad^*_{V(t)}\right) \left(d\overline{L}\left(V(t)\right)\right) = 0 \tag{55}$$

We can use analogy of structure when the convex Gibbs ensemble is homogeneous [106]. We can then apply Euler-Poincaré equation for Lie group thermodynamics. Considering Clairaut's equation:

$$s\left(Q\right) = \langle \beta, Q \rangle - \Phi(\beta) = \left\langle \Theta^{-1}(Q), Q \right\rangle - \Phi\left(\Theta^{-1}(Q)\right) \tag{56}$$

with $Q = \Theta(\beta) = \dfrac{\partial \Phi}{\partial \beta} \in \mathfrak{g}^*$, $\beta = \Theta^{-1}(Q) \in \mathfrak{g}$, a *Souriau-Euler-Poincaré equation* can be elaborated for Souriau Lie group thermodynamics:

$$\frac{dQ}{dt} = ad^*_{\beta} Q \tag{57}$$

or

$$\frac{d}{dt}\left(Ad^*_g Q\right) = 0. \tag{58}$$

The first equation, the Euler-Poincaré equation is a reduction of Euler-Lagrange equations using symmetries and especially the fact that a group is acting homogeneously on the symplectic manifold:

$$\frac{dQ}{dt} = ad^*_{\beta} Q \text{ and } \left\{ \begin{array}{l} s(Q) = \langle \beta, Q \rangle - \Phi(\beta) \\ \beta = \frac{\partial s(Q)}{\partial Q} \in \mathfrak{g}, \ Q = \frac{\partial \Phi(\beta)}{\partial \beta} \in \mathfrak{g}^* \end{array} \right. \tag{59}$$

Back to Koszul model of information geometry, we can then deduce an equivalent of the Euler-Poincaré equation for statistical models

$$\frac{dx^*}{dt} = ad_x^* x^* \text{ and } \begin{cases} \Phi^*(x^*) = \langle x, x^* \rangle - \Phi(x) \\ x = \frac{\partial \Phi^*(x^*)}{\partial x} \in \Omega \, , \, x^* = \frac{\partial \Phi(x)}{\partial x} \in \Omega^* \end{cases} \tag{60}$$

We can use this Euler-Poincaré equation to deduce an associated equation on entropy: $\frac{ds}{dt} = \left\langle \frac{d\beta}{dt}, Q \right\rangle + \left\langle \beta, ad_\beta^* Q \right\rangle - \frac{d\Phi}{dt}$ that reduces to

$$\frac{ds}{dt} = \left\langle \frac{d\beta}{dt}, Q \right\rangle - \frac{d\Phi}{dt} \tag{61}$$

due to $\langle \xi, ad_V X \rangle = -\langle ad_V^* \xi, X \rangle \Rightarrow \left\langle \beta, ad_\beta^* Q \right\rangle = \langle Q, ad_\beta \beta \rangle = 0$.

With these new equation of thermodynamics $\frac{dQ}{dt} = ad_\beta^* Q$ and $\frac{d}{dt}(Ad_g^* Q) = 0$, we can observe that the new important notion is related to co-adjoint orbits, that are associated to a symplectic manifold by Souriau with KKS 2-form.

We will then define the Poincaré-Cartan integral invariant for Lie group thermodynamics. Classically in mechanics, the Pfaffian form $\omega = p \cdot dq - H \cdot dt$ is related to Poincaré-Cartan integral invariant [107]. Dedecker has observed, based on the relation [108]:

$$\omega = \partial_{\dot{q}} L \cdot dq - \left(\partial_{\dot{q}} L \cdot \dot{q} - L \right) \cdot dt = L \cdot dt + \partial_{\dot{q}} L \varpi \text{ with } \varpi = dq - \dot{q} \cdot dt \tag{62}$$

that the property that among all forms $\chi \equiv L \cdot dt \mathrm{mod} \varpi$ the form $\omega = p \cdot dq - H \cdot dt$ is the only one satisfying $d\chi \equiv 0 \mathrm{mod} \varpi$, is a particular case of more general Lepage congruence.

Analogies between geometric mechanics and geometric Lie group thermodynamics, provides the following similarities of structures:

$$\begin{cases} \dot{q} \leftrightarrow \beta \\ p \leftrightarrow Q \end{cases}, \begin{cases} L(\dot{q}) \leftrightarrow \Phi(\beta) \\ H(p) \leftrightarrow s(Q) \\ H = p \cdot \dot{q} - L \leftrightarrow s = \langle Q, \beta \rangle - \Phi \end{cases}$$
$$\text{and } \begin{cases} \dot{q} = \frac{dq}{dt} = \frac{\partial H}{\partial p} \leftrightarrow \beta = \frac{\partial s}{\partial Q} \\ p = \frac{\partial L}{\partial \dot{q}} \leftrightarrow Q = \frac{\partial \Phi}{\partial \beta} \end{cases} \tag{63}$$

We can then consider a similar *Poincaré-Cartan-Souriau Pfaffian form*:

$$\omega = p \cdot dq - H \cdot dt \leftrightarrow \omega = \langle Q, (\beta \cdot dt) \rangle - s \cdot dt = (\langle Q, \beta \rangle - s) \cdot dt = \Phi(\beta) \cdot dt \tag{64}$$

This analogy provides an associated *Poincaré-Cartan-Souriau integral invariant*. Poincaré-Cartan integral invariant $\int\limits_{C_a} p \cdot dq - H.dt = \int\limits_{C_b} p \cdot dq - H \cdot dt$ is given for Souriau thermodynamics by:

$$\int\limits_{C_a} \Phi(\beta) \cdot dt = \int\limits_{C_b} \Phi(\beta) \cdot dt \tag{65}$$

We can then deduce an *Euler-Poincaré-Souriau variational principle* for thermodynamics: The variational principle holds on $\mathfrak{g}$, for variations $\delta\beta = \dot{\eta} + [\beta, \eta]$, where $\eta(t)$ is an arbitrary path that vanishes at the endpoints, $\eta(a) = \eta(b) = 0$:

$$\delta \int_{t_0}^{t_1} \Phi\left(\beta(t)\right) \cdot dt = 0 \tag{66}$$

6. Souriau Affine Representation of Lie Group and Lie Algebra and Comparison with the Koszul Affine Representation

This affine representation of Lie group/algebra used by Souriau has been intensively studied by Marle [7,100,109,110]. Souriau called the mechanics deduced from this model, "affine mechanics". We will explain affine representations and associated notions as cocycles, Souriau moment map and cocycles, equivariance of Souriau moment map, action of Lie group on a symplectic manifold and dual spaces of finite-dimensional Lie algebras. We have observed that these tools have been developed in parallel by Jean-Louis Koszul. We will establish close links and synthetize the comparisons in a table of both approaches.

6.1. Affine Representations and Cocycles

Souriau model of Lie group thermodynamics is linked with affine representation of Lie group and Lie algebra. We will give in the following main elements of this affine representation.

Let G be a Lie group and E a finite-dimensional vector space. A map $A : G \to Aff(E)$ can always be written as:

$$A(g)(x) = R(g)(x) + \theta(g) \text{ with } g \in G, x \in E \tag{67}$$

where the maps $R : G \to GL(E)$ and $\theta : G \to E$ are determined by A. The map A is *an affine representation of* G in E.

The map $\theta : G \to E$ is a one-cocycle of G with values in E, for the linear representation R; it means that θ is a smooth map which satisfies, for all $g, h \in G$:

$$\theta(gh) = R(g)(\theta(h)) + \theta(g) \tag{68}$$

The linear representation R is called the linear part of the affine representation A, and θ is called the one-cocycle of G associated to the affine representation A. A one-coboundary of G with values in E, for the linear representation R, is a map $\theta : G \to E$ which can be expressed as:

$$\theta(g) = R(g)(c) - c\,,\ g \in G \tag{69}$$

where c is a fixed element in E and then there exist an element $c \in E$ such that, for all $g \in G$ and $x \in E$:

$$A(g)(x) = R(g)(x + c) - c \tag{70}$$

Let $\mathfrak{g}$ be a Lie algebra and E a finite-dimensional vector space. A linear map $a : \mathfrak{g} \to aff(E)$ always can be written as:

$$a(X)(x) = r(X)(x) + \Theta(X) \text{ with } X \in \mathfrak{g}, x \in E \tag{71}$$

where the linear maps $r : \mathfrak{g} \to gl(E)$ and $\Theta : \mathfrak{g} \to E$ are determined by a. The map a is an affine representation of G in E. The linear map $\Theta : \mathfrak{g} \to E$ is a one-cocycle of G with values in E, for the linear representation r; it means that Θ satisfies, for all $X, Y \in \mathfrak{g}$:

$$\Theta\left([X, Y]\right) = r(X)\left(\Theta(Y)\right) - r(Y)\left(\Theta(X)\right) \tag{72}$$

Θ is called the one-cocycle of $\mathfrak{g}$ associated to the affine representation a. A one-coboundary of $\mathfrak{g}$ with values in E, for the linear representation r, is a linear map $\Theta : \mathfrak{g} \to E$ which can be expressed as: $\Theta(X) = r(X)(c)\,,\ X \in \mathfrak{g}$ where c is a fixed element in E., and then there exist an element $c \in E$ such that, for all $X \in \mathfrak{g}$ and $x \in E$:

$$a(X)(x) = r(X)(x + c)$$

Let $A : G \to Aff(E)$ be an affine representation of a Lie group $\mathfrak{g}$ in a finite-dimensional vector space E, and $\mathfrak{g}$ be the Lie algebra of G. Let $R : G \to GL(E)$ and $\theta : G \to E$ be, respectively, the linear part and the associated cocycle of the affine representation A. Let $a : \mathfrak{g} \to aff(E)$ be the affine representation of the Lie algebra $\mathfrak{g}$ associated to the affine representation $A : G \to Aff(E)$ of the Lie group G. The linear part of a is the linear representation $r : \mathfrak{g} \to gl(E)$ associated to the linear representation $R : G \to GL(E)$, and the associated cocycle $\Theta : \mathfrak{g} \to E$ is related to the one-cocycle $\theta : G \to E$ by:

$$\Theta(X) = T_e\theta\left(X(e)\right), X \in \mathfrak{g} \tag{73}$$

This is deduced from:

$$\left.\frac{dA\left(\exp(tX)\right)(x)}{dt}\right|_{t=0} = \left.\frac{d\left(R(\exp(tX))(x) + \theta(\exp(tX)\right)}{dt}\right|_{t=0} \Rightarrow a(X)(x) = r(X)(x) + T_e\theta(X) \tag{74}$$

Let G be a connected and simply connected Lie group, $R : G \to GL(E)$ be a linear representation of G in a finite-dimensional vector space E, and $r : \mathfrak{g} \to gl(E)$ be the associated linear representation of its Lie algebra $\mathfrak{g}$. For any one-cocycle $\Theta : \mathfrak{g} \to E$ of the Lie algebra $\mathfrak{g}$ for the linear representation r, there exists a unique one-cocycle $\theta : G \to E$ of the Lie group G for the linear representation R such that:

$$\Theta(X) = T_e\theta\left(X(e)\right) \tag{75}$$

in other words, which has Θ as associated Lie algebra one-cocycle. The Lie group one-cocycle θ is a Lie group one-coboundary if and only if the Lie algebra one-cocycle Θ is a Lie algebra one-coboundary.

$$\left.\frac{d\theta\left(g\exp(tX)\right)}{dt}\right|_{t=0} = \left.\frac{d\left(\theta(g) + R(g)\left(\theta(\exp(tX))\right)\right)}{dt}\right|_{t=0} \Rightarrow T_g\theta\left(TL_g(X)\right) = R(g)\left(\Theta(x)\right) \tag{76}$$

which proves that if it exists, the Lie group one-cocycle θ such that $T_e\theta = \Theta$ is unique.

6.2. Souriau Moment Map and Cocycles

Souriau first introduced the moment map in his book. We will give the link with previous cocycles of affine representation.

There exist J_X linear application from $\mathfrak{g}$ to differential function on M:

$$\begin{aligned} &\mathfrak{g} \to C^\infty(M, R) \\ &X \to J_X \end{aligned} \tag{77}$$

We can then associate a differentiable application J, called moment(um) map for the Hamiltonian Lie group action Φ:

$$\begin{aligned} &J : M \to \mathfrak{g}^* \\ &x \mapsto J(x) \text{ such that } J_X(x) = \langle J(x), X\rangle, \; X \in \mathfrak{g} \end{aligned} \tag{78}$$

Let J moment map, for each $(X, Y) \in \mathfrak{g} \times \mathfrak{g}$, we associate a smooth function $\widetilde{\Theta}\left(X, Y\right) : M \to \Re$ defined by:

$$\widetilde{\Theta}(X, Y) = J_{[X,Y]} - \{J_X, J_Y\} \text{ with } \{.,.\} : \text{Poisson Bracket} \tag{79}$$

It is a Casimir of the Poisson algebra $C^\infty(M, \Re)$, that satisfies:

$$\widetilde{\Theta}([X, Y], Z) + \widetilde{\Theta}([Y, Z], X) + \widetilde{\Theta}([Z, X], Y) = 0 \tag{80}$$

When the Poisson manifold is a connected symplectic manifold, the function $\widetilde{\Theta}\left(X, Y\right)$ is constant on M and the map:

$$\widetilde{\Theta}(X,Y) : \mathfrak{g} \times \mathfrak{g} \to \Re \quad (81)$$

is a skew-symmetric bilinear form, and is called the symplectic Cocycle of Lie algebra $\mathfrak{g}$ associated to the moment map J.

Let $\Theta : \mathfrak{g} \to \mathfrak{g}^*$ be the map such that for all:

$$X, Y \in \mathfrak{g} : \ \langle \Theta(X), Y \rangle = \widetilde{\Theta}(X,Y) \quad (82)$$

The map Θ is therefore the one-cocycle of the Lie algebra $\mathfrak{g}$ with values in $\mathfrak{g}^*$ for the coadjoint representation $X \mapsto ad_X^*$ of $\mathfrak{g}$ associated to the affine action of $\mathfrak{g}$ on its dual:

$$a_\Theta(X)(\xi) = ad_{-X}^*(\xi) + \Theta(X) \ , \ X \in \mathfrak{g} \ , \ \xi \in \mathfrak{g}^* \quad (83)$$

Let G be a Lie group whose Lie algebra is $\mathfrak{g}$. The skew-symmetric bilinear form $\widetilde{\Theta}$ on $\mathfrak{g} = T_e G$ can be extended into a closed differential two-form on G, since the identity on $\widetilde{\Theta}$ means that its exterior differential $d\widetilde{\Theta}$ vanishes. In other words, $\widetilde{\Theta}$ is a 2-cocycle for the restriction of the de Rham cohomology of G to left (or right) invariant differential forms.

6.3. Equivariance of Souriau Moment Map

There exists a unique affine action a such that the linear part is a coadjoint representation:

$$\begin{aligned} a &: G \times \mathfrak{g}^* \to \mathfrak{g}^* \\ a(g,\xi) &= Ad_{g^{-1}}^* \xi + \theta(g) \end{aligned} \quad (84)$$

with $\left\langle Ad_{g^{-1}}^* \xi, X \right\rangle = \langle \xi, Ad_{g^{-1}} X \rangle$ and that induce equivariance of moment J.

6.4. Action of Lie Group on a Symplectic Manifold

Let $\Phi : G \times M \to M$ be an action of Lie group G on differentiable manifold M, the fundamental field associated to an element X of Lie algebra $\mathfrak{g}$ of group G is the vectors field X_M on M:

$$X_M(x) = \frac{d}{dt} \Phi_{\exp(-tX)}(x) \Big|_{t=0} \quad \text{With } \Phi_{g_1}(\Phi_{g_2}(x)) = \Phi_{g_1 g_2}(x) \text{ and } \Phi_e(x) = x \quad (85)$$

Φ is Hamiltonian on a symplectic manifold M, if Φ is symplectic and if for all $X \in \mathfrak{g}$, the fundamental field X_M is globally Hamiltonian.

There is a unique action a of the Lie group G on the dual $\mathfrak{g}^*$ of its Lie algebra for which the moment map J is equivariant, that means satisfies for each $x \in M$

$$J(\Phi_g(x)) = a(g, J(x)) = Ad_{g^{-1}}^*(J(x)) + \theta(g) \quad (86)$$

$\theta : G \to \mathfrak{g}^*$ is called cocycle associated to the differential $T_e\theta$ of 1-cocyle θ associated to J at neutral element e:

$$\langle T_e\theta(X), Y \rangle = \widetilde{\Theta}(X,Y) = J_{[X,Y]} - \{J_X, J_Y\} \quad (87)$$

If instead of J we take the moment map $J'(x) = J(x) + \mu \ , \ x \in M$, where $\mu \in \mathfrak{g}^*$ is constant, the symplectic cocycle θ is replaced by:

$$\theta'(g) = \theta(g) + \mu - Ad_g^* \mu \quad (88)$$

where $\theta' - \theta = \mu - Ad_g^* \mu$ is one-coboundary of G with values in $\mathfrak{g}^*$.

Therefore, the cohomology class of the symplectic cocycle θ only depends on the Hamiltonian action Φ, not on the choice of its moment map J. We have also:

$$\tilde{\Theta}'(X,Y) = \tilde{\Theta}(X,Y) + \langle \mu, [X,Y] \rangle \tag{89}$$

This property is used by Jean-Marie Souriau [10] to offer a very nice cohomological interpretation of the total mass of a classical (nonrelativistic) isolated mechanical system. He [10] proves that the space of all possible motions of the system is a symplectic manifold on which the Galilean group acts by a Hamiltonian action. The dimension of the symplectic cohomology space of the Galilean group (the quotient of the space of symplectic one-cocycles by the space of symplectic one-coboundaries) is equal to 1. The cohomology class of the symplectic cocycle associated to a moment map of the action of the Galilean group on the space of motions of the system is interpreted as the total mass of the system.

For Hamiltonian actions of a Lie group on a connected symplectic manifold, the equivariance of the moment map with respect to an affine action of the group on the dual of its Lie algebra has been proved by Marle [110]. Marle [110] has also developed the notion of symplectic cocycle and has proved that given a Lie algebra symplectic cocycle, there exists on the associated connected and simply connected Lie group a unique corresponding Lie group symplectic cocycle. Marle [104] has also proved that there exists a two-parameter family of deformations of these actions (the Hamiltonian actions of a Lie group on its cotangent bundle obtained by lifting the actions of the group on itself by translations) into a pair of mutually symplectically orthogonal Hamiltonian actions whose moment maps are equivariant with respect to an affine action involving any given Lie group symplectic cocycle. Marle [104] has also explained why a reduction occurs for Euler-Poncaré equation mainly when the Hamiltonian can be expressed as the moment map composed with a smooth function defined on the dual of the Lie algebra; the Euler-Poincaré equation is then equivalent to the Hamilton equation written on the dual of the Lie algebra.

6.5. Dual Spaces of Finite-Dimensional Lie Algebras

Let $\mathfrak{g}$ be a finite-dimensional Lie algebra, and $\mathfrak{g}^*$ its dual space. The Lie algebra $\mathfrak{g}$ can be considered as the dual of $\mathfrak{g}^*$, that means as the space of linear functions on $\mathfrak{g}^*$, and the bracket of the Lie algebra $\mathfrak{g}$ is a composition law on this space of linear functions. This composition law can be extended to the space $C^\infty(\mathfrak{g}^*, \Re)$ by setting:

$$\{f,g\}(x) = \langle x, [df(x), dg(x)] \rangle \ , \ f \text{ and } g \in C^\infty(\mathfrak{g}^*, \Re), \ x \in \mathfrak{g}^* \tag{90}$$

If we apply this formula for Souriau Lie group thermodynamics, and for entropy *s(Q)* depending on geometric heat Q:

$$\{s_1, s_2\}(Q) = \langle Q, [ds_1(Q), ds_2(Q)] \rangle \ , \ s_1 \text{ and } s_2 \in C^\infty(\mathfrak{g}^*, \Re), \ Q \in \mathfrak{g}^* \tag{91}$$

This bracket on $C^\infty(\mathfrak{g}^*, \Re)$ defines a Poisson structure on $\mathfrak{g}^*$, called its canonical Poisson structure. It implicitly appears in the works of Sophus Lie, and was rediscovered by Alexander Kirillov [111], Bertram Kostant and Jean-Marie Souriau.

The above defined canonical Poisson structure on $\mathfrak{g}^*$ can be modified by means of a symplectic cocycle $\tilde{\Theta}$ by defining the new bracket:

$$\{f,g\}_{\tilde{\Theta}}(x) = \langle x, [df(x), dg(x)] \rangle - \tilde{\Theta}(df(x), dg(x)) \tag{92}$$

with $\tilde{\Theta}$ a symplectic cocycle of the Lie algebra $\mathfrak{g}$ being a skew-symmetric bilinear map $\tilde{\Theta} : \mathfrak{g} \times \mathfrak{g} \to \Re$ which satisfies:

$$\tilde{\Theta}([X,Y],Z) + \tilde{\Theta}([Y,Z],X) + \tilde{\Theta}([Z,X],Y) = 0 \tag{93}$$

This Poisson structure is called the modified canonical Poisson structure by means of the symplectic cocycle $\widetilde{\Theta}$. The symplectic leaves of $\mathfrak{g}^*$ equipped with this Poisson structure are the orbits of an affine action whose linear part is the coadjoint action, with an additional term determined by $\widetilde{\Theta}$.

6.6. Koszul Affine Representation of Lie Group and Lie Algebra

Previously, we have developed Souriau's works on the affine representation of a Lie group used to elaborate the Lie group thermodynamics. We will study here another approach of affine representation of Lie group and Lie algebra introduced by Jean-Louis Koszul. We consolidate the link of Jean-Louis Koszul work with Souriau model. This model uses an affine representation of a Lie group and of a Lie algebra in a finite-dimensional vector space, seen as special examples of actions.

Since the work of Henri Poincare and Elie Cartan, the theory of differential forms has become an essential instrument of modern differential geometry [112–115] used by Jean-Marie Souriau for identifying the space of motions as a symplectic manifold. However, as said by Paulette Libermann [116], except Henri Poincaré who wrote shortly before his death a report on the work of Elie Cartan during his application for the Sorbonne University, the French mathematicians did not see the importance of Cartan's breakthroughs. Souriau followed lectures of Elie Cartan in 1945. The second student of Elie Cartan was Jean-Louis Koszul. Koszul introduced the concepts of affine spaces, affine transformations and affine representations [117–124]. More especially, we are interested by Koszul's definition for affine representations of Lie groups and Lie algebras. Koszul studied symmetric homogeneous spaces and defined relation between invariant flat affine connections to affine representations of Lie algebras, and characterized invariant Hessian metrics by affine representations of Lie algebras [117–124]. Koszul provided correspondence between symmetric homogeneous spaces with invariant Hessian structures by using affine representations of Lie algebras, and proved that a simply connected symmetric homogeneous space with invariant Hessian structure is a direct product of a Euclidean space and a homogeneous self-dual regular convex cone [117–124]. Let G be a connected Lie group and let G/K be a homogeneous space on which G acts effectively, Koszul gave a bijective correspondence between the set of G-invariant flat connections on G/K and the set of a certain class of affine representations of the Lie algebra of G [117–124]. The main theorem of Koszul is: let G/K be a homogeneous space of a connected Lie group G and let $\mathfrak{g}$ and k be the Lie algebras of G and K, assuming that G/K is endowed with a G-invariant flat connection, then $\mathfrak{g}$ admits an affine representation (f,q) on the vector space E. Conversely, suppose that G is simply connected and that $\mathfrak{g}$ is endowed with an affine representation, then G/K admits a G-invariant flat connection.

Koszul has proved the following [117–124]. Let Ω be a convex domain in R^n containing no complete straight lines, and an associated convex cone $V(\Omega) = \{(\lambda x, x) \in R^n \times R / x \in \Omega, \lambda \in R^+\}$. Then there exists an affine embedding:

$$\ell : x \in \Omega \mapsto \begin{bmatrix} x \\ 1 \end{bmatrix} \in V(\Omega) \tag{94}$$

If we consider η the group of homomorphism of $A(n,R)$ into $GL(n+1,R)$ given by:

$$s \in A(n,R) \mapsto \begin{bmatrix} \mathbf{f}(s) & \mathbf{q}(s) \\ 0 & 1 \end{bmatrix} \in GL(n+1,R) \tag{95}$$

and associated affine representation of Lie algebra:

$$\begin{bmatrix} f & q \\ 0 & 0 \end{bmatrix} \tag{96}$$

with $A(n,R)$ the group of all affine transformations of R^n. We have $\eta\,(G(\Omega)) \subset G\,(V(\Omega))$ and the pair (η, ℓ) of the homomorphism $\eta : G(\Omega) \to G\,(V(\Omega))$ and the map $\ell : \Omega \to V(\Omega)$ is equivariant.

A Hessian structure (D, g) on a homogeneous space G/K is said to be an invariant Hessian structure if both D and g are G-invariant. A homogeneous space G/K with an invariant Hessian structure (D, g) is called a homogeneous Hessian manifold and is denoted by $(G/K, D, g)$. Another result of Koszul is that a homogeneous self-dual regular convex cone is characterized as a simply connected symmetric homogeneous space admitting an invariant Hessian structure that is defined by the positive definite second Koszul form (we have identified in a previous paper that this second Koszul form is related to the Fisher metric). In parallel, Vinberg [125,126] gave a realization of a homogeneous regular convex domain as a real Siegel domain. Koszul has observed that regular convex cones admit canonical Hessian structures, improving some results of Pyateckii-Shapiro that studied realizations of homogeneous bounded domains by considering Siegel domains in connection with automorphic forms. Koszul defined a characteristic function ψ_Ω of a regular convex cone Ω, and showed that $\psi_\Omega = Dd\log\psi_\Omega$ is a Hessian metric on Ω invariant under affine automorphisms of Ω. If Ω is a homogeneous self dual cone, then the gradient mapping is a symmetry with respect to the canonical Hessian metric, and is a symmetric homogeneous Riemannian manifold. More information on Koszul Hessian geometry can be found in [127–136].

We will now focus our attention to Koszul affine representation of Lie group/algebra. Let G a connex Lie group and E a real or complex vector space of finite dimension, Koszul has introduced an affine representation of G in E such that [117–124]:

$$\begin{aligned} &E \to E \\ &a \mapsto sa \ \forall s \in G \end{aligned} \tag{97}$$

is an affine transformation. We set $A(E)$ the set of all affine transformations of a vector space E, a Lie group called affine transformation group of E. The set $GL(E)$ of all regular linear transformations of E, a subgroup of $A(E)$.

We define a linear representation from G to $GL(E)$:

$$\begin{aligned} &\mathbf{f} : G \to GL(E) \\ &s \mapsto \mathbf{f}(s)a = sa - so \ \forall a \in E \end{aligned} \tag{98}$$

and an application from G to E:

$$\begin{aligned} &\mathbf{q} : G \to E \\ &s \mapsto \mathbf{q}(s) = so \ \forall s \in G \end{aligned} \tag{99}$$

Then we have $\forall s, t \in G$:

$$\mathbf{f}(s)\mathbf{q}(t) + \mathbf{q}(s) = \mathbf{q}(st) \tag{100}$$

deduced from $\mathbf{f}(s)\mathbf{q}(t) + \mathbf{q}(s) = s\mathbf{q}(t) - so + so = s\mathbf{q}(t) = sto = \mathbf{q}(st)$.

On the contrary, if an application q from G to E and a linear representation $\mathbf{f}$ from G to $GL(E)$ verify previous equation, then we can define an affine representation of G in E, written $(\mathbf{f}, \mathbf{q})$:

$$Aff(s) : a \mapsto sa = \mathbf{f}(s)a + \mathbf{q}(s) \ \forall s \in G, \forall a \in E \tag{101}$$

The condition $\mathbf{f}(s)\mathbf{q}(t) + \mathbf{q}(s) = \mathbf{q}(st)$ is equivalent to requiring the following mapping to be an homomorphism:

$$Aff : s \in G \mapsto Aff(s) \in A(E) \tag{102}$$

We write f the linear representation of Lie algebra $\mathfrak{g}$ of G, defined by $\mathbf{f}$ and q the restriction to $\mathfrak{g}$ of the differential to $\mathbf{q}$ (f and q the differential of $\mathbf{f}$ and $\mathbf{q}$ respectively), Koszul has proved that:

$$\begin{gathered} f(X)q(Y) - f(Y)q(X) = q\left([X,Y]\right) \ \forall X, Y \in \mathfrak{g} \\ \text{with } f : \mathfrak{g} \to gl(E) \text{ and } q : \mathfrak{g} \mapsto E \end{gathered} \tag{103}$$

where $gl(E)$ the set of all linear endomorphisms of E, the Lie algebra of $GL(E)$.

Using the computation,

$$q\left(Ad_s Y\right) = \left.\frac{d\mathbf{q}(s \cdot e^{tY} \cdot s^{-1})}{dt}\right|_{t=0} = \mathbf{f}(s)f(Y)\mathbf{q}(s^{-1}) + \mathbf{f}(s)q(Y) \tag{104}$$

We can obtain:

$$q\left([X,Y]\right) = \left.\frac{d\mathbf{q}(Ad_{e^{tX}}Y)}{dt}\right|_{t=0} = f(X)q(Y)\mathbf{q}(e) + \mathbf{f}(e)f(Y)\left(-q(X)\right) + f(X)q(Y) \tag{105}$$

where e is the unit element in G. Since $\mathbf{f}(e)$ is the identity mapping and $\mathbf{q}(e) = 0$, we have the equality: $f(X)q(Y) - f(Y)q(X) = q\left([X,Y]\right)$.

A pair (f,q) of a linear representation f of a Lie algebra $\mathfrak{g}$ on E and a linear mapping q from $\mathfrak{g}$ to E is an affine representation of $\mathfrak{g}$ on E, if it satisfies $f(X)q(Y) - f(Y)q(X) = q\left([X,Y]\right)$.

Conversely, if we assume that $\mathfrak{g}$ admits an affine representation (f,q) on E, using an affine coordinate system $\left\{x^1, ..., x^n\right\}$ on E, we can express an affine mapping $v \mapsto f(X)v + q(Y)$ by an $(n+1) \times (n+1)$ matrix representation:

$$aff(X) = \begin{bmatrix} f(X) & q(X) \\ 0 & 0 \end{bmatrix} \tag{106}$$

where $f(X)$ is a $n \times n$ matrix and $q(X)$ is a n row vector.

$X \mapsto aff(X)$ is an injective Lie algebra homomorphism from $\mathfrak{g}$ in the Lie algebra of all $(n+1) \times (n+1)$ matrices, $gl\ (n+1, R)$:

$$\left| \begin{array}{l} \mathfrak{g} \to gl(n+1, R) \\ X \mapsto aff(X) \end{array} \right. \tag{107}$$

If we denote $\mathfrak{g}_{aff} = aff(\mathfrak{g})$, we write G_{aff} the linear Lie subgroup of $GL(n+1, R)$ generated by $\mathfrak{g}_{aff}$. An element of $s \in G_{aff}$ is expressed by:

$$Aff(s) = \begin{bmatrix} \mathbf{f}(s) & \mathbf{q}(s) \\ 0 & 1 \end{bmatrix} \tag{108}$$

Let M_{aff} be the orbit of G_{aff} through the origin o, then $M_{aff} = \mathbf{q}(G_{aff}) = G_{aff}/K_{aff}$ where $K_{aff} = \left\{s \in G_{aff}/\mathbf{q}(s) = 0\right\} = Ker\left(\mathbf{q}\right)$.

Example. Let Ω be a convex domain in R^n containing no complete straight lines, we define a convex cone $V(\Omega)$ in $R^{n+1} = R^n \times R$ by $V(\Omega) = \left\{(\lambda x, x) \in R^n \times R / x \in \Omega, \lambda \in R^+\right\}$. Then there exists an affine embedding:

$$\ell : x \in \Omega \mapsto \begin{bmatrix} x \\ 1 \end{bmatrix} \in V(\Omega) \tag{109}$$

If we consider η the group of homomorphism of $A(n,R)$ into $GL(n+1,R)$ given by:

$$s \in A(n,R) \mapsto \begin{bmatrix} \mathbf{f}(s) & \mathbf{q}(s) \\ 0 & 1 \end{bmatrix} \in GL(n+1,R) \tag{110}$$

with $A(n,R)$ the group of all affine transformations of R^n. We have $\eta(G(\Omega)) \subset G(V(\Omega))$ and the pair (η,ℓ) of the homomorphism $\eta : G(\Omega) \to G(V(\Omega))$ and the map $\ell : \Omega \to V(\Omega)$ is equivariant:

$$\ell \circ s = \eta(s) \circ \ell \text{ and } d\ell \circ s = \eta(s) \circ d\ell \tag{111}$$

6.7. Comparison of Koszul and Souriau Affine Representation of Lie Group and Lie Algebra

We will compare, in the following Table 1, affine representation of Lie group and Lie algebra from Souriau and Koszul approaches:

Table 1. Table comparing Souriau and Koszul affine representation of Lie group and Lie algebra.

Souriau Model of Affine Representation of Lie Groups and Algebra	Koszul Model of Affine Representation of Lie Groups and Algebra
$A(g)(x) = R(g)(x) + \theta(g)$ with $g \in G, x \in E$ $R : G \to GL(E)$ and $\theta : G \to E$	$Aff(s) : a \mapsto sa = \mathbf{f}(s)a + \mathbf{q}(s) \quad \forall s \in G, \forall a \in E$ $\mathbf{f} : G \to GL(E)$ $s \mapsto \mathbf{f}(s)a = sa - so \quad \forall a \in E$ $\mathbf{q} : G \to E$ $s \mapsto \mathbf{q}(s) = so \quad \forall s \in G$
$\theta(gh) = R(g)(\theta(h)) + \theta(g)$ with $g,h \in G$ $\theta : G \to E$ is a one-cocycle of G with values in E,	$\mathbf{q}(st) = \mathbf{f}(s)\mathbf{q}(t) + \mathbf{q}(s)$
$a(X)(x) = r(X)(x) + \Theta(X)$ with $X \in \mathfrak{g}, x \in E$ The linear map $\Theta : \mathfrak{g} \to E$ is a one-cocycle of G with values in E: $\Theta(X) = T_e\theta(X(e)), X \in \mathfrak{g}$	$v \mapsto f(X)v + q(Y)$ f and q the differential of $\mathbf{f}$ and $\mathbf{q}$ respectively
$\Theta([X,Y]) = r(X)(\Theta(Y)) - r(Y)(\Theta(X))$	$q([X,Y]) = f(X)q(Y) - f(Y)q(X)\ \forall X,Y \in \mathfrak{g}$ with $f : \mathfrak{g} \to gl(E)$ and $q : \mathfrak{g} \mapsto E$
none	$aff(X) = \begin{bmatrix} f(X) & q(X) \\ 0 & 0 \end{bmatrix}$
none	$Aff(s) = \begin{bmatrix} \mathbf{f}(s) & \mathbf{q}(s) \\ 0 & 1 \end{bmatrix}$

6.8. Additional Elements on Koszul Affine Representation of Lie Group and Lie Algebra

Let $\{x^1, x^2, ..., x^n\}$ be a local coordinate system on M, the Christoffel's symbols Γ_{ij}^k of the connection D are defined by:

$$D_{\frac{\partial}{\partial x^i}} \frac{\partial}{\partial x^j} = \sum_{k=1}^{n} \Gamma_{ij}^k \frac{\partial}{\partial x^k} \tag{112}$$

The torsion tensor T of D is given by:

$$T(X,Y) = D_X Y - D_Y X - [X,Y] \tag{113}$$

$$T\left(\frac{\partial}{\partial x^i}, \frac{\partial}{\partial x^j}\right) = \sum_{k=1}^{n} T_{ij}^k \frac{\partial}{\partial x^k} \text{ with } T_{ij}^k = \Gamma_{ij}^k - \Gamma_{ji}^k \tag{114}$$

The curvature tensor R of D is given by:

$$R(X,Y)Z = D_X D_Y Z - D_Y D_X Z - D_{[X,Y]} Z \tag{115}$$

$$R\left(\frac{\partial}{\partial x^k},\frac{\partial}{\partial x^l}\right)\frac{\partial}{\partial x^j}=\sum_i R^i_{jkl}\frac{\partial}{\partial x^i} \text{ with } R^i_{jkl}=\frac{\partial\Gamma^i_{lj}}{\partial x^k}-\frac{\partial\Gamma^i_{kj}}{\partial x^l}+\sum_m\left(\Gamma^m_{lj}\Gamma^i_{km}-\Gamma^m_{kj}\Gamma^i_{lm}\right) \quad (116)$$

The Ricci tensor Ric of D is given by:

$$Ric\,(Y,Z)=Tr\,\{X\to R\,(X,Y)\,Z\} \quad (117)$$

$$R_{jk}=Ric\left(\frac{\partial}{\partial x^j},\frac{\partial}{\partial x^k}\right)=\sum_i R^i_{kij} \quad (118)$$

In the following, we will consider a homogeneous space G/K endowed with a G-invariant flat connection D (homogeneous flat manifold) written $(G/K,\ D)$. Koszul has proved a bijective correspondence between the set of G-invariant flat connections on G/K and the set of affine representations of the Lie algebra of G. Let $(G,\ K)$ be the pair of connected Lie group G and its closed subgroup K. Let $\mathfrak{g}$ the Lie algebra of G and k be the Lie subalgebra of $\mathfrak{g}$ corresponding to K. X^* is defined as the vector field on $M=G/K$ induced by the 1-parameter group of transformation e^{-tX}. We denote $A_{X^*}=L_{X^*}-D_{X^*}$, with L_{X^*} the Lie derivative.

Let V be the tangent space of G/K at $o=\{K\}$ and let consider, the following values at o:

$$f(X)=A_{X^*,o} \quad (119)$$

$$q(X)=X^*_o \quad (120)$$

where $A_{X^*}Y^*=-D_{Y^*}X^*$ (where D is a locally flat linear connection: its torsion and curvature tensors vanish identically), then:

$$f\,([X,Y])=[f(X),f(Y)] \quad (121)$$

$$f(X)q(Y)-f(Y)q(X)=q\,([X,Y]) \quad (122)$$

where ker (k) $=q$, and (f,q) an affine representation of the Lie algebra $\mathfrak{g}$:

$$\forall X\in g,\ X_a=\sum_i\left(\sum_j f(X)^j_i x^i+q(X)^i\right)\frac{\partial}{\partial x^i} \quad (123)$$

The 1-parameter transformation group generated by X_a is an affine transformation group of V, with linear parts given by $e^{-t.f(X)}$ and translation vector parts:

$$\sum_{n=1}^{\infty}\frac{(-t)^n}{n!}f(X)^{n-1}q(X) \quad (124)$$

These relations are proved by using:

$$\begin{cases} A_{X^*}Y^*-A_{Y^*}X^*=[X^*,Y^*] \\ [A_{X^*},A_{Y^*}]=A_{[X^*,Y]^*} \end{cases} \text{ with } A_{X^*}Y^*=-D_{Y^*}X^* \quad (125)$$

based on the property that the connection D is locally flat and there is local coordinate systems on M such that $D_{\frac{\partial}{\partial x^i}}\frac{\partial}{\partial x^j}=0$ with a vanishing torsion and curvature:

$$T\,(X,Y)=0\Rightarrow D_XY-D_YX=[X,Y] \quad (126)$$

$$R\,(X,Y)\,Z=0\Rightarrow D_XD_YZ-D_YD_XZ=D_{[X,Y]}Z \quad (127)$$

deduced from the fact the a locally flat linear connection (vanishing of torsion and curvature).

Let ω be an invariant volume element on G/K in an affine local coordinate system $\{x^1, x^2, ..., x^n\}$ in a neighborhood of o:

$$\omega = \Phi \cdot dx^1 \wedge ... \wedge dx^n \tag{128}$$

We can write $X^* = \sum_i \chi^i \frac{\partial}{\partial x^i}$ and develop the Lie derivative of the volume element ω:

$$L_{X^*}\omega = (L_{X^*}\Phi).dx^1 \wedge ... \wedge dx^n + \sum_j \Phi.dx^1 \wedge \cdots \wedge L_{X^*}dx^j \wedge \cdots \wedge dx^n = \left(X^*\Phi + \left(\sum_j \frac{\partial \chi^j}{\partial x^j} \right) \Phi \right) dx^1 \wedge ... \wedge dx^n \tag{129}$$

Since the volume element ω is invariant by G:

$$L_{X^*}\omega = 0 \Rightarrow X^*\Phi + \left(\sum_j \frac{\partial \chi^j}{\partial x^j} \right) \Phi = 0 \Rightarrow X^*\log\Phi = -\sum_j \frac{\partial \chi^j}{\partial x^j} \tag{130}$$

By using $A_{X^*}Y^* = -D_{Y^*}X^*$, we have:

$$\left(D_{\frac{\partial}{\partial x^i}}(A_{X^*}) \right) \left(\frac{\partial}{\partial x^j} \right) = D_{\frac{\partial}{\partial x^i}} \left(A_{X^*} \left(\frac{\partial}{\partial x^j} \right) \right) - A_{X^*} \left(D_{\frac{\partial}{\partial x^i}} \frac{\partial}{\partial x^j} \right) = -D_{\frac{\partial}{\partial x^i}} D_{\frac{\partial}{\partial x^j}} \left(\sum_k \chi^k \frac{\partial}{\partial x^k} \right) = -\sum_k \frac{\partial^2 \chi^k}{\partial x^i \partial x^j} \frac{\partial}{\partial x^k} \tag{131}$$

But as D is locally flat and X^* is an infinitesimal affine transformation with respect to D:

$$D_{\frac{\partial}{\partial x^i}}(A_{X^*}) = 0 \Rightarrow \frac{\partial^2 \chi^k}{\partial x^i \partial x^j} = 0 \tag{132}$$

The Koszul form and canonical bilinear form are given by:

$$\alpha = \sum_i \frac{\partial \log\Phi}{\partial x^i} dx^i = D\log\Phi \tag{133}$$

$$D\alpha = \sum_{i,j} \frac{\partial^2 \log\Phi}{\partial x^i \partial x^j} dx^i dx^j = Dd\log\Phi \tag{134}$$

$$L_{X^*}\alpha = L_{X^*}D\log\Phi = DL_{X^*}\log\Phi = DX^*\log\Phi = -D\left(\sum_j \frac{\partial \chi^j}{\partial x^j} \right) = -\sum_{,j} \frac{\partial^2 \chi^j}{\partial x^i \partial x^j} dx^i = 0 \tag{135}$$

Then, $L_{X^*}\alpha = 0 \; \forall X \in g$.
By using $X^*\log\Phi = -\sum_j \frac{\partial \chi^j}{\partial x^j}$, we can obtain:

$$\alpha(X^*) = (D\log\Phi)(X^*) \underset{L_{X^*}\alpha = 0}{\Rightarrow} D_{X^*}\log\Phi = -\sum_j \frac{\partial \chi^j}{\partial x^j} \tag{136}$$

By using $A_{X^*}Y^* = -D_{Y^*}X^*$, we can develop:

$$A_{X^*}\left(\frac{\partial}{\partial x^j} \right) = -D_{\frac{\partial}{\partial x^j}} X^* = -\sum_i \frac{\partial \chi^i}{\partial x^j} \frac{\partial}{\partial x^i} \tag{137}$$

As $f(X) = A_{X^*,o}$ and $q(X) = X_o^*$:

$$Tr(f(X)) = Tr(A_{X^*,o}) = -\sum_i \frac{\partial \chi^i}{\partial x^i}(o) = \alpha(X_0^*) = \alpha_0(q(X)) \tag{138}$$

If we use that $L_{X^*}\alpha = 0 \; \forall X \in \mathfrak{g}$, then we obtain:

$$(D\alpha)(X^*, Y^*) = (D_{Y^*}\alpha)(X^*) = -(A_{Y^*}\alpha)(X^*) = -A_{Y^*}(\alpha(X^*)) + \alpha(A_{Y^*}X^*) = \alpha(A_{Y^*}X^*) \tag{139}$$

$$D\alpha_0\left(q(X),q(Y)\right)=\alpha_0\left(f(Y)q(X)\right) \tag{140}$$

To synthetize the result proved by Jean-Louis Koszul, if α_o and $D\alpha_o$ are the values of α and $D\alpha$ at o, then:

$$\alpha_o\left(q(X)\right)=Tr\left(f(X)\right)\ \forall X\in g \tag{141}$$

$$D\alpha_o\left(q(X),q(Y)\right)=\left\langle q(X),q(Y)\right\rangle_o=\alpha_0\left(f(X)q(Y)\right)\ \forall X,Y\in \mathfrak{g} \tag{142}$$

Jean-Louis Koszul has also proved that the inner product $\langle .,. \rangle$ on V, given by the Riemannian metric g_{ij}, satisfies the following conditions:

$$\langle f(X)q(Y),q(Z)\rangle+\langle q(Y),f(X)q(Z)\rangle=\langle f(Y)q(X),q(Z)\rangle+\langle q(X),f(Y)q(Z)\rangle \tag{143}$$

To make the link with Souriau model of thermodynamics, the first Koszul form $\alpha=D\log\Phi=Tr\left(f(X)\right)$ will play the role of the geometric heat Q and the second koszul form $D\alpha=Dd\log\Phi=\langle q(X),q(Y)\rangle_o$ will be the equivalent of Souriau-Fisher metric that is G-invariant.

Koszul theory is wider and integrates "information geometry" in its corpus. Koszul [117–124] has proved general results, for example: on a complex homogeneous space, an invariant volume defines with the complex structure, an invariant Hermitian form. If this space is a bounded domain, then this hermitian form is positive definite and coincides with the classical Bergman metric of this domain. During his stay at Institute for Advanced Study in Princeton, Koszul [117–124] has also demonstrated the reciprocal for a class of complex homogeneous spaces, defined by open orbits of complex affine transformation groups. Koszul and Vey [137,138] have also developed extended results with the following theorem for connected hessian manifolds:

Theorem 3 (Koszul-Vey Theorem). *Let M be a connected hessian manifold with hessian metric g. Suppose that M admits a closed 1-form α such that $D\alpha=g$ and there exists a group G of affine automorphisms of M preserving α:*

- *If M/G* is quasi-compact, then the universal covering manifold of M is affinely isomorphic to a convex domain Ω of an affine space not containing any full straight line.
- *If M/G is compact, then Ω is a sharp convex cone.*

On this basis, Koszul has given a Lie group construction of a homogeneous cone that has been developed and applied in information geometry by Shima and Boyom in the framework of Hessian geometry. The results of Koszul are also fundamental in the framework of Souriau thermodynamics.

7. Souriau Lie Group Model and Koszul Hessian Geometry Applied in the Context of Information Geometry for Multivariate Gaussian Densities

We will enlighten Souriau model with Koszul hessian geometry applied in information geometry [117–124], recently studied in [3,9,139]. We have previously shown that information geometry could be founded on the notion of Koszul-Vinberg characteristic function $\psi_\Omega(x)=\int\limits_{\Omega^*}e^{-\langle x,\xi\rangle}d\xi,\ \forall x\in\Omega$ where Ω is a convex cone and Ω^* the dual cone with respect to Cartan-Killing inner product $\langle x,y\rangle=-B\left(x,\theta(y)\right)$ invariant by automorphisms of Ω, with $B\left(.,.\right)$ the Killing form and $\theta(.)$ the Cartan involution. We can develop the Koszul characteristic function:

$$\psi_\Omega(x+\lambda u)=\psi_\Omega(x)-\lambda\left\langle x^*,u\right\rangle+\frac{\lambda^2}{2}\left\langle K(x)u,u\right\rangle+... \tag{144}$$

$$\text{with } x^*=\frac{d\Phi(x)}{dx},\ \Phi(x)=-\log\psi_\Omega(x) \text{ and } K(x)=\frac{d^2\Phi(x)}{dx^2} \tag{145}$$

This characteristic function is at the cornerstone of modern concept of information geometry, defining Koszul density by solution of maximum Koszul-Shannon entropy [140]:

$$\underset{p}{Max}\left[-\int_{\Omega^*} p_{\hat{\xi}}(\xi)\log p_{\hat{\xi}}(\xi)\cdot d\xi\right] \text{ such that} \int_{\Omega^*} p_{\hat{\xi}}(\xi)d\xi = 1 \text{ and } \int_{\Omega^*} \xi\cdot p_{\hat{\xi}}(\xi)d\xi = \hat{\xi} \tag{146}$$

$$\begin{aligned} &p_{\hat{\xi}}(\xi) = \frac{e^{-\langle\Theta^{-1}(\hat{\xi}),\xi\rangle}}{\int_{\Omega^*} e^{-\langle\Theta^{-1}(\hat{\xi}),\xi\rangle}.d\xi}\ \hat{\xi} = \Theta(\beta) = \frac{\partial\Phi(\beta)}{\partial\beta} \text{ where } \Phi(\beta) = -\log\psi_\Omega(\beta) \\ &\psi_\Omega(\beta) = \int_{\Omega^*} e^{-\langle\beta,\xi\rangle}d\xi\ ,\ S(\hat{\xi}) = -\int_{\Omega^*} p_{\hat{\xi}}(\xi)\log p_{\hat{\xi}}(\xi)\cdot d\xi \text{ and } \beta = \Theta^{-1}(\hat{\xi}) \\ &S(\hat{\xi}) = \langle\hat{\xi},\beta\rangle - \Phi(\beta) \end{aligned} \tag{147}$$

This last relation is a Legendre transform between the logarithm of characteristic function and the entropy:

$$\begin{aligned} &\log p_{\hat{\xi}}(\xi) = -\langle\xi,\beta\rangle + \Phi(\beta) \\ &S(\bar{\xi}) = -\int_{\Omega^*} p_{\hat{\xi}}(\xi)\cdot\log p_{\hat{\xi}}(\xi)\cdot d\xi = -E\left[\log p_{\hat{\xi}}(\xi)\right] \\ &S(\bar{\xi}) = \langle E[\xi],\beta\rangle - \Phi(\beta) = \langle\hat{\xi},\beta\rangle - \Phi(\beta) \end{aligned} \tag{148}$$

The inversion $\Theta^{-1}(\hat{\xi})$ is given by the Legendre transform based on the property that the Koszul-Shannon entropy is given by the Legendre transform of minus the logarithm of the characteristic function:

$$S(\hat{\xi}) = \langle\beta,\hat{\xi}\rangle - \Phi(\beta) \text{ with } \Phi(\beta) = -\log\int_{\Omega^*} e^{-\langle\xi,\beta\rangle}d\xi \quad \forall\beta\in\Omega \text{ and } \forall\xi,\hat{\xi}\in\Omega^* \tag{149}$$

We can observe the fundamental property that $E[S(\xi)] = S(E[\xi])$, $\xi\in\Omega^*$, and also as observed by Maurice Fréchet that "distinguished functions" (densities with estimator reaching the Fréchet-Darmois bound) are solutions of the *Alexis Clairaut equation* introduced by Clairaut in 1734 [141], as illustrated in Figure 8:

$$S(\hat{\xi}) = \left\langle\Theta^{-1}(\hat{\xi}),\hat{\xi}\right\rangle - \Phi\left[\Theta^{-1}(\hat{\xi})\right] \forall\hat{\xi}\in\{\Theta(\beta)/\beta\in\Omega\} \tag{150}$$

(55) $$\mu = \theta\,\mu' - \psi(\mu')$$

c'est-à-dire une équation de Clairaut. La solution $\mu' =$ constante réduirait $f(x, \theta)$, d'après (48) à une fonction indépendante de θ, cas où le problème n'aurait plus de sens. μ est donc donné par la solution singulière de (55), qui est unique et s'obtient en éliminant s entre $\mu = \theta\,s - \psi(s)$ et $\theta = \psi'(s)$ ou encore entre

Figure 8. Clairaut-Legendre equation introduced by Maurice Fréchet in his 1943 paper [141].

Details of Fréchet elaboration for this Clairaut(-Legendre) equation for "distinguished function" is given in Appendix A, and other elements are available on Fréchet's papers [141–144].

In this structure, the Fisher metric $I(x)$ makes appear naturally a *Koszul hessian geometry* [145,146], if we observe that

$$\begin{aligned}
&\log p_{\hat{\xi}}(\xi) = -\langle \xi, \beta \rangle + \Phi(\beta) \\
&S(\bar{\xi}) = -\int_{\Omega^*} p_{\hat{\xi}}(\xi) \cdot \log p_{\hat{\xi}}(\xi) \cdot d\xi = -E\left[\log p_{\hat{\xi}}(\xi)\right] \\
&S(\bar{\xi}) = \langle E[\xi], \beta \rangle - \Phi(\beta) = \langle \hat{\xi}, \beta \rangle - \Phi(\beta)
\end{aligned} \tag{151}$$

Then we can recover the relation with Fisher metric:

$$\begin{aligned}
&I(\beta) = -E\left[\frac{\partial^2 \log p_\beta(\xi)}{\partial \beta^2}\right] = -E\left[\frac{\partial^2 (-\langle \xi, \beta \rangle + \Phi(\beta))}{\partial \beta^2}\right] = -\frac{\partial^2 \Phi(\beta)}{\partial \beta^2} \\
&\hat{\xi} = \frac{\partial \Phi(\beta)}{\partial \beta} \\
&I(\beta) = E\left[\frac{\partial \log p_\beta(\xi)}{\partial \beta} \frac{\partial \log p_\beta(\xi)}{\partial \beta}^T\right] = E\left[(\xi - \hat{\xi})(\xi - \hat{\xi})^T\right] = E[\xi^2] - E[\xi]^2 = Var(\xi)
\end{aligned} \tag{152}$$

with Crouzeix relation established in 1977 [147,148], $\frac{\partial^2 \Phi}{\partial \beta^2} = \left[\frac{\partial^2 S}{\partial \hat{\xi}^2}\right]^{-1}$ giving the dual metric, in dual space, where entropy S and (minus) logarithm of characteristic function, Φ, are dual potential functions.

The first metric of information geometry [149,150], the Fisher metric is given by the hessian of the characteristic function logarithm:

$$I(\beta) = -E\left[\frac{\partial^2 \log p_\beta(\xi)}{\partial \beta^2}\right] = -\frac{\partial^2 \Phi(\beta)}{\partial \beta^2} = \frac{\partial^2 \log \psi_\Omega(\beta)}{\partial \beta^2} \tag{153}$$

$$ds_g^2 = d\beta^T I(\beta) d\beta = \sum_{ij} g_{ij} d\beta_i d\beta_j \text{ with } g_{ij} = [I(\beta)]_{ij} \tag{154}$$

The second metric of information geometry is given by hessian of the Shannon entropy:

$$\frac{\partial^2 S(\hat{\xi})}{\partial \hat{\xi}^2} = \left[\frac{\partial^2 \Phi(\beta)}{\partial \beta^2}\right]^{-1} \text{ with } S(\hat{\xi}) = \langle \hat{\xi}, \beta \rangle - \Phi(\beta) \tag{155}$$

$$ds_h^2 = d\hat{\xi}^T \left[\frac{\partial^2 S(\hat{\xi})}{\partial \hat{\xi}^2}\right] d\hat{\xi} = \sum_{ij} h_{ij} d\hat{\xi}_i d\hat{\xi}_j \text{ with } h_{ij} = \left[\frac{\partial^2 S(\hat{\xi})}{\partial \hat{\xi}^2}\right]_{ij} \tag{156}$$

Both metrics will provide the same distance:

$$ds_g^2 = ds_h^2 \tag{157}$$

From the Cartan inner product, we can generate logarithm of the Koszul characteristic function, and its Legendre transform to define Koszul entropy, Koszul density and Koszul metric, as explained in the following Figure 9:

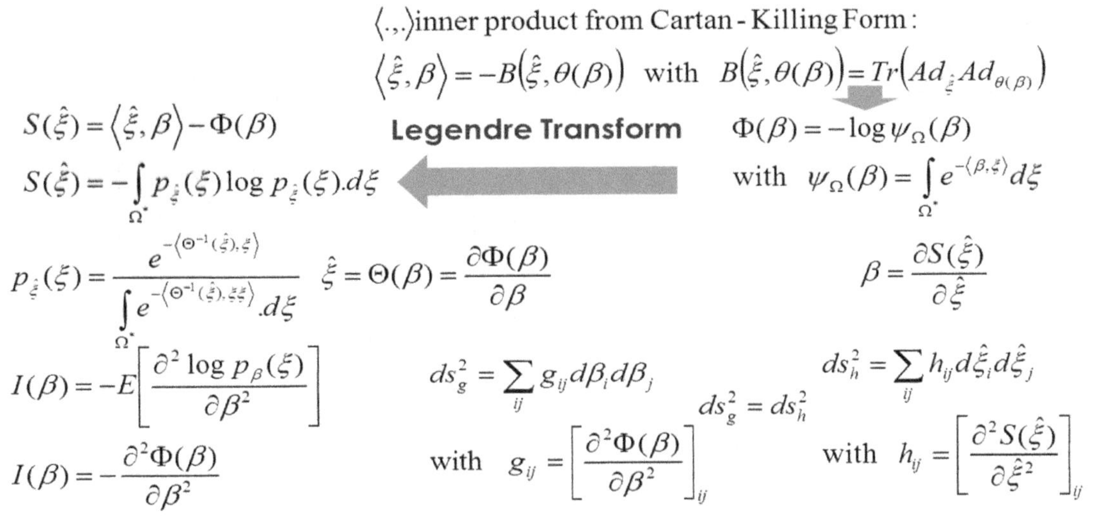

$\langle .,. \rangle$ inner product from Cartan - Killing Form:

$$\left\langle \hat{\xi},\beta \right\rangle = -B\left(\hat{\xi},\theta(\beta)\right) \text{ with } B\left(\hat{\xi},\theta(\beta)\right) = Tr\left(Ad_{\hat{\xi}} Ad_{\theta(\beta)}\right)$$

$$S(\hat{\xi}) = \left\langle \hat{\xi},\beta \right\rangle - \Phi(\beta) \qquad \textbf{Legendre Transform} \qquad \Phi(\beta) = -\log \psi_\Omega(\beta)$$

$$S(\hat{\xi}) = -\int_{\Omega^*} p_{\hat{\xi}}(\xi) \log p_{\hat{\xi}}(\xi).d\xi \qquad \text{with } \psi_\Omega(\beta) = \int_{\Omega^*} e^{-\langle \beta,\xi \rangle} d\xi$$

$$p_{\hat{\xi}}(\xi) = \frac{e^{-\left\langle \Theta^{-1}(\hat{\xi}),\xi \right\rangle}}{\int_{\Omega^*} e^{-\left\langle \Theta^{-1}(\hat{\xi}),\xi \right\rangle}.d\xi} \qquad \hat{\xi} = \Theta(\beta) = \frac{\partial \Phi(\beta)}{\partial \beta} \qquad \beta = \frac{\partial S(\hat{\xi})}{\partial \hat{\xi}}$$

$$I(\beta) = -E\left[\frac{\partial^2 \log p_\beta(\xi)}{\partial \beta^2}\right] \qquad ds_g^2 = \sum_{ij} g_{ij} d\beta_i d\beta_j \qquad ds_g^2 = ds_h^2 \qquad ds_h^2 = \sum_{ij} h_{ij} d\hat{\xi}_i d\hat{\xi}_j$$

$$I(\beta) = -\frac{\partial^2 \Phi(\beta)}{\partial \beta^2} \qquad \text{with } g_{ij} = \left[\frac{\partial^2 \Phi(\beta)}{\partial \beta^2}\right]_{ij} \qquad \text{with } h_{ij} = \left[\frac{\partial^2 S(\hat{\xi})}{\partial \hat{\xi}^2}\right]_{ij}$$

Figure 9. Generation of Koszul elements from Cartan inner product.

This information geometry has been intensively studied for structured matrices [151–166] and in statistics [167] and is linked to the seminal work of Siegel [168] on symmetric bounded domains.

We can apply this Koszul geometry framework for cones of symmetric positive definite matrices. Let the inner product $\langle \eta, \xi \rangle = Tr\left(\eta^T \xi\right), \forall \eta, \xi \in Sym(n)$ given by Cartan-Killing form, Ω be the set of symmetric positive definite matrices is an open convex cone and is self-dual $\Omega^* = \Omega$.

$$\begin{aligned} &\langle \eta, \xi \rangle = Tr\left(\eta^T \xi\right), \forall \eta, \xi \in Sym(n) \\ &\psi_\Omega(\beta) = \int_{\Omega^*} e^{-\langle \beta, \xi \rangle} d\xi = \det(\beta)^{-\frac{n+1}{2}} \psi_\Omega(I_d) \\ &\hat{\xi} = \frac{\partial \Phi(\beta)}{\partial \beta} = \frac{\partial(-\log \psi_\Omega(\beta))}{\partial \beta} = \frac{n+1}{2}\beta^{-1} \end{aligned} \tag{158}$$

$$\begin{aligned} &p_{\hat{\xi}}(\xi) = e^{-\langle \Theta^{-1}(\hat{\xi}),\xi \rangle + \Phi(\Theta^{-1}(\hat{\xi}))} = \psi_\Omega(I_d) \cdot \left[\det\left(\alpha \hat{\xi}^{-1}\right)\right] \cdot e^{-Tr(\alpha \hat{\xi}^{-1} \xi)} \\ &\text{with } \alpha = \frac{n+1}{2} \end{aligned} \tag{159}$$

We will in the following illustrate information geometry for multivariate Gaussian density [169]:

$$p_{\hat{\xi}}(\xi) = \frac{1}{(2\pi)^{n/2} \det(R)^{1/2}} e^{-\frac{1}{2}(z-m)^T R^{-1}(z-m)} \tag{160}$$

If we develop:

$$\begin{aligned} \frac{1}{2}(z-m)^T R^{-1}(z-m) &= \frac{1}{2}\left[z^T R^{-1} z - m^T R^{-1} z - z^T R^{-1} m + m^T R^{-1} m\right] \\ &= \frac{1}{2} z^T R^{-1} z - m^T R^{-1} z + \tfrac{1}{2} m^T R^{-1} m \end{aligned} \tag{161}$$

We can write the density as a Gibbs density:

$$\begin{aligned} &p_{\hat{\xi}}(\xi) = \frac{1}{(2\pi)^{n/2} \det(R)^{1/2} e^{\frac{1}{2} m^T R^{-1} m}} e^{-\left[-m^T R^{-1} z + \frac{1}{2} z^T R^{-1} z\right]} = \frac{1}{Z} e^{-\langle \xi, \beta \rangle} \\ &\xi = \begin{bmatrix} z \\ zz^T \end{bmatrix} \text{ and } \beta = \begin{bmatrix} -R^{-1} m \\ \frac{1}{2} R^{-1} \end{bmatrix} = \begin{bmatrix} a \\ H \end{bmatrix} \\ &\text{with } \langle \xi, \beta \rangle = a^T z + z^T H z = Tr\left[z a^T + H^T z z^T\right] \end{aligned} \tag{162}$$

We can then rewrite density with canonical variables:

$$p_{\hat{\xi}}(\xi)=\frac{1}{\int\limits_{\Omega^*}e^{-\langle\xi,\beta\rangle}.d\xi}e^{-\langle\xi,\beta\rangle}=\frac{1}{Z}e^{-\langle\xi,\beta\rangle}\text{ with }\log(Z)=n\log(2\pi)+\frac{1}{2}\log\det(R)+\frac{1}{2}m^TR^{-1}m$$

$$\xi=\begin{bmatrix}z\\zz^T\end{bmatrix},\ \hat{\xi}=\begin{bmatrix}E[z]\\E\left[zz^T\right]\end{bmatrix}=\begin{bmatrix}m\\R+mm^T\end{bmatrix},\ \beta=\begin{bmatrix}a\\H\end{bmatrix}=\begin{bmatrix}-R^{-1}m\\\frac{1}{2}R^{-1}\end{bmatrix} \tag{163}$$

$$\text{with }\langle\xi,\beta\rangle=Tr\left[za^T+H^Tzz^T\right]$$

$$R=E\left[(z-m)(z-m)^T\right]=E\left[zz^T-mz^T-zm^T+mm^T\right]=E\left[zz^T\right]-mm^T$$

The first potential function (free energy/logarithm of characteristic function) is given by:

$$\psi_\Omega(\beta)=\int\limits_{\Omega^*}e^{-\langle\xi,\beta\rangle}\cdot d\xi$$

$$\text{and }\Phi(\beta)=-\log\psi_\Omega(\beta)=\frac{1}{2}\left[-Tr\left[H^{-1}aa^T\right]+\log\left[(2)^n\det H\right]-n\log(2\pi)\right] \tag{164}$$

We verify the relation between the first potential function and moment:

$$\frac{\partial\Phi(\beta)}{\partial\beta}=\frac{\partial\left[-\log\psi_\Omega(\beta)\right]}{\partial\beta}=\int\limits_{\Omega^*}\xi\frac{e^{-\langle\xi,\beta\rangle}}{\int\limits_{\Omega^*}e^{-\langle\xi,\beta\rangle}\cdot d\xi}\cdot d\xi=\int\limits_{\Omega^*}\xi\cdot p_{\hat{\xi}}(\xi)\cdot d\xi=\hat{\xi}$$

$$\frac{\partial\Phi(\beta)}{\partial\beta}=\begin{bmatrix}\dfrac{\partial\Phi(\beta)}{\partial a}\\\dfrac{\partial\Phi(\beta)}{\partial H}\end{bmatrix}=\begin{bmatrix}m\\R+mm^T\end{bmatrix}=\hat{\xi} \tag{165}$$

The second potential function (Shannon entropy) is given as a Legendre transform of the first one:

$$S(\hat{\xi})=\langle\hat{\xi},\beta\rangle-\Phi(\beta)\text{ with }\frac{\partial\Phi(\beta)}{\partial\beta}=\hat{\xi}\text{ and }\frac{\partial S(\hat{\xi})}{\partial\hat{\xi}}=\beta$$

$$S(\hat{\xi})=-\int\limits_{\Omega^*}\frac{e^{-\langle\xi,\beta\rangle}}{\int\limits_{\Omega^*}e^{-\langle\xi,\beta\rangle}\cdot d\xi}\log\frac{e^{-\langle\xi,\beta\rangle}}{\int\limits_{\Omega^*}e^{-\langle\xi,\beta\rangle}\cdot d\xi}\cdot d\xi=-\int\limits_{\Omega^*}p_{\hat{\xi}}(\xi)\log p_{\hat{\xi}}(\xi)\cdot d\xi \tag{166}$$

$$S(\hat{\xi})=-\int\limits_{\Omega^*}p_{\hat{\xi}}(\xi)\log p_{\hat{\xi}}(\xi)\cdot d\xi=\frac{1}{2}\left[\log(2)^n\det\left[H^{-1}\right]+n\log(2\pi\cdot e)\right]=\frac{1}{2}\left[\log\det[R]+n\log(2\pi\cdot e)\right] \tag{167}$$

This remark was made by Jean-Souriau in his book [10] as soon as 1969. He has observed, as illustrated in Figure 10 that if we take vector with tensor components $\xi=\begin{pmatrix}z\\z\otimes z\end{pmatrix}$, components of $\hat{\xi}$ will provide moments of the first and second order of the density of probability $p_{\hat{\xi}}(\xi)$. He used this change of variable $z'=H^{1/2}z+H^{-1/2}a$, to compute the logarithm of the characteristic function $\Phi(\beta)$:

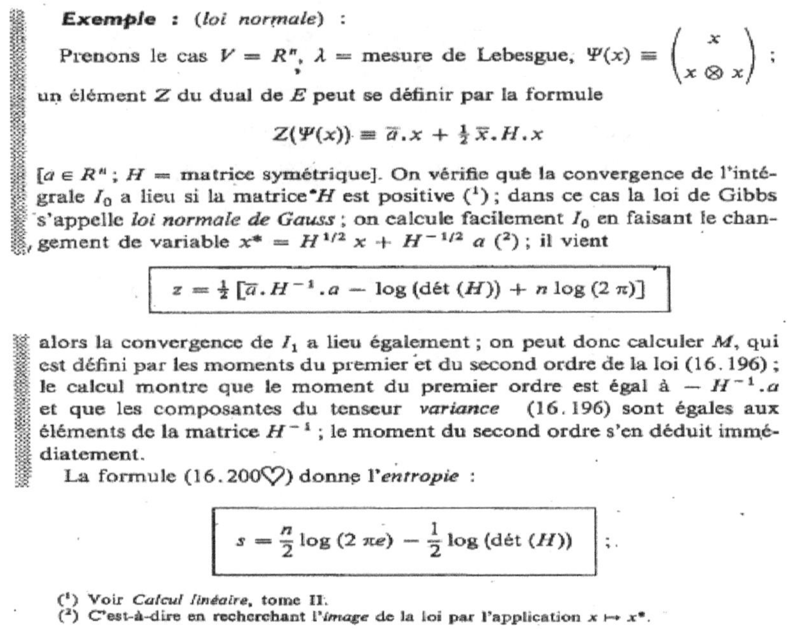

Exemple : (*loi normale*) :

Prenons le cas $V = R^n$, λ = mesure de Lebesgue, $\Psi(x) = \begin{pmatrix} x \\ x \otimes x \end{pmatrix}$;
un élément Z du dual de E peut se définir par la formule

$$Z(\Psi(x)) \equiv \bar{a}.x + \tfrac{1}{2}\bar{x}.H.x$$

[$a \in R^n$; H = matrice symétrique]. On vérifie que la convergence de l'intégrale I_0 a lieu si la matrice H est positive ([1]) ; dans ce cas la loi de Gibbs s'appelle *loi normale de Gauss* ; on calcule facilement I_0 en faisant le changement de variable $x^* = H^{1/2}x + H^{-1/2}a$ ([2]) ; il vient

$$z = \tfrac{1}{2}\left[\bar{a}.H^{-1}.a - \log(\text{dét}(H)) + n\log(2\pi)\right]$$

alors la convergence de I_1 a lieu également ; on peut donc calculer M, qui est défini par les moments du premier et du second ordre de la loi (16.196) ; le calcul montre que le moment du premier ordre est égal à $-H^{-1}.a$ et que les composantes du tenseur *variance* (16.196) sont égales aux éléments de la matrice H^{-1} ; le moment du second ordre s'en déduit immédiatement.

La formule (16.200♡) donne l'*entropie* :

$$s = \frac{n}{2}\log(2\pi e) - \frac{1}{2}\log(\text{dét}(H))$$;.

([1]) Voir *Calcul linéaire*, tome II.
([2]) C'est-à-dire en recherchant l'*image* de la loi par l'application $x \mapsto x^*$.

Figure 10. Introduction of potential function for multivariate Gaussian law in Souriau book [10].

We can finally compute the metric from the matrix g_{ij}:

$$ds^2 = \sum_{ij} g_{ij}d\theta_i d\theta_j = dm^T R^{-1} dm + \frac{1}{2}Tr\left[\left(R^{-1}dR\right)^2\right] \tag{168}$$

and from classical expression of the Euler-Lagrange equation:

$$\sum_{i=1}^{n} g_{ik}\ddot{\theta}_i + \sum_{i,j=1}^{n} \Gamma_{ijk}\dot{\theta}_i\dot{\theta}_j = 0\ ,\ k = 1, \ldots, n \text{ with } \Gamma_{ijk} = \frac{1}{2}\left[\frac{\partial g_{jk}}{\partial\theta_i} + \frac{\partial g_{jk}}{\partial\theta_j} + \frac{\partial g_{ij}}{\partial\theta_k}\right] \tag{169}$$

That is explicitely given by [170]:

$$\begin{cases} \ddot{R} + \dot{m}\dot{m}^T - \dot{R}R^{-1}\dot{R} = 0 \\ \ddot{m} - \dot{R}R^{-1}\dot{m} = 0 \end{cases} \tag{170}$$

We cannot integrate this Euler-Lagrange equation. We will see that Lie group theory will provide new reduced equation, Euler-Poincaré equation, using Souriau theorem.

We make reference to the book of Deza that gives a survey about distance and metric space [171].

The case of Natural Exponential families that are invariant by an affine group has been studied by Casalis (in 1999 paper and in her Ph.D. thesis) [172–178] and by Letac [179–181]. We give the details of Casalis' development in Appendix C. Barndorff-Nielsen has also studied transformation models for exponential families [182–186]. In this section, we will only consider the case of multivariate Gaussian densities.

8. Affine Group Action for Multivariate Gaussian Densities and Souriau's Moment Map: Computation of Geodesics by Geodesic Shooting

To more deeply understand Koszul and Souriau Lie group models of information geometry, we will illustrate their tools for multivariate Gaussian densities.

Consider the general linear group $GL(n)$ consisting of the invertible $n \times n$ matrices, that is a topological group acting linearly on R^n by:

$$\begin{aligned} GL(n) \times R^n &\to R^n \\ (A,x) &\mapsto Ax \end{aligned} \tag{171}$$

The group $GL(n)$ is a Lie group, is a subgroup of the general affine group $GA(n)$, composed of all pairs (A,v) where $A \in GL(n)$ and $v \in R^n$, the group operation given by:

$$(A_1,v_1)(A_2,v_2) = (A_1A_2, A_1v_2 + v_1) \tag{172}$$

$GL(n)$ is an open subset of R^{n^2}, and may be considered as n^2-dimensional differential manifold with the same differentiable structure than R^{n^2}. Multiplication and inversion are infinitely often differentiable mappings. Consider the vector space $gl(n)$ of real $n \times n$ matrices and the commutator product:

$$\begin{aligned} gl(n) \times gl(n) &\to gl(n) \\ (A,B) &\mapsto AB - BA = [A,B] \end{aligned} \tag{173}$$

This is a Lie product making $gl(n)$ into a Lie algebra. The exponential map is then the mapping defined by:

$$\begin{aligned} \exp{:}gl(n) &\to GL(n) \\ A &\mapsto \exp(A) = \sum_{n=0}^{\infty} \tfrac{A^n}{n!} \end{aligned} \tag{174}$$

Restricting A to have positive determinant, one obtains the positive general affine group $GA_+(n)$ that acts transitively on R^n by:

$$((A,v),x) \mapsto Ax + v \tag{175}$$

In case of symmetric positive definite matrices $Sym^+(n)$, we can use the Cholesky decomposition:

$$R = LL^T \tag{176}$$

where L is a lower triangular matrix with real and positive diagonal entries, and L^T denotes the transpose of L, to define the square root of R.

Given a positive semidefinite matrix R, according to the spectral theorem, the continuous functional calculus can be applied to obtain a matrix $R^{1/2}$ such that $R^{1/2}$ is itself positive and $R^{1/2}R^{1/2} = R$. The operator $R^{1/2}$ is the unique non-negative square root of R.

$N_n = \{\aleph(\mu,\Sigma)/\mu \in R^n, \Sigma \in Sym^+{}_n\}$ the class of regular multivariate normal distributions, where μ is the mean vector and Σ is the (symmetric positive definite) covariance matrix, is invariant under the transitive action of $GA(n)$. The induced action of $GA(n)$ on $R^n \times Sym^+{}_n$ is then given by:

$$\begin{aligned} GA(n) \times (R^n \times Sym^+n) &\to R^n \times Sym^+n \\ ((A,v),(\mu,\Sigma)) &\mapsto (A\mu + v, A\Sigma A^T) \end{aligned} \tag{177}$$

and

$$\begin{aligned} GA(n) \times R^n &\to R^n \\ ((A,v),x) &\mapsto Ax + v \end{aligned} \tag{178}$$

As the isotropy group of $(0, I_n)$ is equal to $O(n)$, we can observe that:

$$N_n = GA(n)/O(n) \tag{179}$$

N_n is an open subset of the vector space $T_n = \{(\eta,\Omega)/\eta \in R^n, \Omega \in Sym_n\}$ and is a differentiable manifold, where the tangent space at any point may be identified with T_n.

The Fisher information defines a metric given to N_n a Riemannian manifold structure. The inner product of two tangent vectors $(\eta_1, \Omega_1) \in T_n$, $(\eta_2, \Omega_2) \in T_n$ at the point $(\mu, \Sigma) \in N_n$ is given by:

$$g_{(\mu,\Sigma))}\left((\eta_1, \Omega_1), (\eta_1, \Omega_1)\right) = \eta_1^T \Sigma^{-1} \eta_2 + \frac{1}{2} Tr\left(\Sigma^{-1}\Omega_1\Sigma^{-1}\Omega_2\right) \quad (180)$$

Niels Christian Bang Jesperson has proved that the transformation model on R^n with parameter set $R^n \times Sym^+{}_n$ are exactly those of the form $p_{\mu,\Sigma} = f_{\mu,\Sigma}\lambda$ where λ is the Lebesque measure, where $f_{\mu,\Sigma}(x) = h\left((x-\mu)^T \Sigma^{-1}(x-\mu)\right)/\det(\Sigma)^{1/2}$ and $h : [0, +\infty[\rightarrow R^+$ is a continuous function with $\int\limits_0^{+\infty} h(s)s^{\frac{n}{2}-1}ds < +\infty$. Distributions with densities of this form are called elliptic distributions.

To improve understanding of tools, we will consider $GA(n)$ as a sub-group of affine group, that could be defined by a matrix Lie group G_{aff}, that acts for multivariate Gaussian laws, as illustrated in Figure 11:

$$\begin{bmatrix} Y \\ 1 \end{bmatrix} = \begin{bmatrix} R^{1/2} & m \\ 0 & 1 \end{bmatrix} \begin{bmatrix} X \\ 1 \end{bmatrix} = \begin{bmatrix} R^{1/2}X + m \\ 1 \end{bmatrix}, \begin{cases} (m, R) \in R^n \times Sym^+(n) \\ M = \begin{bmatrix} R^{1/2} & m \\ 0 & 1 \end{bmatrix} \in G_{aff} \end{cases} \quad (181)$$

$$X \approx \aleph(0, I) \rightarrow Y \approx \aleph(m, R)$$

We can verify that M is a Lie group with classical properties, that product of M preserves the structure, the associativity, the non-commutativity, and the existence of neutral element:

$$\left.\begin{array}{l} M_1 \cdot M_2 = \begin{bmatrix} R_1^{1/2} & m_1 \\ 0 & 1 \end{bmatrix} \begin{bmatrix} R_2^{1/2} & m_2 \\ 0 & 1 \end{bmatrix} = \begin{bmatrix} R_1^{1/2}R_2^{1/2} & R_1^{1/2}m_2 + m_1 \\ 0 & 1 \end{bmatrix} \\ M_2 \cdot M_1 = \begin{bmatrix} R_2^{1/2} & m_2 \\ 0 & 1 \end{bmatrix} \begin{bmatrix} R_1^{1/2} & m_1 \\ 0 & 1 \end{bmatrix} = \begin{bmatrix} R_2^{1/2}R_1^{1/2} & R_2^{1/2}m_1 + m_2 \\ 0 & 1 \end{bmatrix} \end{array}\right\}$$
$$\Rightarrow \begin{cases} M_1 \cdot M_2 \in G_{aff} \\ M_2 \cdot M_1 \in G_{aff} \\ M_1 \cdot M_2 \neq M_2 \cdot M_1 \\ M_1 \cdot (M_2 \cdot M_3) = (M_1 \cdot M_2) \cdot M_3 \\ M_1 \cdot I = M_1 \end{cases} \quad (182)$$

We can also observe that the inverse preserves the structure:

$$M = \begin{bmatrix} R^{1/2} & m \\ 0 & 1 \end{bmatrix} \Rightarrow M_R^{-1} = M_L^{-1} = M^{-1} = \begin{bmatrix} R^{-1/2} & -R^{-1/2}m \\ 0 & 1 \end{bmatrix} \in G_{aff} \quad (183)$$

To this Lie group we can associate a Lie algebra whose underlying vector space is the tangent space of the Lie group at the identity element and which completely captures the local structure of the group. This Lie group acts smoothly on the manifold, and acts on the vector fields. Any tangent vector at the identity of a Lie group can be extended to a left (respectively right) invariant vector field by left (respectively right) translating the tangent vector to other points of the manifold. This identifies the tangent space at the identity $\mathfrak{g} = T_I(G)$ with the space of left invariant vector fields, and therefore makes the tangent space at the identity into a Lie algebra, called the Lie algebra of G.

$$L_G : \begin{cases} G_{aff} \rightarrow G_{aff} \\ M \mapsto L_M N = M \cdot N \end{cases} \text{ and } R_G : \begin{cases} G_{aff} \rightarrow G_{aff} \\ M \mapsto R_M N = N \cdot M \end{cases} \quad (184)$$

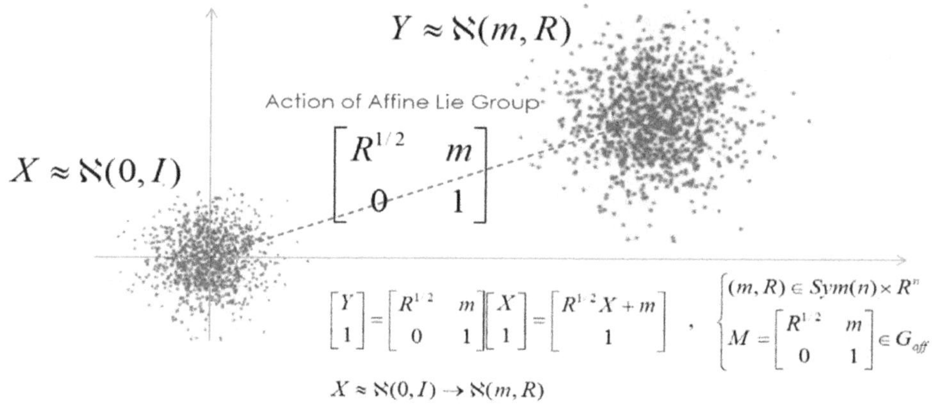

Figure 11. Affine Lie group action for multivariate Gaussian law.

Considering the curve $\gamma(t)$ and its derivative $\dot{\gamma}(t)$:

$$\gamma(t)=\begin{bmatrix} R^{1/2}(t) & m(t) \\ 0 & 1 \end{bmatrix} \text{ and } \dot{\gamma}(t)=\begin{bmatrix} \dot{R}^{1/2}(t) & \dot{m}(t) \\ 0 & 0 \end{bmatrix} \tag{185}$$

We can consider the curve with the point $\gamma(0)$ moved at the identity element on the left or on the right. Then, the tangent plan at identity element provides the Lie algebra:

$$\Gamma_L(t)=L_{M^{-1}}\left(\gamma(t)\right)=\begin{bmatrix} R^{-1/2}R^{1/2}(t) & R^{-1/2}\left(m(t)-m\right) \\ 0 & 1 \end{bmatrix} \tag{186}$$

$$\dot{\Gamma}_L(t)\Big|_{t=0}=\begin{bmatrix} R^{-1/2}\dot{R}^{1/2}(0) & R^{-1/2}\dot{m}(0) \\ 0 & 1 \end{bmatrix}=\frac{d}{dt}\left(L_{M^{-1}}(\gamma(t))\right)\Big|_{t=0}=dL_{M^{-1}}\dot{\gamma}(0)=dL_{M^{-1}}\dot{M} \tag{187}$$

Lie algebra on the right and on the left is the defined by:

$$dL_{M^{-1}}:T_M(G)\rightarrow \mathfrak{g}_L$$

$$\dot{M}\mapsto \Omega_L=dL_{M^{-1}}\dot{M}=M^{-1}\dot{M}=\begin{bmatrix} R^{-1/2}\dot{R}^{1/2} & R^{-1/2}\dot{m} \\ 0 & 0 \end{bmatrix} \tag{188}$$

$$dR_{M^{-1}}:T_M(G)\rightarrow \mathfrak{g}_R$$

$$\dot{M}\mapsto \Omega_R=dR_{M^{-1}}\dot{M}=\dot{M}M^{-1}=\begin{bmatrix} R^{-1/2}\dot{R}^{1/2} & \dot{m}-R^{-1/2}\dot{R}^{1/2}m \\ 0 & 0 \end{bmatrix} \tag{189}$$

We can then observe the velocities in two different ways, either by placing in a fixed outside frame, either by putting in place of the element in the process of moving by placing in the reference frame of the element.

$$\begin{bmatrix} X(t) \\ 1 \end{bmatrix}=M\begin{bmatrix} x \\ 1 \end{bmatrix}\Rightarrow\begin{bmatrix} \dot{X}(t) \\ 0 \end{bmatrix}=\Omega_R\begin{bmatrix} X(t) \\ 1 \end{bmatrix} \text{ with } x \text{ fixed} \tag{190}$$

$$\begin{bmatrix} x(t) \\ 1 \end{bmatrix}=M^{-1}\begin{bmatrix} X \\ 1 \end{bmatrix}\Rightarrow\begin{bmatrix} \dot{x}(t) \\ 0 \end{bmatrix}=-\Omega_L\begin{bmatrix} X \\ 1 \end{bmatrix} \text{ with } X \text{ fixed} \tag{191}$$

In the following, we will complete the global view by the operators which will allow to link algebra (from the left or the right) between them and also connect to their dual. We will first consider the automorphisms, the action by conjugation of the Lie group on itself that allows this operator to carry a member of the group.

$$\begin{aligned} & AD : G \times G \to G \\ & \quad M, N \mapsto AD_M N = M.N.M^{-1} \end{aligned} \tag{192}$$

$$\begin{cases} M_1 = \begin{bmatrix} R_1^{1/2} & m_1 \\ 0 & 1 \end{bmatrix}, M_2 = \begin{bmatrix} R_2^{1/2} & m_2 \\ 0 & 1 \end{bmatrix} \\ AD_{M_1} M_2 = \begin{bmatrix} R_2^{1/2} & -R_2^{1/2} m_1 + R_1^{1/2} m_2 + m_1 \\ 0 & 1 \end{bmatrix} \end{cases} \tag{193}$$

If now we consider a curve *N(t)* curve on the manifold via the identity at *t* = 0. Its image by the previous operator will be then curve $\gamma = M \cdot N(t) \cdot M^{-1}$ passing through identity element at *t = 0*. As $\dot{N}(0)$ is an element of the Lie algebra and its image by previous conjugation operator is called the Adjoint operator:

$$\begin{aligned} & Ad : G \times \mathfrak{g} \to \mathfrak{g} \\ & \quad M, n \mapsto Ad_M n = M.n.M^{-1} = \left.\frac{d}{dt}\right|_{t=0} (AD_M N(t)) \text{ with } \begin{cases} N(0) = I \\ \dot{N}(0) = n \in g \end{cases} \end{aligned} \tag{194}$$

We can then compute the Adjoint operator for the previous Lie group:

$$\begin{cases} n_{2L} = \begin{bmatrix} R_2^{-1/2} \dot{R}_2^{1/2} & R_2^{-1/2} \dot{m}_2 \\ 0 & 0 \end{bmatrix}, n_{2R} = \begin{bmatrix} R_2^{-1/2} \dot{R}_2^{1/2} & -R_2^{-1/2} \dot{R}_2^{1/2} m_2 + \dot{m}_2 \\ 0 & 0 \end{bmatrix} \\ Ad_{M_1} n_{2L} = n_{2R} \text{ and } Ad_{M2} n_{2R} = \begin{bmatrix} R_2^{-1/2} \dot{R}_2^{1/2} & -R_2^{-1/2} \dot{R}_2^{1/2} m_2 + \dot{R}_2^{1/2} m_2 + R_2^{1/2} \dot{m}_2 \\ 0 & 0 \end{bmatrix}, Ad_{M_1^{-1}} n_{2R} = n_{2L} \end{cases} \tag{195}$$

We recall that the Lie algebra has been defined as the tangent space at the identity of a Lie group. We will then introduce a Lie bracket $[.,.]$, the expression of the operator associated with the combined action of the Lie algebra on itself, called an adjoint operator. The adjoint operator represents the action by conjugation of the Lie algebra on itself and is defined by:

$$\begin{aligned} & ad : \mathfrak{g} \times \mathfrak{g} \to \mathfrak{g} \\ & \quad n, m \mapsto ad_m n = m \cdot n - n \cdot m = \left.\frac{d}{dt}\right|_{t=0} (Ad_M n(t)) = [m, n] \text{ with } \begin{cases} \dot{N}(0) = n \in g \\ \dot{M}(0) = m \in g \end{cases} \end{aligned} \tag{196}$$

We can then compute this operator for our use case:

$$n_{1L} = \begin{bmatrix} R_1^{-1/2} \dot{R}_1^{1/2} & R_1^{-1/2} \dot{m}_1 \\ 0 & 0 \end{bmatrix}, n_{2L} = \begin{bmatrix} R_2^{-1/2} \dot{R}_2^{1/2} & R_2^{-1/2} \dot{m}_2 \\ 0 & 0 \end{bmatrix} \tag{197}$$

$$ad_{n_{1L}} n_{2L} = [n_{1L}, n_{2L}] = \begin{bmatrix} 0 & R_1^{-1/2} \left(\dot{R}_1^{1/2} \dot{m}_2 - \dot{R}_2^{1/2} \dot{m}_1 \right) R_2^{-1/2} \\ 0 & 0 \end{bmatrix} \tag{198}$$

$$ad_{n_{1R}} n_{2R} = [n_{1R}, n_{2R}] = \begin{bmatrix} 0 & R_1^{-1/2} \dot{R}_1^{1/2} \left(-R_2^{-1/2} \dot{R}_2^{1/2} m_2 + \dot{m}_2 \right) - R_2^{-1/2} \dot{R}_2^{1/2} \left(-R_1^{-1/2} \dot{R}_1^{1/2} m_1 + \dot{m}_1 \right) \\ 0 & 0 \end{bmatrix} \tag{199}$$

To study the geodesic trajectories of the group, we consider the Lagrangian from the total kinetic energy (a quadratic form on speeds). It may therefore in particular be written in the left algebra "left", with the scalar product associated with the metric.

$$E_L = \frac{1}{2}\langle n_L, n_L \rangle = \frac{1}{2} Tr\left[n_L^T n_L\right] \quad (200)$$

If we consider as scalar product:

$$\begin{aligned} &\langle .,. \rangle : \mathfrak{g}^* \times \mathfrak{g} \to R \\ &\quad k, n \mapsto \langle k, n \rangle = Tr\left(k^T n\right) \end{aligned} \quad (201)$$

and left algebra:

$$n_L = \begin{bmatrix} R^{-1/2}\dot{R}^{1/2} & R^{-1/2}\dot{m} \\ 0 & 0 \end{bmatrix} \quad (202)$$

we obtain for the total kinetic energy

$$E_L = \frac{1}{2}\left(Tr\left(R^{-1}\dot{R}\right) + \dot{m}^T R^{-1}\dot{m}\right) \quad (203)$$

We will then introduce the coadjoint operator that will enable us to work on the elements of the dual algebra of the Lie algebra defined above. Like algebra, which is physically the space of instantaneous speeds, the dual algebra is the space of moments. For the dual of left algebra, the moment is given by:

$$\Pi_L = \frac{\partial E_L}{\partial n_L} = n_L \quad (204)$$

Where E_L is the kinetic energy of the system and is currently associated with Π_L is an element of the left algebra. The moment space is the dual algebra, denoted $\mathfrak{g}^*$, associated with the Lie algebra $\mathfrak{g}$. This value is deduced from the computation:

$$\begin{aligned} &\left\langle \frac{\partial E_L}{\partial n_L}, \delta U \right\rangle = \lim_{\varepsilon \to 0} \frac{E_L(n_L + \varepsilon \cdot \delta U) - E_L(n_L)}{\varepsilon} \\ &\text{with } E_L(n_L + \varepsilon \cdot \delta U) = \frac{1}{2}\langle n_L + \varepsilon.\delta U, n_L + \varepsilon \cdot \delta U \rangle = \frac{1}{2}(n_L + \varepsilon \cdot \delta U)^T (n_L + \varepsilon \cdot \delta U) \\ &\left\langle \frac{\partial E_L}{\partial n_L}, \delta U \right\rangle = 2 \cdot \frac{1}{2} tr\left(\eta_L^T \delta U\right) = \langle n_L, \delta U \rangle \Rightarrow \frac{\partial E_L}{\partial n_L} = n_L \end{aligned} \quad (205)$$

Then the moment map is given by:

$$\begin{aligned} &\alpha_M : \mathfrak{g} \to \mathfrak{g}^* \\ &\quad n_L \mapsto \Pi_L = \eta_L \end{aligned} \quad (206)$$

We can observe that the application that turns left algebra into dual algebra is the identity application but, physically, the first are moments and the seconds are instantaneous speeds.

We can also define the moment Π_R associated to the right algebra η_R by:

$$\langle \Pi_L, n_L \rangle = \left\langle \Pi_L, M^{-1} n_R M \right\rangle = \langle \Pi_R, n_R \rangle \quad (207)$$

But as $\Pi_L = n_L$, we can deduce that:

$$\langle n_L, M^{-1} n_R M\rangle = \langle \Pi_R, n_R\rangle$$
$$\text{with } M = \begin{bmatrix} R^{1/2} & m \\ 0 & 1 \end{bmatrix},\ n_L = \begin{bmatrix} R^{-1/2}\dot{R}^{1/2} & R^{-1/2}\dot{m} \\ 0 & 0 \end{bmatrix} \text{ and } \eta_R = \begin{bmatrix} R^{-1/2}\dot{R}^{1/2} & \dot{m} - R^{-1/2}\dot{R}^{1/2}\dot{m} \\ 0 & 0 \end{bmatrix} \tag{208}$$
$$\Rightarrow \Pi_R = \begin{bmatrix} R^{-1/2}\dot{R}^{1/2} + R^{-1}\dot{m}m^T & R^{-1}\dot{m} \\ 0 & 0 \end{bmatrix}$$

Then, the operator that transform the right algebra to its dual algebra is given by:

$$\beta_M : \mathfrak{g} \to \mathfrak{g}^*$$
$$n_R = \begin{bmatrix} \eta_{R1} & \eta_{R2} \\ 0 & 0 \end{bmatrix} \mapsto \Pi_R = \begin{bmatrix} \eta_{R1}\left(1 + m^T R^{-1} m\right) + \eta_{R2} m^T R^{-1} & \eta_{R1} R^{-1} m + R^{-1}\eta_{R2} \\ 0 & 0 \end{bmatrix} \tag{209}$$

There is an operator to change the view of algebra. Therefore, there is an operator that did the same to the dual algebra. This is called the co-adjoint operator and it is the conjugate action of the Lie group on its dual algebra:

$$\begin{cases} Ad^* : G \times \mathfrak{g}^* \to \mathfrak{g} \\ M, \eta \mapsto Ad^*_M \eta \end{cases} \text{ with } \langle Ad^*_M \eta, n\rangle = \langle \eta, Ad_M n\rangle \text{ where } n \in \mathfrak{g} \tag{210}$$

We can then develop this expression for our use in the case of an affine sup-group. We find:

$$\begin{cases} M = \begin{bmatrix} A & b \\ 0 & 1 \end{bmatrix} \in G \\ \eta = \begin{bmatrix} \eta_1 & \eta_2 \\ 0 & 0 \end{bmatrix} \in \mathfrak{g}^* \\ n = \begin{bmatrix} n_1 & n_2 \\ 0 & 0 \end{bmatrix} \in \mathfrak{g} \end{cases} \Rightarrow \begin{cases} \langle Ad^*_M \eta, n\rangle = \langle \eta, Ad_M n\rangle = \langle \eta, MnM^{-1}\rangle \\ \langle Ad^*_M \eta, n\rangle = \left\langle \begin{bmatrix} \eta_1 - \eta_2 b^T & A\eta_2 \\ 0 & 0 \end{bmatrix}, n \right\rangle \end{cases} \Rightarrow Ad^*_M \eta = \begin{bmatrix} \eta_1 - \eta_2 b^T & A\eta_2 \\ 0 & 0 \end{bmatrix} \tag{211}$$

and we can also observe that:

$$Ad^*_{M^{-1}} \eta = \begin{bmatrix} \eta_1 + A\eta_2 b^T & A\eta_2 \\ 0 & 0 \end{bmatrix} \tag{212}$$

Similarly there exists the following relation between the left and the right algebras:

$$Ad^*_M \Pi_R = \Pi_L \text{ and } Ad^*_{M^{-1}} \Pi_L = \Pi_R \tag{213}$$

As we have defined a commutator on the Lie algebra, it is possible to define one on its dual algebra. This commutator on the dual algebra can also be defined using the operator expressing the combined action of the algebra of its dual algebra. This operator is called the co-adjoint operator:

$$\begin{cases} ad^* : \mathfrak{g} \times \mathfrak{g}^* \to \mathfrak{g}^* \\ n, \eta \mapsto ad^*_n \eta \end{cases} \text{ with } \langle ad^*_n \eta, \kappa\rangle = \langle \eta, ad_n \kappa\rangle \text{ where } \kappa \in \mathfrak{g} \tag{214}$$

We can develop this co-adjoint operator on its dual algebra for our use-case:

$$\begin{cases} \kappa = \begin{bmatrix} \kappa_1 & \kappa_2 \\ 0 & 0 \end{bmatrix} \in G \\ \eta = \begin{bmatrix} \eta_1 & \eta_2 \\ 0 & 0 \end{bmatrix} \in \mathfrak{g}^* \\ n = \begin{bmatrix} n_1 & n_2 \\ 0 & 0 \end{bmatrix} \in \mathfrak{g} \end{cases} \Rightarrow \begin{cases} \langle ad_n^* \eta, \kappa \rangle = \langle \eta, ad_n \kappa \rangle = \langle \eta, n\kappa - \kappa n \rangle \\ \langle ad_n^* \eta, \kappa \rangle = \left\langle \begin{bmatrix} -\eta_2 n_2^T & n_1 \eta_2 \\ 0 & 0 \end{bmatrix}, \kappa \right\rangle \end{cases} \Rightarrow \begin{cases} ad_n^* \eta = \begin{bmatrix} -\eta_2 n_2^T & n_1 \eta_2 \\ 0 & 0 \end{bmatrix} \\ ad_n^* \eta = \{n, \eta\} \end{cases} \tag{215}$$

This co-adjoint operator will give the Euler-Poincaré equation. While the Euler-Lagrange equations is defined on the tangent bundle (union of the tangent spaces at each point) of the manifold and give the geodesics, the Euler-Poincaré equation gives a differential system on the dual Lie algebra of the group associated with the manifold.

We can also complete these maps by using additional ones. First, $p \in T_M^* G$ the moment associated with $\dot{M} \in T_M G$ in tangent space of G at M and also two other moments map the element of the dual algebra in dual tangent space, respectively on the left and on the right:

$$\begin{cases} \langle \Pi_L, n_L \rangle = \left\langle dL_{M^{-1}}^* \Pi_L, \dot{M} \right\rangle \\ \left\langle \Pi_L, dL_{M^{-1}} \dot{M} \right\rangle = \left\langle \Pi_L, M^{-1} \dot{M} \right\rangle \end{cases} \Rightarrow p = \left(M^{-1} \right)^T \Pi_L \tag{216}$$

where

$$\begin{array}{c} dL_{M^{-1}}^* : \mathfrak{a}_L^* \to T_M^* G \\ \Pi_L \mapsto p = \left(M^{-1} \right)^T \Pi_L \end{array} \quad \text{and} \quad \begin{array}{c} dR_{M^{-1}}^* : \mathfrak{g}_R^* \to T_M^* G \\ \Pi_R \mapsto p = \Pi_R \left(M^{-1} \right)^T \end{array} \tag{217}$$

From these relations, we can also observe that:

$$\begin{array}{l} \Pi_L = n_L = M^{-1} \dot{M} \\ \Rightarrow \begin{cases} p = \left(M^{-1} \right)^T M^{-1} \dot{M} \\ p = \Xi_M \cdot \dot{M} \text{ with } \Xi_M = \left(M^{-1} \right)^T M^{-1} \end{cases} \end{array} \tag{218}$$

All these maps could be summarized in the following Figure 12:

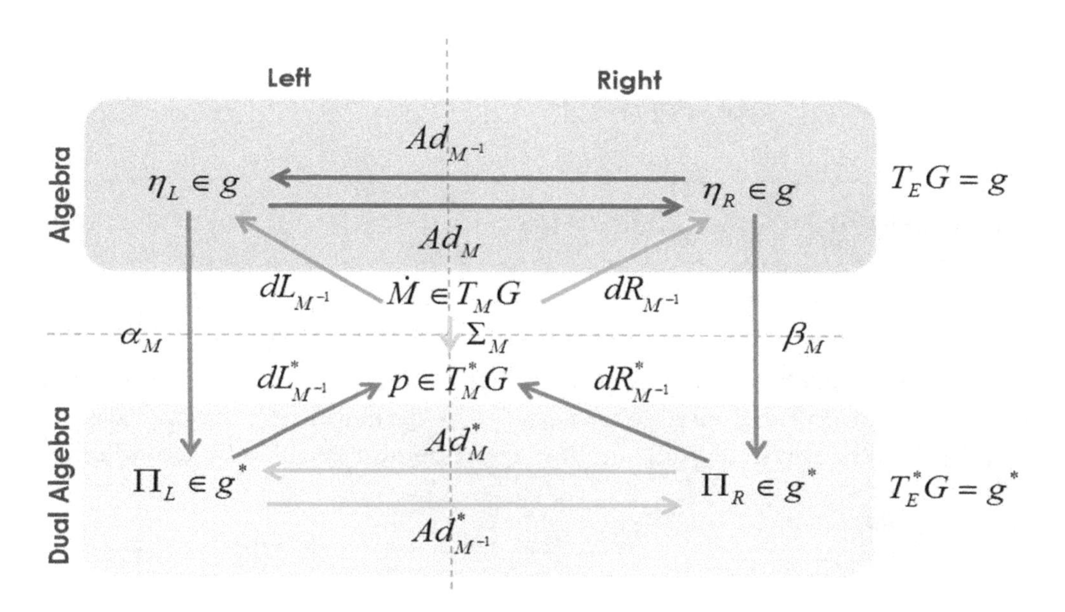

Figure 12. Maps between algebras.

Heni Poincaré proved that when a Lie algebra acts locally and transitively on the configuration space of a Lagrangian mechanical system, the Euler-Lagrange equations are equivalent to a new system of differential equations defined on the product of the configuration space with the Lie algebra.

If we consider that the following function is stationary for a Lagragian $l(.)$ invariant with respect to the action of a group on the left:

$$S(\eta_L) = \int_a^b l(\eta_L)dt \text{ with } \delta S(\eta_L) = 0 \text{ and } l : \mathfrak{g} \to R \tag{219}$$

The solution is given by the Euler-Poincaré equation:

$$\begin{gathered} \frac{d}{dt}\frac{\delta l}{\delta \eta_L} = ad^*_{\eta_L}\frac{\delta l}{\delta \eta_L} \\ \delta\eta_L = \dot{\Gamma} + ad_{\eta_L}\Gamma \text{ where } \Gamma(t) \in \mathfrak{g} \end{gathered} \tag{220}$$

If we take for the function $l(.)$, the total kinetic energy E_L, using $\Pi_L = M^{-1}\dot{M} = \frac{\partial E_L}{\partial n_L} \in \mathfrak{g}_L$, then the Euler-Poincaré equation is given by:

$$\frac{d\Pi_L}{dt} = ad^*_{n_L}\Pi_L \text{ with } \frac{\delta l}{\delta \eta_L} = \frac{\partial E_L}{\partial n_L} = \Pi_L \in \mathfrak{g}_L \tag{221}$$

The following quantities are conserved:

$$\frac{d\Pi_R}{dt} = 0 \tag{222}$$

With this second theorem, it is possible to write the geodesic not from its coordinate system but from the quantity of motion, and in addition to determine explicitly what the conserved quantities along the geodesic are (conservations are related to the symmetries of the variety and hence the invariance of the Lagrangian under the action of the group).

For our use-case, the Euler-Poincaré equation is given by:

$$\left\{ \begin{array}{l} \dot{\eta}_{L1} = -\eta_{L2}\eta_{L2}^T \\ \dot{\eta}_{L2} = \eta_{L2}\eta_{L1} \end{array} \right. \text{ with } \left\{ \begin{array}{l} \eta_{L1} = R^{-1/2}\dot{R}^{1/2} \\ \eta_{L2} = R^{-1/2}\dot{m} \end{array} \right. \Rightarrow \left\{ \begin{array}{l} \left(R^{-1/2}\dot{R}^{1/2}\right)^{\bullet} = -R^{-1/2}\dot{m}\dot{m}^T R^{-1/2} \\ \left(R^{-1/2}\dot{m}\right)^{\bullet} = \dot{R}^{-1/2}\dot{R}^{1/2}R^{-1/2}\dot{m} \end{array} \right. \tag{223}$$

If we remark that we have $R^{-1/2}\dot{R}^{1/2} = R^{-1/2}\left(R^{-1/2}\dot{R}\right) = R^{-1}\dot{R}$, then the conserved Souriau moment could be given by:

$$\Pi_R = \begin{bmatrix} R^{-1/2}\dot{R}^{1/2} + R^{-1}\dot{m}m^T & R^{-1}\dot{m} \\ 0 & 0 \end{bmatrix} = \begin{bmatrix} R^{-1}\dot{R} + R^{-1}\dot{m}m^T & R^{-1}\dot{m} \\ 0 & 0 \end{bmatrix} \tag{224}$$

Components of the Souriau moment give the conserved quantities that are the classical elements given by Emmy Noether Theorem (Souriau moment is a geometrization of Emmy Noether Theorem):

$$\frac{d\Pi_R}{dt} = \begin{bmatrix} \frac{d\left(R^{-1}\dot{R}+R^{-1}\dot{m}m^T\right)}{dt} & \frac{d\left(R^{-1}\dot{m}\right)}{dt} \\ 0 & 0 \end{bmatrix} = 0 \Rightarrow \left\{ \begin{array}{c} R^{-1}\dot{R} + R^{-1}\dot{m}m^T = B = cste \\ R^{-1}\dot{m} = b = cste \end{array} \right. \tag{225}$$

From this constant, we can obtain a reduced equation of geodesic:

$$\left\{ \begin{array}{l} \dot{m} = Rb \\ \dot{R} = R\left(B - bm^T\right) \end{array} \right. \tag{226}$$

This is the Euler-Poincaré equation of geodesic. We can observe that we have obtained a reduction of the following Euler-Lagrange equation [27,156,187]: $\begin{cases} \ddot{R} + \dot{m}\dot{m}^T - \dot{R}R^{-1}\dot{R} = 0 \\ \ddot{m} - \dot{R}R^{-1}\dot{m} = 0 \end{cases}$ associated to the information geometry metric $ds^2 = dm^T R^{-1} dm + \frac{1}{2} Tr\left(\left(R^{-1}dR\right)^2\right)$.

The Fisher information defines a metric turning $N_n = \{(m,R) \in R^n \times Sym^+(n)\}$ into a Riemannian manifold. The inner product of two tangent vectors $(m_1, R_1) \in T_n$ and $(m_2, R_2) \in T_n$ at the point $(\mu, \Sigma) \in N_n$ is given by:

$$g_{(\mu,\Sigma)}\left((m_1, R_1), (m_2, R_2)\right) = m_1^T \Sigma^{-1} m_2 + \frac{1}{2} tr\left(\Sigma^{-1} R_1 \Sigma^{-1} R_2\right) \tag{227}$$

and the geodesic is given by:

$$l(\chi) = \int_{t_0}^{t_1} \sqrt{g_{\chi(t)}\left(\dot{\chi}(t), \dot{\chi}(t)\right)} dt \tag{228}$$

We can also observe that the manifold of multivariate Gaussian is homogeneous with respect to positive affine group $GA^+(n)$:

$$ds_Y^2 = ds_X^2 \text{ for } Y = \Sigma^{1/2} X + \mu \text{ with } \mathrm{GA}^+(n) = \{(\mu, \Sigma) \in R \times GL(R) / \det(\Sigma) > 0\} \tag{229}$$

characterized by the action of the group $(m,R) \mapsto \rho.(m,R) = \left(\Sigma^{1/2} m + \mu, \Sigma^{1/2} R \Sigma^{1/2T}\right), \rho \in GA^+(n)$

$$\text{with} \begin{bmatrix} Y \\ 1 \end{bmatrix} = \begin{bmatrix} \Sigma^{1/2} & \mu \\ 0 & 1 \end{bmatrix} \begin{bmatrix} X \\ 1 \end{bmatrix} \tag{230}$$

$$\begin{aligned} ds_Y^2 &= d\left(\Sigma^{1/2} m + \mu\right)^T \left(\Sigma^{1/2} R \Sigma^{1/2T}\right)^{-1} d\left(\Sigma^{1/2} m + \mu\right) + \frac{1}{2} Tr\left(\left(\left(\Sigma^{1/2} R \Sigma^{1/2T}\right)^{-1} d\left(\Sigma^{1/2} R \Sigma^{1/2T}\right)\right)^2\right) \\ ds_Y^2 &= dm^T R^{-1} dm + \frac{1}{2} Tr\left(\left(R^{-1} dR\right)^2\right) = ds_X^2 \end{aligned} \tag{231}$$

Since the special orthogonal group $SO(n) = \{\delta \in GL(R) / \det(\delta) = 1\}$ is the stabilizer subgroup of $(0, I_n)$, we have the following isomorphism:

$$\begin{aligned} &GA^+(n)/SO(n) \to N_n = \{(m,R) \in R^n \times Sym^+(n)\} \\ &\rho = (\mu, \Sigma) \mapsto \rho.(0, I_n) = \left(\mu, \Sigma^{1/2} \Sigma^{1/2T}\right) = (\mu, \Sigma) \end{aligned} \tag{232}$$

We can then restrict the computation of the geodesic from $(0, I_n)$ and then we can partially integrate the system of equations:

$$\begin{cases} \dot{m} = Rb \\ \dot{R} = R\left(B - bm^T\right) \end{cases} \tag{233}$$

where $\left(R^{-1}(0)\dot{m}(0), R^{-1}(0)\left(\dot{R}(0) + \dot{m}(0)m(0)^T\right)\right) = (b, B) \in R^n \times Sym_n(R)$ are the integration constants.

From this Euler-Poincaré equation, we can compute geodesics by geodesic shooting [188–191] using classical Eriksen equations [192–195], by the following change of parameters:

$$\begin{cases} \Delta(t) = R^{-1}(t) \\ \delta(t) = R^{-1}(t) m(t) \end{cases} \Rightarrow \begin{cases} \dot{\Delta} = -B\Delta + bm^T \\ \dot{\delta} = -B\delta + \left(1 + \delta^T \Delta^{-1} \delta\right) b \\ \Delta(0) = I_p, \delta(0) = 0 \end{cases} \text{ with } \begin{cases} \dot{\Delta}(0) = -B \\ \dot{\delta}(0) = b \end{cases} \tag{234}$$

The initial speed of the geodesic is given by $\left(\dot{\delta}(0),\dot{\Delta}(0)\right)$. The geodesic shooting is given by the exponential map:

$$\Lambda(t)=\exp(tA)=\sum_{n=0}^{\infty}\frac{(tA)^n}{n!}=\begin{pmatrix}\Delta & \delta & \Phi\\ \delta^T & \varepsilon & \gamma^T\\ \Phi^T & \gamma & \Gamma\end{pmatrix}\text{ with }A=\begin{pmatrix}-B & b & 0\\ b^T & 0 & -b^T\\ 0 & -b & B\end{pmatrix}\tag{235}$$

This equation can be interpreted by group theory. A could be considered as an element of Lie algebra $so(n+1,n)$ of the special Lorentz group $SO_O(n+1,n)$ and more specifically as the element p of Cartan Decomposition $\mathrm{l}+\mathrm{p}$ where l is the Lie algebra of a maximal compact sub-group $K=S\left(O(n+1)\times O(n)\right)$ of the group $G=SO_O(n+1,n)$. We know that its exponential map defines a geodesic on Riemannian Symetric space G/K.

This equation can be established by the following developments:

$$\dot{\Lambda}(t)=A.\Lambda(t)\Rightarrow\begin{pmatrix}\dot{\Delta} & \dot{\delta} & \dot{\Phi}\\ \dot{\delta}^T & \dot{\varepsilon} & \dot{\gamma}^T\\ \dot{\Phi}^T & \dot{\gamma} & \dot{\Gamma}\end{pmatrix}=\begin{pmatrix}-B & b & 0\\ b^T & 0 & -b^T\\ 0 & -b & B\end{pmatrix}.\begin{pmatrix}\Delta & \delta & \Phi\\ \delta^T & \varepsilon & \gamma^T\\ \Phi^T & \gamma & \Gamma\end{pmatrix}\tag{236}$$

We can then deduce that:

$$\begin{cases}\dot{\Delta}=-B\Delta+b\delta^T\\ \dot{\delta}=-B\delta+\varepsilon b\end{cases}\tag{237}$$

If $\varepsilon=1+\delta^T\Delta^{-1}\delta$, then (Δ,δ) is solution to the geodesic equation previously defined. Since $\varepsilon(0)=1$, it suffices to demonstrate that $\dot{\varepsilon}=\dot{\tau}$ where $\tau=\delta^T\Delta^{-1}\delta$.

From $\dot{\Lambda}(t)=\Lambda(t).A$, using that $\dot{\delta}^T=b^T\Delta-b^T\Phi^T$, we can deduce:

$$\begin{cases}\dot{\varepsilon}=b^T\delta-b^T\gamma\\ \dot{\tau}=b^T\delta-b^T\left((\tau-\varepsilon)\Delta^{-1}\delta+\Phi^T\Delta^{-1}\delta\right)\end{cases}\tag{238}$$

Then $\dot{\varepsilon}=\dot{\tau}$, if $\gamma=(\tau-\varepsilon)\Delta^{-1}\delta+\Phi\Delta^{-1}\delta$, that could be verified using relation $\Lambda.\Lambda^{-1}=I$, by observing that:

$$\Lambda^{-1}=\exp(-tA)=\Lambda(-t)=\begin{bmatrix}\Gamma & \gamma & \Phi^T\\ \gamma^T & \varepsilon & \delta^T\\ \Phi & \delta & \Delta\end{bmatrix}\tag{239}$$

$$\Lambda.\Lambda^{-1}=I\Rightarrow\begin{cases}\Delta\gamma+\varepsilon\delta+\Phi\delta=0\\ \Delta\Phi^T+\delta\delta^T+\Phi\Delta=0\end{cases}\Rightarrow\begin{cases}\gamma=-\varepsilon\Delta^{-1}\delta-\Delta^{-1}\Phi\delta\\ \Phi^T\Delta^{-1}+\Delta^{-1}\delta\delta^T\Delta^{-1}+\Delta^{-1}\Phi=0\end{cases}\Rightarrow\begin{cases}\gamma=-\varepsilon\Delta^{-1}\delta-\Delta^{-1}\Phi\delta\\ \Phi^T\Delta^{-1}\delta+\tau\Delta^{-1}\delta+\Delta^{-1}\Phi\delta=0\end{cases}\tag{240}$$

We can then compute γ from two last equations:

$$\gamma=(\tau-\varepsilon)\Delta^{-1}\delta+\Phi^T\Delta^{-1}\delta\tag{241}$$

As $\dot{\tau}=b^T\delta-b^T\left((\tau-\varepsilon)\Delta^{-1}\delta+\Phi^T\Delta^{-1}\delta\right)$ then we can deduce that $\dot{\tau}=b^T\delta-b^T\gamma$ and then $\dot{\tau}=\dot{\varepsilon}$.

To interpret elements of Λ, $(\Gamma(t),\gamma(t))=(\Delta(-t),\delta(-t))$, opposite points to $(\Delta(t),\delta(t))$, and $\varepsilon=1+\delta^T\Delta^{-1}\delta=1+\gamma^T\Gamma^{-1}\gamma$.

Then the geodesic that goes through the origin $(0,I_n)$ with initial tangent vector $(b,-B)$ is the curve given by $(\delta(t),\Delta(t))$. Then the distance computation is reduced to estimate the initial tangent vector space related by $\left(R^{-1}(0)\dot{m}(0), R^{-1}(0)\left(\dot{R}(0)+\dot{m}(0)m(0)^T\right)\right) = (b, B) \in R^n \times Sym_n(R)$The distance will be then given by the initial tangent vector:

$$d = \sqrt{\dot{m}(0)^T R^{-1}(0)\dot{m}(0) + \frac{1}{2}Tr\left[\left(R^{-1}(0)\dot{R}(0)\right)^2\right]} \tag{242}$$

This initial tangent vector will be identified by "Geodesic Shooting". Let $V = \log_A B$:

$$\begin{cases} \frac{dV_m}{dt} = \frac{1}{2}\left(\frac{dR}{dt}\right)R^{-1}V_m + \frac{1}{2}V_R R^{-1}\left(\frac{dm}{dt}\right) \\ \frac{dV_R}{dt} = \frac{1}{2}\left(\left(\frac{dR}{dt}\right)R^{-1}V_m + V_R R^{-1}\left(\frac{dR}{dt}\right)\right) - \frac{1}{2}\left(\left(\frac{dm}{dt}\right)V_m^T + V_m^T\left(\frac{dm}{dt}\right)\right) \end{cases} \tag{243}$$

Geodesic Shooting is corrected by using Jacobi Field J and parallel transport: $J(t) = \left.\frac{\partial \chi_\alpha(t)}{\partial \alpha}\right|_{t=0}$ solution to $\frac{d^2J(t)}{dt^2} + R\left(J(t), \dot{\chi}(t)\right)\dot{\chi}(t) = 0$ with R the Riemann Curvarture tensor.

We consider a geodesic χ between θ_0 and θ_1 with an initial tangent vector V, and we suppose that V is perturbated by W, to $V+W$. The variation of the final point θ_1 can be determined thanks to the Jacobi field with $J(0)=0$ and $\dot{J}(0)=W$. In term of the exponential map, this could be written:

$$J(t) = \left.\frac{d}{d\alpha}\exp_{\theta_0}\left(t\left(V+\alpha W\right)\right)\right|_{\alpha=0} \tag{244}$$

This could be illustrated in the Figure 13:

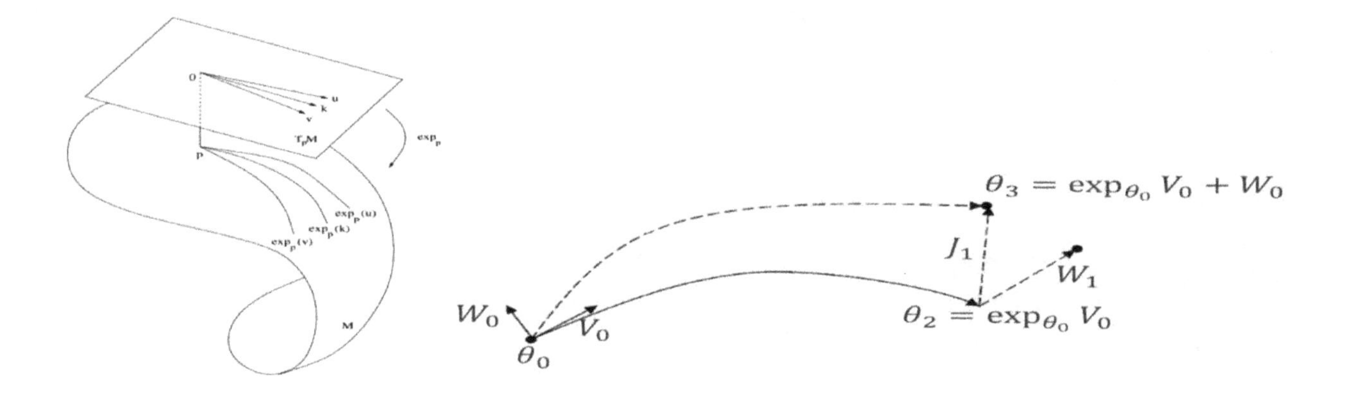

Figure 13. Geodesic shooting principle.

We give some illustration, in Figure 14, of geodesic shooting to compute the distance between multivariate Gaussian density for the case $n = 2$:

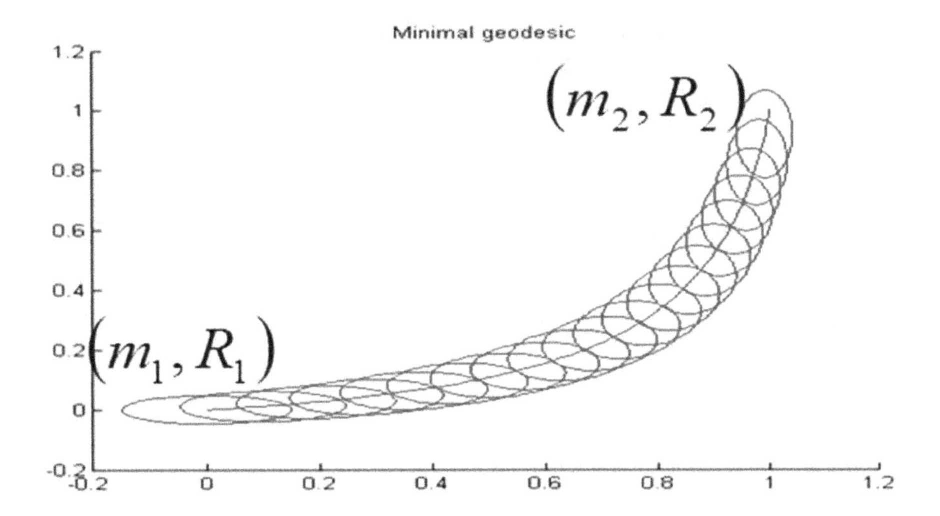

Figure 14. GeodesicsShooting between two multivariate Gaussian in case $n = 2$.

9. Souriau Riemannian Metric for Multivariate Gaussian Densities

To illustrate the Souriau-Fisher metric, we will consider the family of multivariate Gaussian densities and will develop some elements that we have previously developed purely theoretically.

For the families of multivariate Gaussian densities, that we have identified as homogeneous manifold with the associated sub-group of the affine group $\begin{bmatrix} R^{1/2} & m \\ 0 & 1 \end{bmatrix}$, we have seen that if we consider them as elements of exponential families, we can write $\hat{\xi}$ (element of the dual Lie algebra) that play the role of geometric heat Q in Souriau Lie group thermodynamics, and β the geometric (Planck) temperature.

$$\hat{\xi} = \begin{bmatrix} E[z] \\ E[zz^T] \end{bmatrix} = \begin{bmatrix} m \\ R + mm^T \end{bmatrix}, \ \beta = \begin{bmatrix} -R^{-1}m \\ \frac{1}{2}R^{-1} \end{bmatrix} \tag{245}$$

These elements are homeomorphic to the matrix elements in matrix Lie algebra and dual Lie algebra:

$$\hat{\xi} = \begin{bmatrix} R + mm^T & m \\ 0 & 0 \end{bmatrix} \in \mathfrak{g}^*, \ \beta = \begin{bmatrix} \frac{1}{2}R^{-1} & -R^{-1}m \\ 0 & 0 \end{bmatrix} \in \mathfrak{g} \tag{246}$$

If we consider $M = \begin{bmatrix} R'^{1/2} & m' \\ 0 & 1 \end{bmatrix}$, then we can compute the co-adjoint operator:

$$Ad^*_M \hat{\xi} = \begin{bmatrix} R + mm^T - mm'^T & R'^{1/2}m \\ 0 & 0 \end{bmatrix} \tag{247}$$

We can also compute the adjoint operator:

$$\begin{gathered} Ad_M \beta = M \cdot \beta \cdot M^{-1} = \begin{bmatrix} R'^{1/2} & m' \\ 0 & 1 \end{bmatrix} \begin{bmatrix} \frac{1}{2}R^{-1} & -R^{-1}m \\ 0 & 0 \end{bmatrix} \begin{bmatrix} R'^{-1/2} & -R'^{-1/2}m' \\ 0 & 1 \end{bmatrix} \\ Ad_M \beta = \begin{bmatrix} \frac{1}{2}R'^{1/2}R^{-1}R'^{-1/2} & -\frac{1}{2}R'^{1/2}R^{-1}R'^{-1/2}m' - R'^{1/2}R^{-1}m \\ 0 & 0 \end{bmatrix} \end{gathered} \tag{248}$$

We can rewrite $Ad_M\beta$ with the following identification:

$$Ad_M\beta = \begin{bmatrix} \frac{1}{2}\Omega^{-1} & -\Omega^{-1}n \\ 0 & 0 \end{bmatrix}$$
$$\text{with } \Omega = R'^{1/2}RR'^{-1/2} \text{ and } n = \left(\frac{1}{2}m' + R'^{1/2}m\right) \tag{249}$$

We have then to develop $\hat{\xi}\left(Ad_M(\beta)\right)$, that is to say $\hat{\xi}(\beta)$ after action of the group on the Lie algebra for β, given by $Ad_M(\beta)$. By analogy of structure between $\hat{\xi}(\beta)$ and β, we can write:

$$\left.\begin{array}{l} \beta = \begin{bmatrix} \frac{1}{2}R^{-1} & -R^{-1}m \\ 0 & 0 \end{bmatrix} \\ \hat{\xi}(\beta) = \begin{bmatrix} R + mm^T & m \\ 0 & 0 \end{bmatrix} \end{array}\right\} \Rightarrow \left\{\begin{array}{l} Ad_M\beta = \begin{bmatrix} \frac{1}{2}\Omega^{-1} & -\Omega^{-1}n \\ 0 & 0 \end{bmatrix} \\ \hat{\xi}\left(Ad_M(\beta)\right) = \begin{bmatrix} \Omega + nn^T & n \\ 0 & 0 \end{bmatrix} \end{array}\right. \tag{250}$$

We have then to identify the cocycle $\theta(M)$ from $\hat{\xi}\left(Ad_M(\beta)\right) = Ad^*_M(\hat{\xi}) + \theta(M)$ $\Rightarrow \theta(M) = \hat{\xi}\left(Ad_M(\beta)\right) - Ad^*_M\hat{\xi}$ where:

$$Ad^*_M\hat{\xi} = \begin{bmatrix} R + mm^T - mm'^T & R'^{1/2}m \\ 0 & 0 \end{bmatrix} \tag{251}$$

$$\hat{\xi}\left(Ad_M(\beta)\right) = \begin{bmatrix} R'^{1/2}RR'^{-1/2} + \left(\frac{1}{2}m' + R'^{1/2}m\right)\left(\frac{1}{2}m' + R'^{1/2}m\right)^T & \left(\frac{1}{2}m' + R'^{1/2}m\right) \\ 0 & 0 \end{bmatrix} \tag{252}$$

The cocycle is then given by:

$$\begin{array}{l} \theta(M) = \begin{bmatrix} R'^{1/2}RR'^{-1/2} + \left(\frac{1}{2}m' + R'^{1/2}m\right)\left(\frac{1}{2}m' + R'^{1/2}m\right)^T & \left(\frac{1}{2}m' + R'^{1/2}m\right) \\ 0 & 0 \end{bmatrix} - \begin{bmatrix} R + mm^T - mm'^T & R'^{1/2}m \\ 0 & 0 \end{bmatrix} \\ \theta(M) = \begin{bmatrix} \left(R'^{1/2}RR'^{-1/2} - R\right) + \left(R'^{1/2}mm^TR'^{1/2T} - mm^T\right) + \left(\frac{1}{2}m'm^TR'^{1/2T} + \frac{1}{2}R'^{1/2}mm'^T - mm'^T\right) & \frac{1}{2}m' \\ 0 & 0 \end{bmatrix} \end{array} \tag{253}$$

From $\theta(M) = \hat{\xi}\left(Ad_M(\beta)\right) - Ad^*_M\hat{\xi}$, we can compute cocycle in Lie algebra

$$\Theta = T_e\theta \tag{254}$$

used to define the tensor:

$$\begin{array}{c} \widetilde{\Theta}(X, Y) : \mathfrak{g} \times \mathfrak{g} \rightarrow \Re \\ X, Y \mapsto \langle \Theta(X), Y \rangle \end{array} \tag{255}$$

In this second part, we will compute the Souriau-Fisher metric given by:

$$g_\beta\left([\beta, Z_1], [\beta, Z_2]\right) = \widetilde{\Theta}_\beta\left(Z_1, [\beta, Z_2]\right) \tag{256}$$

with

$$\widetilde{\Theta}_\beta(Z_1, Z_2) = \widetilde{\Theta}(Z_1, Z_2) + \langle \hat{\xi}, ad_{Z_1}Z_2 \rangle = \langle \Theta(Z_1), Z_2 \rangle + \langle \hat{\xi}, [Z_1, Z_2] \rangle \tag{257}$$

$$\begin{aligned} g_\beta\left([\beta, Z_1], [\beta, Z_2]\right) &= \widetilde{\Theta}_\beta\left(Z_1, [\beta, Z_2]\right) = \widetilde{\Theta}\left(Z_1, [\beta, Z_2]\right) + \langle \hat{\xi}, [Z_1, [\beta, Z_2]] \rangle \\ &= \langle \Theta(Z_1), [\beta, Z_2] \rangle + \langle \hat{\xi}, [Z_1, [\beta, Z_2]] \rangle \end{aligned} \tag{258}$$

where

$$\beta = \begin{bmatrix} \frac{1}{2}R^{-1} & -R^{-1}m \\ 0 & 0 \end{bmatrix} \text{ and } \hat{\xi} = \begin{bmatrix} R + mm^T & m \\ 0 & 0 \end{bmatrix} \tag{259}$$

$$\text{If we set } Z_1 = \begin{bmatrix} \frac{1}{2}\Omega_1^{-1} & -\Omega_1^{-1}n_1 \\ 0 & 0 \end{bmatrix} \text{ and } Z_2 = \begin{bmatrix} \frac{1}{2}\Omega_2^{-1} & -\Omega_2^{-1}n_2 \\ 0 & 0 \end{bmatrix} \tag{260}$$

With $\langle ..., ... \rangle$ the inner product given by

$$\langle \xi, \beta \rangle = Tr\left[ba^T + H^T L\right] \text{ with } \xi = \begin{bmatrix} L & b \\ 0 & 0 \end{bmatrix}, \beta = \begin{bmatrix} H & a \\ 0 & 0 \end{bmatrix} \tag{261}$$

$$\begin{gathered}
[\beta, Z_2] = \beta Z_2 - Z_2 \beta = \begin{bmatrix} \frac{1}{2}R^{-1} & -R^{-1}m \\ 0 & 0 \end{bmatrix} \begin{bmatrix} \frac{1}{2}\Omega_2^{-1} & -\Omega_2^{-1}n_2 \\ 0 & 0 \end{bmatrix} - \begin{bmatrix} \frac{1}{2}\Omega_2^{-1} & -\Omega_2^{-1}n_2 \\ 0 & 0 \end{bmatrix} \begin{bmatrix} \frac{1}{2}R^{-1} & -R^{-1}m \\ 0 & 0 \end{bmatrix} \\
[\beta, Z_2] = \begin{bmatrix} \frac{1}{4}\left(R^{-1}\Omega_2^{-1} - \Omega_2^{-1}R^{-1}\right) & -\frac{1}{2}\left(R^{-1}\Omega_2^{-1}n_2 - \Omega_2^{-1}R^{-1}m\right) \\ 0 & 0 \end{bmatrix}
\end{gathered} \tag{262}$$

$$\begin{aligned}
[Z_1, [\beta, Z_2]] &= \begin{bmatrix} \frac{1}{2}\Omega_1^{-1} & -\Omega_1^{-1}n_1 \\ 0 & 0 \end{bmatrix} \begin{bmatrix} \frac{1}{4}\left(R^{-1}\Omega_2^{-1} - \Omega_2^{-1}R^{-1}\right) & -\frac{1}{2}\left(R^{-1}\Omega_2^{-1}n_2 - \Omega_2^{-1}R^{-1}m\right) \\ 0 & 0 \end{bmatrix} \\
&\quad - \begin{bmatrix} \frac{1}{4}\left(R^{-1}\Omega_2^{-1} - \Omega_2^{-1}R^{-1}\right) & -\frac{1}{2}\left(R^{-1}\Omega_2^{-1}n_2 - \Omega_2^{-1}R^{-1}m\right) \\ 0 & 0 \end{bmatrix} \begin{bmatrix} \frac{1}{2}\Omega_1^{-1} & -\Omega_1^{-1}n_1 \\ 0 & 0 \end{bmatrix} \\
&= \begin{bmatrix} \frac{1}{8}\left(\Omega_1^{-1}\left(R^{-1}\Omega_2^{-1} - \Omega_2^{-1}R^{-1}\right) - \left(R^{-1}\Omega_2^{-1} - \Omega_2^{-1}R^{-1}\right)\Omega_1^{-1}\right) & -\frac{1}{4}\left(\Omega_1^{-1}\left(R^{-1}\Omega_2^{-1}n_2 - \Omega_2^{-1}R^{-1}m\right) - \left(R^{-1}\Omega_2^{-1} - \Omega_2^{-1}R^{-1}\right)\Omega_1^{-1}n_1\right) \\ 0 & 0 \end{bmatrix}
\end{aligned} \tag{263}$$

We can then compute:

$$\begin{aligned}
\langle \hat{\xi}, [Z_1, [\beta, Z_2]] \rangle &= Tr\left[\frac{1}{4}m\left(\left(R^{-1}\Omega_2^{-1} - \Omega_2^{-1}R^{-1}\right)\Omega_1^{-1}n_1 - \Omega_1^{-1}\left(R^{-1}\Omega_2^{-1}n_2 - \Omega_2^{-1}R^{-1}m\right)\right)^T\right] \\
&+ Tr\left[\left(\frac{1}{8}\left(\Omega_1^{-1}\left(R^{-1}\Omega_2^{-1} - \Omega_2^{-1}R^{-1}\right) - \left(R^{-1}\Omega_2^{-1} - \Omega_2^{-1}R^{-1}\right)\Omega_1^{-1}\right)\right)\left(R + mm^T\right)\right]
\end{aligned} \tag{264}$$

The Souriau-Fisher metric is defined in Lie algebra $g_\beta\left([\beta, Z_1], [\beta, Z_2]\right)$ where:

$$\begin{gathered}
[\beta, Z_1] = \begin{bmatrix} \frac{1}{4}\left(R^{-1}\Omega_1^{-1} - \Omega_1^{-1}R^{-1}\right) & -\frac{1}{2}\left(R^{-1}\Omega_1^{-1}n_1 - \Omega_1^{-1}R^{-1}m\right) \\ 0 & 0 \end{bmatrix} = \begin{bmatrix} \frac{1}{2}G_1^{-1} & -G_1^{-1}g_1 \\ 0 & 0 \end{bmatrix} \\
\text{with } G_1 = 2\left(\Omega_1 R - R\Omega_1\right) \text{ and } g_1 = (I - R\Omega_1 R^{-1}\Omega_1^{-1})n_1 + (\Omega_1 R\Omega_1^{-1}R^{-1} - I)m \\
[\beta, Z_2] = \begin{bmatrix} \frac{1}{4}\left(R^{-1}\Omega_2^{-1} - \Omega_2^{-1}R^{-1}\right) & -\frac{1}{2}\left(R^{-1}\Omega_2^{-1}n_2 - \Omega_2^{-1}R^{-1}m\right) \\ 0 & 0 \end{bmatrix} = \begin{bmatrix} \frac{1}{2}G_2^{-1} & -G_2^{-1}g_2 \\ 0 & 0 \end{bmatrix} \\
\text{with } G_2 = 2\left(\Omega_2 R - R\Omega_2\right) \text{ and } g_2 = (I - R\Omega_2 R^{-1}\Omega_2^{-1})n_2 + (\Omega_2 R\Omega_2^{-1}R^{-1} - I)m
\end{gathered} \tag{265}$$

and

$$\beta = \begin{bmatrix} \frac{1}{2}R^{-1} & -R^{-1}m \\ 0 & 0 \end{bmatrix} \tag{266}$$

Another approach to develop the Souriau-Fisher metric $g_\beta\left([\beta, Z_1], [\beta, Z_2]\right)$ is to compute the tensor $\widetilde{\Theta}(X, Y)$ from the moment map J:

$$\widetilde{\Theta}(X, Y) = J_{[X,Y]} - \{J_X, J_Y\} \text{ with } \{.,.\} \text{ Poisson Bracket and } J \text{ the Moment Map} \tag{267}$$

$$\widetilde{\Theta}(X, Y) : \mathfrak{g} \times \mathfrak{g} \to \Re \tag{268}$$

We can then write the Souriau-Fisher metric as:

$$\widetilde{\Theta}_\beta (Z_1, Z_2) = J_{[Z_1,Z_2]} - \{J_{Z_1}, J_{Z_2}\} + \langle \hat{\xi}, [Z_1, Z_2] \rangle \quad (269)$$

Where the associated differentiable application J, called moment map is:

$$\begin{array}{l} J : M \to g^* \qquad \text{such that } J_X(x) = \langle J(x), X \rangle, \; X \in g \\ \quad x \mapsto J(x) \end{array} \quad (270)$$

This moment map could be identified with the operator that transforms the right algebra to an element of its dual algebra given by:

$$\begin{array}{l} \beta_M : \mathfrak{g} \to \mathfrak{g}^* \\ Z = \begin{bmatrix} \mathrm{N} & \eta \\ 0 & 0 \end{bmatrix} \mapsto J = \begin{bmatrix} \mathrm{N}\left(1 + m^T R^{-1} m\right) + \eta m^T R^{-1} & \mathrm{N} R^{-1} m + R^{-1} \eta \\ 0 & 0 \end{bmatrix} \end{array} \quad (271)$$

10. Conclusions

In this paper, we have developed a Souriau model of Lie group thermodynamics that recovers the symmetry broken by lack of covariance of Gibbs density in classical statistical mechanics with respect to dynamic groups action in physics (Galileo and Poincaré groups, sub-group of affine group). The ontological model of Souriau gives geometric status to (Planck) temperature (element of Lie alebra), heat (element of dual Lie algebra) and entropy. Souriau said in one of his papers [30] on this new "Lie group thermodynamics" that "*these formulas are universal, in that they do not involve the symplectic manifold, but only group G, the symplectic cocycle. Perhaps this Lie group thermodynamics could be of interest for mathematics*".

For this new covariant thermodynamics, the fundamental notion is the coadjoint orbit that is linked to positive definite KKS (Kostant–Kirillov–Souriau) 2-form [196]:

$$\omega_w(X, Y) = \langle w, [U, V] \rangle \text{ with } X = ad_w U \in T_w\mathrm{M} \text{ and } Y = ad_w V \in T_w\mathrm{M} \quad (272)$$

that is the Kähler-form of a G-invariant kähler structure compatible with the canonical complex structure of M, and determines a canonical symplectic structure on M. When the cocycle is equal to zero, the KKS and Souriau-Fisher metric are equal. This 2-form introduced by Jean-Marie Souriau is linked to the coadjoint action and the coadjoint orbits of the group on its moment space. Souriau provided a classification of the homogeneous symplectic manifolds with this moment map. The coadjoint representation of a Lie group G is the dual of the adjoint representation. If $\mathfrak{g}$ denotes the Lie algebra of G, the corresponding action of G on $\mathfrak{g}^*$, the dual space to $\mathfrak{g}$, is called the coadjoint action. Souriau proved based on the moment map that a symplectic manifold is always a coadjoint orbit, affine of its group of Hamiltonian transformations, deducing that coadjoint orbits are the universal models of symplectic manifolds: a symplectic manifold homogeneous under the action of a Lie group, is isomorphic, up to a covering, to a coadjoint orbit. So the link between Souriau-Fisher metric and KKS 2-form will provide a symplectic structure and foundation to information manifolds. For Souriau thermodynamics, the Souriau-Fisher metric is the canonical structure linked to KKS 2-form, modified by the cocycle (its symplectic leaves are the orbits of the affine action that makes equivariant the moment map). This last property allows us to determine all homogeneous spaces of a Lie group admitting an invariant symplectic structure by the action of this group: for example, there are the orbits of the coadjoint representation of this group or of a central extension of this group (the central extension allowing suppressing the cocycle). For affine coadjoint orbits, we make reference to Alice Tumpach Ph.D. [197–199] who has developed previous works of Neeb [200], Biquard and Gauduchon [201–204].

Other promising domains of research are theory of generating maps [205–208] and the link with Poisson geometry through affine Poisson group. As observed by Pierre Dazord [209] in his paper *"Groupe de Poisson Affines"*, the extension of a Poisson group to an affine Poisson group due to Drinfel'd [210] includes the affine structures of Souriau on dual Lie algebra. For an affine Poisson group, its universal covering could be identified to a vector space with an associated affine structure. If this vector space is an abelian affine Poisson group, we can find the affine structure of Souriau. For the abelian group $(R^3,+)$, affine Poisson groups are the affine structures of Souriau.

Souriau model of Lie group thermodynamics could be a promising way to achieve René Thom's dream to replace thermodynamics by geometry [211,212], and could be extended to the second order extension of the Gibbs state [213,214].

We could explore the links between *"stochastic mechanics"* (*mécanique alétoire*) developed by Jean-Michel Bismut based on Malliavin Calculus (stochastic calculus of variations) and Souriau "Lie group thermodynamics", especially to extend covariant Souriau Gibbs density on the stochastic symplectic manifold (e.g., to model centrifuge with random vibrating axe and the Gibbs density).

We have seen that Souriau has replaced classical Maximum Entropy approach by replacing Lagrange parameters by only one geometric "temperature vector" as element of Lie algebra. In parallel, as refered in [15], Ingarden has introduced [213,214] second and higher order temperature of the Gibbs state that could be extended to Souriau theory of thermodynamics. Ingarden higher order temperatures could be defined in the case when no variational is considered, but when a probability distribution depending on more than one parameter. It has been observed that Ingarden can fail if the following assumptions are not fulfilled: the number of components of the sum goes to infinity and the components of the sum are stochastically independent. Gibbs hypothesis can also fail if stochastic interactions with the environment are not sufficiently weak. In all these cases, we never observe absolute thermal equilibrium of Gibbs type but only flows or turbulence. Nonequilibrium thermodynamics could be indirectly addressed by means of the concept of high order temperatures. Momentum $Q = \frac{\partial \Phi(\beta)}{\partial \beta}$ should be replaced by higher order moments given by the relation $Q_k =$

$$\frac{\partial \Phi(\beta_1, ..., \beta_n)}{\partial \beta_k} = \frac{\int\limits_M U^k(\xi) \cdot e^{-\sum\limits_{k=1}^{n} \langle \beta_k, U^k(\xi) \rangle} d\omega}{\int\limits_M e^{-\sum\limits_{k=1}^{n} \langle \beta_k, U^k(\xi) \rangle} d\omega}$$ defined by extended Massieu characteristic function

$\Phi(\beta_1, ..., \beta_n) = -\log \int\limits_M e^{-\sum\limits_{k=1}^{n} \langle \beta_k, U^k(\xi) \rangle} d\omega$. Entropy is defined by Legendre transform of this Massieu characteristic function $S(Q_1, ..., Q_n) = \sum\limits_{k=1}^{n} \langle \beta_k, Q_k \rangle - \Phi(\beta_1, ..., \beta_n)$ where $\beta_k = \frac{\partial S(Q_1, ..., Q_n)}{\partial Q_k}$. We are able also to define high order thermal capacities given by $K_k = -\frac{\partial Q_k}{\partial \beta_k}$. The Gibbs density could be then extended with respect to high order temperatures by $p_{Gibbs}(\xi) = e^{\sum\limits_{k=1}^{n} \langle \beta_k, U^k(\xi) \rangle - \Phi(\beta_1, ..., \beta_n)} = \frac{e^{-\sum\limits_{k=1}^{n} \langle \beta_k, U^k(\xi) \rangle}}{\int\limits_M e^{-\sum\limits_{k=1}^{n} \langle \beta_k, U^k(\xi) \rangle} d\omega}$.

We also have to make reference to the works of Streater [16], Nencka [215] and Burdet [216]. Nencka and Streater [215], for certain unitary representations of a Lie algebra $\mathfrak{g}$, define the statistical manifold $\mathcal{M}$ of states as the convex cone of $X \in \mathfrak{g}$ for which the partition function $Z = Tr\left[\exp(-X)\right]$ is finite. The Hessian of $\log Z$ defines a Riemannian metric g on dual Lie algebra $\mathfrak{g}^*$. They observe that $\mathfrak{g}^*$ foliates into the union of coadjoint orbits, each of which can be given a complex Kostant structure (that of Kostant).

To conclude, we will make reference to Alain Berthoz [217] at College de France who has studied brain coding of movement. The most recent studies on this topic, by Alexandre Afgoustidis Ph.D. [218] "*Invariant Harmonic Analysis and Geometry in the Workings of the Brain*" supervised by Daniel Bennequin, Afgoustidis [218] consolidate the idea that brain vestibular channels and otolithes code Lie algebra of the homogeneous Galileo group as illustrated in the following Figure 15.

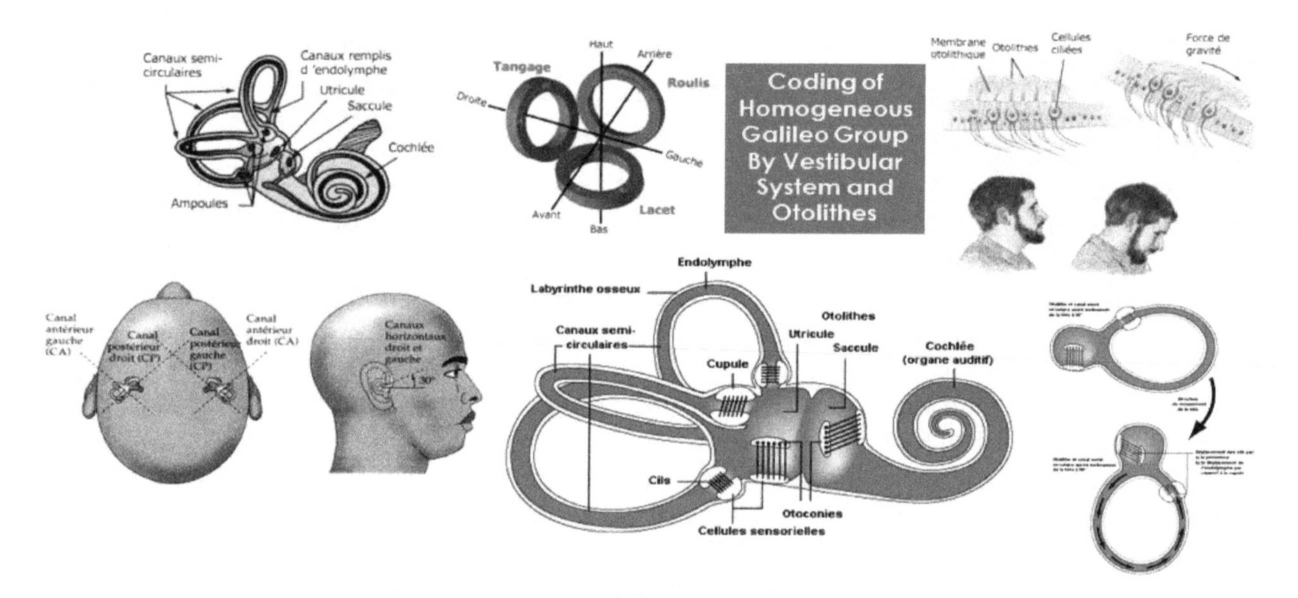

Figure 15. Coding of homogeneous Galileo algebra by vestibular system and otolithes.

Souriau gave the same ideas in this direction regarding how the brain could code invariants [219]:

> *Lorsque il y un tremblement de terre, nous assistons à la mort de l'Espace. ... Nous vivons avec nos habitudes que nous pensons universelles. ... La neuroscience s'occupe rarement de la géométrie ... Pour les singes qui vivent dans les arbres, certaines propriétés du groupe d'Euclide sont mieux câblées dans leurs cerveaux (When there is an earthquake, we are witnessing the death of Space ... We live with our habits that we think are universal.... Neuroscience rarely is interested in geometry ... For the monkeys that live in trees, some properties of the Euclid group are better coded in their brains).*

Souriau added anecdotes from a discussion with a student of Bohr that [220]:

> *L'élève demanda à Bohr qu'il ne comprenait pas le principe de correspondance. Bohr lui demanda de s'assoir et il tourna autour de lui. Bohr lui dit tu dois commencer à avoir mal au cœur, c'est que tu commences à comprendre ce qu'est le principe de correspondance (The student said to Bohr that* he *did not understand the principle of correspondence. Bohr asked him to sit and he turned around. Bohr said, you should start to be seasick, it is then that you begin to understand what the correspondence principle is.).*

Acknowledgments: I would like to thank Charles-Michel Marle and Gery de Saxcé for the fruitful discussions on Souriau model of statistical physics that help me to understand the fundamental notion of affine representation of Lie group and algebra, moment map and coadjoint orbits. I would also like to thank Michel Boyom that introduce me to Jean-Louis Koszul works on affine representation of Lie group and Lie algebra.

> *Si on ajoute que la critique qui accoutume l'esprit, surtout en matière de faits, à recevoir de simples probabilités pour des preuves, est, par cet endroit, moins propre à le former, que ne le doit être la géométrie qui lui fait contracter l'habitude de n'acquiescer qu'à l'évidence; nous répliquerons qu'à la rigueur on pourrait conclure de cette différence même, que la critique donne, au contraire, plus d'exercice à l'esprit que la géométrie: parce que l'évidence, qui est une et absolue, le fixe au premier aspect sans lui laisser ni la liberté de douter, ni le mérite de choisir; au lieu que les probabilités étant susceptibles du plus et du moins, il faut, pour se mettre en état de prendre un parti, les comparer ensemble, les discuter et les peser. Un genre d'étude qui rompt, pour ainsi dire,*

l'esprit à cette opération, est certainement d'un usage plus étendu que celui où tout est soumis à l'évidence; parce que les occasions de se déterminer sur des vraisemblances ou probabilités, sont plus fréquentes que celles qui exigent qu'on procède par démonstrations: pourquoi ne dirions –nous pas que souvent elles tiennent aussi à des objets beaucoup plus importants?

—Joseph de Maistre in L'Espit de Finesse [221]

Le cadavre qui s'acoutre se méconnait et imaginant l'éternité s'en approrie l'illusion ... C'est pourquoi j'abandonnerai ces frusques et jetant le masque de mes jours, je fuirai le temps où, de concert avec les autres, je m'éreinte à me trahir.

—Emile Cioran in Précis de decomposition [222]

Appendix A. Clairaut(-Legendre) Equation of Maurice Fréchet Associated to "Distinguished Functions" as Fundamental Equation of Information Geometry

Before Rao [223,224], in 1943, Maurice Fréchet [141] wrote a seminal paper introducing what was then called the Cramer-Rao bound. This paper contains in fact much more that this important discovery. In particular, Maurice Fréchet introduces more general notions relative to "distinguished functions", densities with estimator reaching the bound, defined with a function, solution of Clairaut's equation. The solutions "envelope of the Clairaut's equation" are equivalent to standard Legendre transform without convexity constraints but only smoothness assumption. This Fréchet's analysis can be revisited on the basis of Jean-Louis Koszul's works as a seminal foundation of "information geometry".

We will use Maurice Fréchet notations, to consider the estimator:

$$T = H(X_1, ..., X_n) \tag{A1}$$

and the random variable

$$A(X) = \frac{\partial \log p_\theta(X)}{\partial \theta} \tag{A2}$$

that are associated to:

$$U = \sum_i A(X_i) \tag{A3}$$

The normalizing constraint $\int_{-\infty}^{+\infty} p_\theta(x)dx = 1$ implies that: $\int_{-\infty}^{+\infty} ... \int_{-\infty}^{+\infty} \prod_i p_\theta(x_i)dx_i = 1$

If we consider the derivative if this last expression with respect to θ, then

$$\int_{-\infty}^{+\infty} ... \int_{-\infty}^{+\infty} \left[\sum_i A(x_i)\right] \prod_i p_\theta(x_i)dx_i = 0 \text{ gives : } E_\theta[U] = 0 \tag{A4}$$

Similarly, if we assume that $E_\theta[T] = \theta$, then $\int_{-\infty}^{+\infty} ... \int_{-\infty}^{+\infty} H(x_1, ..., x_n) \prod_i p_\theta(x_i)dx_i = \theta$, and we obtain by derivation with respect to θ:

$$E[(T - \theta)U] = 1 \tag{A5}$$

But as $E[T] = \theta$ and $E[U] = 0$, we immediately deduce that:

$$E[(T - E[T])(U - E[U])] = 1 \tag{A6}$$

From Schwarz inequality, we can develop the following relations:

$$\begin{gathered} [E(ZT)]^2 \leq E[Z^2] E[T^2] \\ 1 \leq E\left[(T - E[T])^2\right] E\left[(U - E[U])^2\right] = (\sigma_T \sigma_U)^2 \end{gathered} \tag{A7}$$

U being the summation of independent variables, Bienaymé equality could be applied:

$$(\sigma_U)^2 = \sum_i \left[\sigma_{A(X_i)}\right]^2 = n\,(\sigma_A)^2 \tag{A8}$$

From which, Fréchet deduced the bound, rediscovered by Cramer and Rao 2 years later:

$$(\sigma_T)^2 \geq \frac{1}{n\,(\sigma_A)^2} \tag{A9}$$

Fréchet [141] observed that it is a remarkable inequality where the second member is independent of the choice of the function H defining the "empirical value" T, where the first member can be taken to any empirical value $T = H\,(X_1, ..., X_n)$ subject to the unique condition $E_\theta\,[T] = \theta$ regardless is θ.

The classic condition that the Schwarz inequality becomes an equality helps us to determine when σ_T reaches its lower bound $\frac{1}{\sqrt{n}\sigma_n}$.

The previous inequality becomes an equality if there are two numbers α and β (not random and not both zero) such that $\alpha\,(H' - \theta) + \beta U = 0$, with H' being a particular function among eligible H such that we have an equality. This equality is rewritten $H' = \theta + \lambda' U$ with λ' being a non-random number.

If we use the previous equation, then:

$$E\,[(T - E\,[T])\,(U - E\,[U])] = 1 \Rightarrow E\,\left[(H' - \theta)\,U\right] = \lambda' E_\theta\left[U^2\right] = 1 \tag{A10}$$

We obtain:

$$U = \sum_i A\,(X_i) \Rightarrow \lambda' n E_\theta\left[A^2\right] = 1 \tag{A11}$$

From which we obtain λ' and the form of the associated estimator H':

$$\lambda' = \frac{1}{nE\,[A^2]} \Rightarrow H' = \theta + \frac{1}{nE\,[A^2]}\sum_i \frac{\partial \log p_\theta(X_i)}{\partial\theta} \tag{A12}$$

It is therefore deduced that the estimator that reaches the terminal is of the form:

$$H' = \theta + \frac{\sum_i \frac{\partial \log p_\theta(X_i)}{\partial\theta}}{n\int\limits_{-\infty}^{+\infty}\left[\frac{\partial p_\theta(x)}{\partial\theta}\right]^2 \frac{dx}{p_\theta(x)}} \tag{A13}$$

with $E\,[H'] = \theta + \lambda' E\,[U] = \theta$.

H' would be one of the eligible functions, if H' would be independent of θ. Indeed, if we consider $E_{\theta_0}\,[H'] = \theta_0$, $E\left[(H' - \theta_0)^2\right] \leq E_{\theta_0}\left[(H - \theta_0)^2\right]$ $\forall H$ such that $E_{\theta_0}\,[H] = \theta_0$.

$H = \theta_0$ satisfies the equation and inequality shows that it is almost certainly equal to θ_0.

So to look for θ_0, we should know beforehand θ_0.

At this stage, Fréchet [141] looked for "distinguished functions" ("densités distinguées" in French), as any probability density $p_\theta(x)$ such that the function:

$$h(x) = \theta + \frac{\frac{\partial \log p_\theta(x)}{\partial\theta}}{\int\limits_{-\infty}^{+\infty}\left[\frac{\partial p_\theta(x)}{\partial\theta}\right]^2 \frac{dx}{p_\theta(x)}} \tag{A14}$$

is independent of θ. The objective of Fréchet is then to determine the minimizing function $T = H'\,(X_1, ..., X_n)$ that reaches the bound. We can deduce from previous relations that:

$$\lambda(\theta)\frac{\partial \log p_\theta(x)}{\partial\theta} = h(x) - \theta \tag{A15}$$

But as $\lambda(\theta) > 0$, *we can consider* $\frac{1}{\lambda(\theta)}$ *as the second derivative of a function* $\Phi(\theta)$ such that:

$$\frac{\partial \log p_\theta(x)}{\partial \theta} = \frac{\partial^2 \Phi(\theta)}{\partial \theta^2}\left[h(x) - \theta\right] \tag{A16}$$

From which we deduce that:

$$\ell(x) = \log p_\theta(x) - \frac{\partial \Phi(\theta)}{\partial \theta}\left[h(x) - \theta\right] - \Phi(\theta) \tag{A17}$$

Is an independent quantity of θ. *A distinguished function* will be then given by:

$$p_\theta(x) = e^{\frac{\partial \Phi(\theta)}{\partial \theta}[h(x)-\theta]+\Phi(\theta)+\ell(x)} \tag{A18}$$

With the normalizing constraint $\int\limits_{-\infty}^{+\infty} p_\theta(x)dx = 1$.

These two conditions are sufficient. Indeed, reciprocally, let three functions $\Phi(\theta)$, $h(x)$ and $\ell(x)$ that we have, for any

$$\theta : \int\limits_{-\infty}^{+\infty} e^{\frac{\partial \Phi(\theta)}{\partial \theta}[h(x)-\theta]+\Phi(\theta)+\ell(x)} dx = 1 \tag{A19}$$

Then the function is distinguished:

$$\theta + \frac{\frac{\partial \log p_\theta(x)}{\partial \theta}}{\int\limits_{-\infty}^{+\infty} \left[\frac{\partial p_\theta(x)}{\partial \theta}\right]^2 \frac{dx}{p_\theta(x)}} = \theta + \lambda(x)\frac{\partial^2 \Phi(\theta)}{\partial \theta^2}\left[h(x) - \theta\right] \tag{A20}$$

$$\text{If } \lambda(x)\frac{\partial^2 \Phi(\theta)}{\partial \theta^2} = 1, \text{ when } \frac{1}{\lambda(x)} = \int\limits_{-\infty}^{+\infty} \left[\frac{\partial \log p_\theta(x)}{\partial \theta}\right]^2 p_\theta(x)dx = (\sigma_A)^2 \tag{A21}$$

The function is reduced to $h(x)$ and then is not dependent of θ.

We have then the following relation:

$$\frac{1}{\lambda(x)} = \int\limits_{-\infty}^{+\infty} \left(\frac{\partial^2 \Phi(\theta)}{\partial \theta^2}\right)^2 \left[h(x) - \theta\right]^2 e^{\frac{\partial \Phi(\theta)}{\partial \theta}(h(x)-\theta)+\Phi(\theta)+\ell(x)} dx \tag{A22}$$

The relation is valid for any θ, we can derive prefious equation with respect with θ:

$$\int\limits_{-\infty}^{+\infty} e^{\frac{\partial \Phi(\theta)}{\partial \theta}(h(x)-\theta)+\Phi(\theta)+\ell(x)} \left(\frac{\partial^2 \Phi(\theta)}{\partial \theta^2}\right)\left[h(x) - \theta\right] dx = 0 \tag{A23}$$

We can divide by $\frac{\partial^2 \Phi(\theta)}{\partial \theta^2}$ because it does not depend on x.

If we derive again with respect to θ, we will have:

$$\int\limits_{-\infty}^{+\infty} e^{\frac{\partial \Phi(\theta)}{\partial \theta}(h(x)-\theta)+\Phi(\theta)+\ell(x)} \left(\frac{\partial^2 \Phi(\theta)}{\partial \theta^2}\right)\left[h(x) - \theta\right]^2 dx = \int\limits_{-\infty}^{+\infty} e^{\frac{\partial \Phi(\theta)}{\partial \theta}(h(x)-\theta)+\Phi(\theta)+\ell(x)} dx = 1 \tag{A24}$$

Combining this relation with that of $\frac{1}{\lambda(x)}$, we can deduce that $\lambda(x)\frac{\partial^2\Phi(\theta)}{\partial\theta^2} = 1$ and as $\lambda(x) > 0$ then $\frac{\partial^2\Phi(\theta)}{\partial\theta^2} > 0$.

Fréchet emphasizes at this step [141], another way to approach the problem. We can select arbitrarily $h(x)$ and $l(x)$ and then $\Phi(\theta)$ is determined by:

$$\int_{-\infty}^{+\infty} e^{\frac{\partial\Phi(\theta)}{\partial\theta}[h(x)-\theta]+\Phi(\theta)+\ell(x)}dx = 1 \tag{A25}$$

That could be rewritten:

$$e^{\theta.\frac{\partial\Phi(\theta)}{\partial\theta}-\Phi(\theta)} = \int_{-\infty}^{+\infty} e^{\frac{\partial\Phi(\theta)}{\partial\theta}h(x)+\ell(x)}dx \tag{A26}$$

If we then fixed arbitrarily $h(x)$ and $l(x)$ and let s an arbitrary variable, the following function will be an explicit positive function given by $e^{\Psi(s)}$:

$$\int_{-\infty}^{+\infty} e^{s.h(x)+\ell(x)}dx = e^{\Psi(s)} \tag{A27}$$

Fréchet obtained finally the function $\Phi(\theta)$ *as solution of the equation* [141]:

$$\Phi(\theta) = \theta \cdot \frac{\partial\Phi(\theta)}{\partial\theta} - \Psi\left(\frac{\partial\Phi(\theta)}{\partial\theta}\right) \tag{A28}$$

Fréchet noted that this is the Alexis Clairaut equation [141].

The case $\frac{\partial\Phi(\theta)}{\partial\theta} = cste$ would reduce the density to a function that would be independent of θ, and so $\Phi(\theta)$ is given by a singular solution of this Clairaut equation, which is unique and could be computed by eliminating the variable s between:

$$\Phi = \theta \cdot s - \Psi(s) \text{ and } \theta = \frac{\partial\Psi(s)}{\partial s} \tag{A29}$$

Or between:

$$e^{\theta\cdot s-\Phi(\theta)} = \int_{-\infty}^{+\infty} e^{s\cdot h(x)+\ell(x)}dx \text{ and } \int_{-\infty}^{+\infty} e^{s\cdot h(x)+\ell(x)}[h(x)-\theta]\,dx = 0 \tag{A30}$$

$\Phi(\theta) = -\log \int_{-\infty}^{+\infty} e^{s\cdot h(x)+\ell(x)}dx + \theta \cdot s$ where s is given implicitly by $\int_{-\infty}^{+\infty} e^{s\cdot h(x)+\ell(x)}[h(x)-\theta]\,dx = 0$.

Then we know the distinguished function, H' among functions $H(X_1, ..., X_n)$ verifying $E_\theta[H] = \theta$ and such that σ_H reaches for each value of θ, an absolute minimum, equal to $\frac{1}{\sqrt{n}\sigma_A}$.

For the previous equation:

$$h(x) = \theta + \frac{\frac{\partial \log p_\theta(x)}{\partial\theta}}{\int_{-\infty}^{+\infty}\left[\frac{\partial p_\theta(x)}{\partial\theta}\right]^2 \frac{dx}{p_\theta(x)}} \tag{A31}$$

We can rewrite the estimator as:

$$H'(X_1, ..., X_n) = \frac{1}{n}\left[h(X_1) + ... + h(X_n)\right] \tag{A32}$$

and compute the associated empirical value:

$$t = H'(x_1, ..., x_n) = \frac{1}{n}\sum_i h(x_i) = \theta + \lambda(\theta)\sum_i \frac{\partial \log p_\theta(x_i)}{\partial \theta}$$

If we take $\theta = t$, we have as $\lambda(\theta) > 0$:

$$\sum_i \frac{\partial \log p_t(x_i)}{\partial t} = 0 \tag{A33}$$

When $p_\theta(x)$ is a distinguished function, the empirical value t of θ corresponding to a sample $x_1, ..., x_n$ is a root of previous equation in t. This equation has a root and only one when X is a distinguished variable. Indeed, as we have:

$$p_\theta(x) = e^{\frac{\partial \Phi(\theta)}{\partial \theta}[h(x)-\theta]+\Phi(\theta)+\ell(x)} \tag{A34}$$

$$\sum_i \frac{\partial \log p_t(x_i)}{\partial t} = \frac{\partial^2 \Phi(t)}{\partial t^2}\left[\frac{\sum_i h(x_i)}{n} - t\right] \text{ with } \frac{\partial^2 \Phi(t)}{\partial t^2} > 0 \tag{A35}$$

We can then recover the unique root: $t = \frac{\sum_i h(x_i)}{n}$.

This function $T \equiv H'(X_1, ..., X_n) = \frac{1}{n}\sum_i h(X_i)$ can have an arbitrary form, that is a sum of functions of each only one of the quantities and it is even the arithmetic average of N values of a same auxiliary random variable $Y = h(X)$. The dispersion is given by:

$$(\sigma_{T_n})^2 = \frac{1}{n(\sigma_A)^2} = \frac{1}{n\int_{-\infty}^{+\infty}\left[\frac{\partial p_\theta(x)}{\partial \theta}\right]^2 \frac{dx}{p_\theta(x)}} = \frac{1}{n\frac{\partial^2 \Phi(\theta)}{\partial \theta^2}} \tag{A36}$$

and T_n follows the probability density:

$$p_\theta(t) = \sqrt{n}\frac{1}{\sigma_A\sqrt{2\pi}}e^{-\frac{n(t-\theta)^2}{2\cdot\sigma_A^2}} \text{ with } (\sigma_A)^2 = \frac{\partial^2 \Phi(\theta)}{\partial \theta^2} \tag{A37}$$

Clairaut Equation and Legendre Transform

We have just observed that Fréchet shows that distinguished functions depend on a function $\Phi(\theta)$, solution of the Clairaut equation:

$$\Phi(\theta) = \theta \cdot \frac{\partial \Phi(\theta)}{\partial \theta} - \Psi\left(\frac{\partial \Phi(\theta)}{\partial \theta}\right) \tag{A38}$$

Or given by the Legendre transform:

$$\Phi = \theta \cdot s - \Psi(s) \text{ and } \theta = \frac{\partial \Psi(s)}{\partial s} \tag{A39}$$

Fréchet also observed that this function $\Phi(\theta)$ could be rewritten:
$\Phi(\theta) = -\log \int_{-\infty}^{+\infty} e^{s\cdot h(x)+\ell(x)}dx + \theta \cdot s$ where s is given implicitly by $\int_{-\infty}^{+\infty} e^{s\cdot h(x)+\ell(x)}\left[h(x) - \theta\right]dx = 0$.

This equation is the fundamental equation of information geometry.

The "Legendre" transform was introduced by Adrien-Marie Legendre in 1787 [225] to solve a minimal surface problem Gaspard Monge in 1784. Using a result of Jean Baptiste Meusnier, a student of Monge, it solves the problem by a change of variable corresponding to the transform which now entitled with his name. Legendre wrote: *"I have just arrived by a change of variables that can be useful in other occasions."* About this transformation, Darboux [226] in his book gives an interpretation of Chasles: *"This comes after a comment by Mr. Chasles, to substitute its polar reciprocal on the surface compared to a paraboloïd."* The equation of Clairaut was introduced 40 years earlier in 1734 by Alexis Clairaut [225]. Solutions "envelope of the Clairaut equation" are equivalent to the Legendre transform with unconditional convexity, but only under differentiability constraint. Indeed, for a non-convex function, Legendre transformation is not defined where the Hessian of the function is canceled, so that the equation of Clairaut only makes the hypothesis of differentiability. The portion of the strictly convex function g in Clairaut equation $y = px - g(p)$ to the function f giving the envelope solutions by the formula $y = f(x)$ is precisely the Legendre transformation. The approach of Fréchet may be reconsidered in a more general context on the basis of the work of Jean-Louis Koszul.

Appendix B. Balian Gauge Model of Thermodynamics and its Compliance with Souriau Model

Supported by Industial group TOTAL (previously Elf-Aquitaine), Roger Balian has introduced a Gauge theory of thermodynamics [103] and has also developed information geometry in statistical physics and quantum physics [103,227–235]. Balian has observed that the entropy S (we use Balian notation, contrary with previous section where we use $-S$ as neg-entropy) can be regarded as an extensive variable $q^0 = S(q^1, ..., q^n)$, with $q^i (i = 1, ..., n)$, n independent quantities, usually extensive and conservative, characterizing the system. The n intensive variables γ_i are defined as the partial derivatives:

$$\gamma_i = \frac{\partial S(q^1, ..., q^n)}{\partial q^i} \tag{B1}$$

Balian has introduced a non-vanishing gauge variable p_0, without physical relevance, which multiplies all the intensive variables, defining a new set of variables:

$$p_i = -p_0 . \gamma_i \, , \; i = 1, ..., n \tag{B2}$$

The $2n + 1$-dimensional space is thereby extended into a $2n + 2$-dimensional thermodynamic space T spanned by the variables $p_i \, , q^i$ with $i = 0, 1, ..., n$, where the physical system is associated with a $n + 1$-dimensional manifold M in T, parameterized for instance by the coordinates $q^1, ..., q^n$ and p_0. A gauge transformation which changes the extra variable p_0 while keeping the ratios $p_i / p_0 = -\gamma_i$ invariant *is not observable*, so that a state of the system is represented by any point of a one-dimensional ray lying in M, along which the physical variables $q^0, ..., q^n, \gamma_1, ..., \gamma_n$ are fixed. Then, the relation between contact and canonical transformations is a direct outcome of this gauge invariance: the contact structure $\tilde{\omega} = dq^0 - \sum_{i=1}^{n} \gamma_i \cdot dq^i$ in $n + 1$ dimension can be embedded into a symplectic structure in $2n + 2$ dimension, with 1-form:

$$\omega = \sum_{i=0}^{n} p_i \cdot dq^i \tag{B3}$$

as *symplectization*, with geometric interpretation in the theory of fiber bundles.

The $n + 1$-dimensional thermodynamic manifolds M are characterized by the vanishing of this form $\omega = 0$. The 1-form induces then *a symplectic structure on* T:

$$d\omega = \sum_{i=0}^{n} dp_i \wedge dq^i \tag{B4}$$

Any thermodynamic manifold M belongs to the set of the so-called Lagrangian manifolds in T, which are the integral submanifolds of $d\omega$ with maximum dimension $(n+1)$. Moreover, M *is gauge invariant*, which is implied by $\omega = 0$. The extensivity of the entropy function $S\left(q^1, \ldots, q^n\right)$ is expressed by the Gibbs-Duhem relation $S = \sum_{i=1}^{n} q^i \frac{\partial S}{\partial q^i}$, rewritten with previous relation $\sum_{i=0}^{n} p_i q^i = 0$, defining a $2n+1$-dimensional extensivity sheet in T, where the thermodynamic manifolds M should lie. Considering an infinitesimal canonical transformation, generated by the Hamiltonian $h(q^0, q^1, \ldots, q^n, p_0, p_1, \ldots, p_n)$, $\dot{q}_i = \frac{\partial h}{\partial p_i}$ and $\dot{p}_i = \frac{\partial h}{\partial q^i}$, the Hamilton's equations are given by Poisson bracket:

$$\dot{g} = \{g, h\} = \sum_{i=0}^{n} \frac{\partial g}{\partial q^i}\frac{\partial h}{\partial p_i} - \frac{\partial h}{\partial q_i}\frac{\partial g}{\partial p_i} \tag{B5}$$

The concavity of the entropy $S\left(q^1, \ldots, q^n\right)$, as function of the extensive variables, expresses the stability of equilibrium states. This property produces constraints on the physical manifolds M in the $2n+2$-dimensional space. It entails the existence of a metric structure in the n-dimensional space q_i relying on the quadratic form:

$$ds^2 = -d^2 S = -\sum_{i,j=1}^{n} \frac{\partial^2 S}{\partial q^i \partial q^j} dq^i dq^j \tag{B6}$$

which defines a distance between two neighboring thermodynamic states.

$$\text{As } d\gamma_i = \sum_{j=1}^{n} \frac{\partial^2 S}{\partial q^i \partial q^j} dq^j, \text{ then: } ds^2 = -\sum_{i=1}^{n} d\gamma_i dq_i = \frac{1}{p_0}\sum_{i=0}^{n} dp_i dq^i \tag{B7}$$

The factor $1/p_0$ *ensures gauge invariance*. In a continuous transformation generated by h, the metric evolves according to:

$$\frac{d}{d\tau}(ds^2) = \frac{1}{p_0}\frac{\partial h}{\partial q^0} ds^2 + \frac{1}{p_0}\sum_{i,j=0}^{n}\left(\frac{\partial^2 h}{\partial q^i \partial p_j} dp_i dp_j - \frac{\partial^2 h}{\partial q^i \partial q^j} dq^i dq^j\right) \tag{B8}$$

We can observe that *this gauge theory of thermodynamics is compatible with Souriau Lie groupTthermodynamics*, where we have to consider the Souriau vector $\beta = \begin{bmatrix} \gamma_1 \\ \vdots \\ \gamma_n \end{bmatrix}$, transformed in a new vector:

$$p_i = -p_0 . \gamma_i,\ p = \begin{bmatrix} -p_0\gamma_1 \\ \vdots \\ -p_0\gamma_n \end{bmatrix} = -p_0 \cdot \beta \tag{B9}$$

Appendix C. Casalis-Letac Affine Group Invariance for Natural Exponential Families

The characterization of the natural exponential families of R^d which are preserved by a group of affine transformations has been examined by Muriel Casalis in her Ph.D. [173] and her different papers [172,174–178]. Her method has consisted of translating the invariance property of the family into a property concerning the measures which generate it, and to characterize such measures.

Let E a vector space of finite size, E^* its dual. $\langle \theta, x \rangle$ duality bracket with $(\theta, x) \in E^* \times E$. μ positive Radon measure on E, Laplace transform is:

$$L_\mu : E^* \to [0, \infty] \text{ with } \theta \mapsto L_\mu(\theta) = \int_E e^{\langle \theta, x \rangle} \mu(dx) \tag{C1}$$

Let transformation $k_\mu(\theta)$ defined on $\Theta(u)$ interior of $D_\mu = \{\theta \in E^*, L_\mu < \infty\}$:

$$k_\mu(\theta) = \log L_\mu(\theta) \tag{C2}$$

natural exponential families are given by:

$$F(\mu) = \left\{P(\theta,\mu)(dx) = e^{\langle\theta,x\rangle - k_\mu(\theta)}\mu(dx), \theta \in \Theta(\mu)\right\} \tag{C3}$$

with injective function (domain of means):

$$k'_\mu(\theta) = \int_E xP(\theta,\mu)\,\mu(dx) \tag{C4}$$

the inverse function:

$$\psi_\mu : M_F \to \Theta(\mu) \text{ with } M_F = \mathrm{Im}\left(k'_\mu\left(\Theta(\mu)\right)\right) \tag{C5}$$

and the Covariance operator:

$$V_F(m) = k''_\mu\left(\psi_\mu(m)\right) = \left(\psi'_\mu(m)\right)^{-1}, \; m \in M_F \tag{C6}$$

Measure generetad by a family F is then given by:

$$F(\mu) = F(\mu') \Leftrightarrow \exists (a,b) \in E^* \times R, \text{such that } \mu'(dx) = e^{\langle a,x\rangle + b}\mu(dx) \tag{C7}$$

Let F an exponential family of E generated by μ and $\varphi : x \mapsto g_\varphi x + v_\varphi$ with $g_\varphi \in GL(E)$ automorphisms of E and $v_\varphi \in E$, then the family $\varphi(F) = \{\varphi(P(\theta,\mu)), \theta \in \Theta(\mu)\}$ is an exponential familly of E generated by $\varphi(\mu)$

Definition C1. *An exponential family F is invariant by a group G (affine group of E), if*

$$\forall \varphi \in G, \varphi(F) = F : \; \forall \mu, F(\varphi(\mu)) = F(\mu) \tag{C8}$$

(the contrary could be false)

Then Muriel Casalis has established the following theorem:

Theorem C1 (Casalis). *Let $F = F(\mu)$ an exponential family of E and G affine group of E, then F is invariant by G if and only:*

$$\exists a : G \to E^*, \exists b : G \to R, \text{ such that:}$$

$$\forall (\varphi,\varphi') \in G^2, \begin{cases} a(\varphi\varphi') = {}^t_g{}^{-1}_\varphi a(\varphi') + a(\varphi) \\ b(\varphi\varphi') = b(\varphi) + b(\varphi') - \left\langle a(\varphi'), g_\varphi^{-1} v_\varphi \right\rangle \end{cases} \tag{C9}$$

$$\forall \varphi \in G, \varphi(\mu)(dx) = e^{\langle a(\varphi),x\rangle + b(\varphi)}\mu(dx)$$

When G is a linear subgroup, b is a character of G and a could be obtained by the help of cohomology of Lie groups.

If we define action of G on E^* by:

$$g \cdot x = {}^t_g{}^{-1} x, g \in G, x \in E^* \tag{C10}$$

It can be verified that:

$$a(g_1 g_2) = g_1 \cdot a(g_2) + a(g_1) \tag{C11}$$

the action a is an inhomogeneous 1-cocycle:

$\forall n > 0$, let the set of all functions from G^n to E^*, $\Im(G^n, E^*)$ called inhomogenesous n-cochains, then we can define the operators $d^n : \Im(G^n, E^*) \rightarrow \Im(G^{n+1}, E^*)$ by:

$$d^n F(g_1, \cdots, g_{n+1}) = g_1.F(g_2, \cdots, g_{n+1}) + \sum_{i=1}^{n} (-1)^i F(g_1, g_2, \cdots, g_i g_{i+1}, \cdots, g_n) + (-1)^{n+1} F(g_1, g_2, \cdots, g_n) \tag{C12}$$

Let $Z^n(G, E^*) = Ker(d^n), B(G, E^*) = Im(d^{n-1})$, with Z^n inhomogneous n-cocycles, the quotient:

$$H^n(G, E^*) = Z^n(G, E^*) / B^n(G, E^*) \tag{C13}$$

is the Cohomology group of G with value in E^*. We have:

$$\begin{aligned} d^0 : E^* &\rightarrow \Im(G, E^*) \\ x &\mapsto (g \mapsto g \cdot x - x) \end{aligned} \tag{C14}$$

$$Z^0 = \{x \in E^*; g \cdot x = x, \forall g \in G\} \tag{C15}$$

$$\begin{aligned} d^1 &: \Im(G, E^*) \rightarrow \Im(G^2, E^*) \\ F &\mapsto d^1 F,\ d^1 F(g_1, g_2) = g_1 \cdot F(g_2) - F(g_1 g_2) + F(g_1) \end{aligned} \tag{C16}$$

$$Z^1 = \left\{F \in \Im(G, E^*); F(g_1 g_2) = g_1 \cdot F(g_2) + F(g_1), \forall (g_1, g_2) \in G^2\right\} \tag{C17}$$

$$B^1 = \{F \in \Im(G, E^*); \exists x \in E^*, F(g) = g \cdot x - x\} \tag{C18}$$

When the Cohomology group $H^1(G, E^*) = 0$ then:

$$Z^1(G, E^*) = B^1(G, E^*) \tag{C19}$$

Then if $F = F(\mu)$ is an exponential family invariant by G, μ verifies:

$$\forall g \in G, g(\mu)(dx) = e^{\langle c,x\rangle - \langle c, g^{-1}x\rangle + b(g)} \mu(dx) \tag{C20}$$

$$\forall g \in G, g\left(e^{\langle c,x\rangle} \mu(dx)\right) = e^{b(g)} e^{\langle c,x\rangle} \mu(dx) \text{ with } \mu_0(dx) = e^{\langle c,x\rangle} \mu(dx) \tag{C21}$$

For all compact group, $H^1(G, E^*) = 0$ and we can express a:

$$\begin{aligned} A &: G \rightarrow GA(E) \\ g &\mapsto A_g,\ A_g(\theta) = {}^t g^{-1} \theta + a(g) \end{aligned} \tag{C22}$$

$$\begin{gathered} \forall (g, g') \in G^2, A_{gg'} = A_g A_{g'} \\ A(G) \text{ compact sub} - \text{group of } GA(E) \end{gathered} \tag{C23}$$

$$\exists \text{fixed point} \Rightarrow \forall g \in G, A_g(c) = {}^t g^{-1} c + a(g) = c \Rightarrow a(g) = \left(I_d - {}^t g^{-1}\right) c \tag{C24}$$

References

1. Bernard, C. Introduction à l'Étude de la Médecine Expérimentale. Available online: http://classiques.uqac.ca/classiques/bernard_claude/intro_etude_medecine_exp/intro_medecine_exper.pdf (accessed on 17 October 2016).
2. Thom, R. *Logos et Théorie des Catastrophes*; Editions Patiño: Genève, Switzerland, 1988.

3. Barbaresco, F. Symplectic structure of information geometry: Fisher metric and Euler-Poincaré equation of souriau lie group thermodynamics. In *Geometric Science of Information, Second International Conference GSI 2015*; Nielsen, F., Barbaresco, F., Eds.; Springer: Berlin/Heidelberg, Germany, 2015; Volume 9389, pp. 529–540.
4. De Saxcé, G.; Vallée, C. *Galilean Mechanics and Thermodynamics of Continua*; Wiley-ISTE: London, UK, 2016.
5. Vallée, C. Relativistic thermodynamics of continua. *Int. J. Eng. Sci.* **1981**, *19*, 589–601. [CrossRef]
6. Vallée, C.; Lerintiu, C. Convex analysis and entropy calculation in statistical mechanics. *Proc. A Razmadze Math. Inst.* **2005**, *137*, 111–129.
7. Marle, C.M. From Tools in Symplectic and Poisson Geometry to J.-M. Souriau's Theories of Statistical Mechanics and Thermodynamics. *Entropy* **2016**, *18*, 370. [CrossRef]
8. De Saxcé, G. Link between lie group statistical mechanics and thermodynamics of continua. In *Special Issue MDPI Entropy "Differential Geometrical Theory of Statistics"*; MDPI: Basel, Switzerland, 2016; Volume 18, p. 254.
9. Barbaresco, F. Koszul information geometry and souriau geometric temperature/capacity of lie group thermodynamics. *Entropy* **2014**, *16*, 4521–4565. [CrossRef]
10. Souriau, J.M. *Structure des Systèmes Dynamiques*; Editions Jacques Gabay: Paris, France, 1970. (In French)
11. Souriau, J.M. Structure of Dynamical Systems, volume 149 of Progress in Mathematics. In *A Symplectic View of Physics*; Birkhäuser: Basel, Switzerland, 1997.
12. Nielsen, F.; Barbaresco, F. *Geometric Science of Information*; Springer: Berlin/Heidelberg, Germany, 2015.
13. Kosmann-Schwarzbach, Y. La géométrie de Poisson, création du XXième siècle. In *Siméon-Denis Poisson*; Ecole Polytechnique: Paris, France, 2013; pp. 129–172.
14. Bismut, J.M. *Mécanique Aléatoire*; Springer: Berlin/Heidelberg, Germany, 1981; Volume 866.
15. Casas-Vázquez, J.; Jou, D. Temperature in non-equilibrium states: A review of open problems and current proposals. *Rep. Prog. Phys.* **2003**, *66*, 1937–2023. [CrossRef]
16. Streater, R.F. The information manifold for relatively bounded potentials. *Tr. Mat. Inst. Steklova* **2000**, *228*, 217–235.
17. Arnold, V.I. Sur la géométrie différentielle des groupes de Lie de dimension infinie et ses applications à l'hydrodynamique des fluides parfaits. *Ann. Inst. Fourier* **1966**, *16*, 319–361. [CrossRef]
18. Arnold, V.I.; Givental, A.B. Symplectic geometry. In *Dynamical Systems IV: Symplectic Geometry and Its Applications, Encyclopaedia of Mathematical Sciences*; Arnol'd, V.I., Novikov, S.P., Eds.; Springer: Berlin, Germany, 1990; Volume 4, pp. 1–136.
19. Marle, C.M.; de Saxcé, G.; Vallée, C. L'oeuvre de Jean-Marie Souriau, Gazette de la SMF, Hommage à Jean-Marie Souriau. 2012. Published by SMF, Paris.
20. Patrick Iglesias, Itinéraire d'un Mathématicien: Un entretien Avec Jean-Marie Souriau, Le Journal de Maths des Elèves. Available online: http://www.lutecium.fr/jp-petit/science/gal_port/interview_Souriau.pdf (accessed on 27 October 2016). (In French)
21. Iglesias, P. *Symétries et Moment*; Hermann: Paris, France, 2000.
22. Kosmann-Schwarzbach, Y. *Groupes et Symmetries*; Ecole Polytechnique: Paris, France, 2006.
23. Kosmann-Schwarzbach, Y. En homage à Jean-Marie Souriau, quelques souvenirs. *Gazette des Mathématiciens* **2012**, *133*, 105–106.
24. Ghys, E. Actions localement libres du groupe affine. *Invent. Math.* **1985**, *82*, 479–526. [CrossRef]
25. Rais, M. La representation coadjointe du groupe affine. *Ann. Inst. Fourier* **1978**, *28*, 207–237. (In French) [CrossRef]
26. Souriau, J.M. Mécanique des états condensés de la matière. In Proceedings of the 1st International Seminar of Mechanics Federation of Grenoble, Grenoble, France, 19–21 May 1992. (In French)
27. Souriau, J.M. Géométrie de l'espace de phases. *Commun. Math. Phys.* **1966**, *374*, 1–30. (In French)
28. Souriau, J.M. Définition covariante des équilibres thermodynamiques. *Nuovo Cimento* **1966**, *1*, 203–216. (In French)
29. Souriau, J.M. *Mécanique Statistique, Groupes de Lie et Cosmologie*; Colloques Internationaux du CNRS Numéro 237: Paris, France, 1974; pp. 59–113. (In French)
30. Souriau, J.M. *Géométrie Symplectique et Physique Mathématique*; Éditions du C.N.R.S.: Paris, France, 1975. (In French)
31. Souriau, J.M. *Thermodynamique Relativiste des Fluides*; Centre de Physique Théorique: Marseille, France, 1977. (In French)

32. Souriau, J.M. *Interpretation Géometrique des Etatsquantiques*; Springer: Berlin/Heidelberg, Germany, 1977; Volume 570. (In French)
33. Souriau, J.M. Thermodynamique et géométrie. In *Differential Geometrical Methods in Mathematical Physics II*; Bleuler, K., Reetz, A., Petry, H.R., Eds.; Springer: Berlin/Heidelberg, Germany, 1978; pp. 369–397. (In French)
34. Souriau, J.M. *Dynamic Systems Structure*; Chapters 16–19; Unpublished work, 1980.
35. Souriau, J.M.; Iglesias, P. *Heat Cold and Geometry. Differential Geometry and Mathematical Physics, Mathematical Physics Studies Volume*; Springer: Amsterdam, The Netherlands, 1983; pp. 37–68.
36. Souriau, J.M. Mécanique classique et géométrie symplectique. CNRS Marseille. *Cent. Phys. Théor.* Report ref. CPT-84/PE-1695 1984. (In French)
37. Souriau, J.M. On Geometric Mechanics. *Discret. Cont. Dyn. Syst. J.* **2007**, *19*, 595–607. [CrossRef]
38. Laplace, P.S. Mémoire sur la probabilité des causes sur les évènements. In *Mémoires de Mathématique et de Physique*; De l'Imprimerie Royale: Paris, France, 1774. (In French)
39. Gibbs, J.W. Elementary principles in statistical mechanics. In *The Rational Foundation of Thermodynamics*; Scribner: New York, NY, USA, 1902.
40. Ruelle, D.P. *Thermodynamic Formalism*; Addison-Wesley: New York, NY, USA, 1978.
41. Ruelle, D.P. Extending the definition of entropy to nonequilibrium steady states. *Proc. Natl. Acad. Sci. USA* **2003**, *100*, 3054–3058. [CrossRef] [PubMed]
42. Jaynes, E.T. Information theory and statistical mechanics. *Phys. Rev.* **1957**, *106*, 620–630. [CrossRef]
43. Jaynes, E.T. Information theory and statistical mechanics II. *Phys. Rev.* **1957**, *108*, 171–190. [CrossRef]
44. Jaynes, E.T. Prior probabilities. *IEEE Trans. Syst. Sci. Cybern.* **1968**, *4*, 227–241. [CrossRef]
45. Jaynes, E.T. The well-posed problem. *Found. Phys.* **1973**, *3*, 477–493. [CrossRef]
46. Jaynes, E.T. Where do we stand on maximum entropy? In *The Maximum Entropy Formalism*; Levine, R.D., Tribus, M., Eds.; MIT Press: Cambridge, MA, USA, 1979; pp. 15–118.
47. Jaynes, E.T. The minimum entropy production principle. *Annu. Rev. Phys. Chem.* **1980**, *31*, 579–601. [CrossRef]
48. Jaynes, E.T. On the rationale of maximum entropy methods. *IEEE Proc.* **1982**, *70*, 939–952. [CrossRef]
49. Jaynes, E.T. *Papers on Probability, Statistics and Statistical Physics*; Reidel: Dordrecht, The Netherlands, 1982.
50. Ollivier, Y. Aspects de l'entropie en Mathématiques et en Physique (Théorie de l'information, Systèmes Dynamiques, Grandes Déviations, Irréversibilité). Available online: http://www.yann-ollivier.org/entropie/entropie.pdf (accessed on 7 August 2015). (In French)
51. Villani, C. (Ir)rréversibilité et Entropie. Available online: http://www.bourbaphy.fr/villani.pdf (accessed on 5 August 2015). (In French)
52. Godement, R. *Introduction à la Théorie des Groupes de Lie*; Springer: Berlin/Heidelberg, Germany, 2004.
53. Guichardet, A. *Cohomologie des Groups Topologiques et des Algèbres de Lie*; Cedic/Fernand Nathan: Paris, France, 1980.
54. Guichardet, A. La method des orbites: Historiques, principes, résultats. In *Leçons de Mathématiques D'aujourd'hui*; Cassini: Paris, France, 2010; Volume 4, pp. 33–59.
55. Guichardet, A. *Le Problème de Kepler, Histoire et Théorie*; Ecole Polytechnique: Paris, France, 2012.
56. Dubois, J.G.; Dufour, J.P. La théorie des catastrophes. V. Transformées de Legendre et thermodynamique. In *Annales de l'IHP Physique Théorique*; Institut Henri Poincaré: Paris, France, 1978; Volume 29, pp. 1–50.
57. Monge, G. *Sur le Calcul Intégral des Equations aux Différences Partielles*; Mémoires de l'Académie des Sciences: Paris, France, 1784; pp. 118–192. (In French)
58. Moreau, J.J. Fonctions convexes duales et points proximaux dans un espace hilbertien. *C. R. Acad. Sci.* **1962**, *255*, 2897–2899. (In French)
59. Plastino, A.; Plastino, A.R. On the Universality of thermodynamics' Legendre transform structure. *Phys. Lett. A* **1997**, *226*, 257–263. [CrossRef]
60. Friedrich, T. Die fisher-information und symplectische strukturen. *Math. Nachr.* **1991**, *153*, 273–296. (In German) [CrossRef]
61. Massieu, F. Sur les Fonctions caractéristiques des divers fluides. *C. R. Acad. Sci.* **1869**, *69*, 858–862. (In French)
62. Massieu, F. Addition au précédent Mémoire sur les Fonctions caractéristiques. *C. R. Acad. Sci.* **1869**, *69*, 1057–1061. (In French)

63. Massieu, F. *Exposé des Principes Fondamentaux de la Théorie Mécanique de la Chaleur (note Destinée à Servir D'introduction au Mémoire de L'auteur sur les Fonctions Caractéristiques des Divers Fluides et la Théorie des Vapeurs)*; Académie des Sciences: Paris, France, 1873; p. 31. (In French)
64. Massieu, F. *Thermodynamique: Mémoire sur les Fonctions Caractéristiques des Divers Fluides et sur la Théorie des Vapeurs*; Académie des Sciences: Paris, France, 1876; p. 92. (In French)
65. Massieu, F. Sur les Intégrales Algébriques des Problèmes de Mécanique. Suivie de Sur le Mode de Propagation des Ondes Planes et la Surface de L'onde Elémentaire dans les Cristaux Biréfringents à Deux Axes. Ph.D. Thesis, Faculté des Sciences de Paris, Paris, France, 1861.
66. Nivoit, E. Notice sur la vie et les Travaux de M. Massieu, Inspecteur Général des Mines. Available online: http://facultes19.ish-lyon.cnrs.fr/fiche.php?indice=1153 (accessed 27 October).
67. Gibbs, J.W. Graphical Methods in the Thermodynamics of Fluids. In *The Scientific Papers of J. Willard Gibbs*; Dover: New York, NY, USA, 1961.
68. Brillouin, L. *Science and Information Theory*; Academic Press: New York, NY, USA, 1956.
69. Brillouin, L. Maxwell's demon cannot operate: Information and entropy. *J. Appl. Phys.* **1951**, *22*, 334–337. [CrossRef]
70. Brillouin, L. Physical entropy and information. *J. Appl. Phys.* **1951**, *22*, 338–343. [CrossRef]
71. Brillouin, L. Negentropy principle of information. *J. Appl. Phys.* **1953**, *24*, 1152–1163. [CrossRef]
72. Duhem, P. Sur les équations générales de la thermodynamique. In *Annales scientifiques de l'École Normale Supérieure*; Ecole Normale Supérieure: Paris, France, 1891; Volume 8, pp. 231–266. (In French)
73. Duhem, P. Commentaire aux principes de la Thermodynamique—Première partie. *J. Math. Appl.* **1892**, *8*, 269–330. (In French)
74. Duhem, P. Commentaire aux principes de la Thermodynamique—Troisième partie. *J. Math. Appl.* **1894**, *10*, 207–286. (In French)
75. Duhem, P. Les théories de la chaleur. *Revue des deux Mondes* **1895**, *130*, 851–868.
76. Carathéodory, C. Untersuchungen über die Grundlagen der Thermodynamik (Examination of the foundations of thermodynamics). *Math. Ann.* **1909**, *67*, 355–386. [CrossRef]
77. Carnot, S. *Réflexions sur la Puissance Motrice du feu*; Dover: New York, NY, USA, 1960.
78. Clausius, R. *On the Mechanical Theory of Heat*; Browne, W.R., Translator; Macmillan: London, UK, 1879.
79. Darrigol, O. The Origins of the Entropy Concept. Available online: http://www.bourbaphy.fr/darrigol.pdf (accessed on 5 August 2015). (In French)
80. Gromov, M. In a Search for a Structure, Part 1: On Entropy. Available online: http://www.ihes.fr/~gromov/PDF/structre-serch-entropy-july5-2012.pdf (accessed on 6 August 2015).
81. Gromov, M. Six Lectures on Probability, Symmetry, Linearity. Available online: http://www.ihes.fr/~gromov/PDF/probability-huge-Lecture-Nov-2014.pdf (accessed on 6 August 2015).
82. Gromov, M. Metric Structures for Riemannian and Non-Riemannian Spaces (Modern Birkhäuser Classics), 3rd ed.Lafontaine, J., Pansu, P., Eds.; Birkhäuser: Basel, Switzerland, 2006.
83. Kozlov, V.V. Heat equilibrium by Gibbs and poincaré. *Dokl. RAN* **2002**, *382*, 602–606. (In French)
84. Poincaré, H. Sur les tentatives d'explication mécanique des principes de la thermodynamique. *C. R. Acad. Sci.* **1889**, *108*, 550–553.
85. Poincaré, H. *Thermodynamique, Cours de Physique Mathématique*. Available online: http://gallica.bnf.fr/ark:/12148/bpt6k2048983 (accessed on 24 October 2016). (In French)
86. Poincaré, H. *Calcul des Probabilités*; Gauthier-Villars: Paris, France, 1896. (In French)
87. Poincaré, H. Réflexions sur la théorie cinétique des gaz. *J. Phys. Theor. Appl.* **1906**, *5*, 369–403. [CrossRef]
88. Fourier, J. *Théorie Analytique de la Chaleur*; Chez Firmin Didot: Paris, France, 1822. (In French)
89. Clausius, R. *Théorie Mécanique de la Chaleur*; Lacroix: Paris, France, 1868. (In French)
90. Poisson, S.D. *Théorie Mathématique de la Chaleur*; Bachelier: Paris, France, 1835. (In French)
91. Kosmann-Schwarzbach, Y. *Siméon-Denis Poisson: Les Mathématiques au Service de la Science*; Ecole Polytechnique: Paris, France, 2013. (In French)
92. Smale, S. Topology and Mechanics. *Invent. Math.* **1970**, *10*, 305–331. [CrossRef]
93. Cushman, R.; Duistermaat, J.J. The quantum mechanical spherical pendulum. *Bull. Am. Math. Soc.* **1988**, *19*, 475–479. [CrossRef]
94. Guillemin, V.; Sternberg, S. The moment map and collective motion. *Ann. Phys.* **1980**, *1278*, 220–253. [CrossRef]

95. De Saxcé, G.; Vallée, C. Bargmann group, momentum tensor and Galilean invariance of Clausius-Duhem Inequality. *Int. J. Eng. Sci.* **2012**, *50*, 216–232. [CrossRef]
96. De Saxcé, G. Entropy and structure for the thermodynamic systems. In *Geometric Science of Information, Second International Conference GSI 2015 Proceedings*; Nielsen, F., Barbaresco, F., Eds.; Lecture Notes in Computer Science; Springer: Berlin/Heidelberg, Germany, 2015; Volume 9389, pp. 519–528.
97. Kapranov, M. Thermodynamics and the moment map. 2011; arXiv:1108.3472v1.
98. Pavlov, V.P.; Sergeev, V.M. Thermodynamics from the differential geometry standpoint. *Theor. Math. Phys.* **2008**, *157*, 1484–1490. [CrossRef]
99. Cartier, P.; DeWitt-Morette, C. *Functional Integration. Action and Symmetries*; Cambridge University Press: Cambridge, UK, 2004.
100. Libermann, P.; Marle, C.M. *Symplectic Geometry and Analytical Mechanics*; Springer Science & Business Media: Berlin/Heidelberg, Germany, 1987.
101. Lichnerowicz, A. Espaces homogènes Kähleriens. In *Colloque de Géométrie Différentielle*; CNRSP: Paris, France, 1953; pp. 171–184. (In French)
102. Lichnerowicz, A. *Représentation Coadjointe Quotient et Espaces Homogènes de Contact, cours du Collège de France*; Springer: Berlin/Heidelberg, Germany, 1986. (In French)
103. Balian, R.; Valentin, P. Hamiltonian structure of thermodynamics with gauge. *Eur. Phys. J. B* **2001**, *21*, 269–282. [CrossRef]
104. Marle, C.M. On Henri Poincaré's note "Sur une forme nouvelle des équations de la mécanique". *J. Geom. Symmetry Phys.* **2013**, *29*, 1–38.
105. Poincaré, H. Sur une forme nouvelle des équations de la Mécanique. *C. R. Acad. Sci.* **1901**, *7*, 369–371. (In French)
106. Sternberg, S. Symplectic homogeneous spaces. *Trans. Am. Math. Soc.* **1975**, *212*, 113–130. [CrossRef]
107. Bourguignon, J.P. *Calcul Variationnel*; Ecole Polytechnique: Paris, France, 2007. (In French)
108. Dedecker, P. A property of differential forms in the calculus of variations. *Pac. J. Math.* **1957**, *7*, 1545–1549. [CrossRef]
109. Marle, C.M. On mechanical systems with a Lie group as configuration space. In *Jean Leray '99 Conference Proceedings*; De Gosson, M., Ed.; Springer: Berlin/Heidelberg, Germany, 2003; pp. 183–203.
110. Marle, C.M. Symmetries of Hamiltonian systems on symplectic and poisson manifolds. In *Similarity and Symmetry Methods*; Springer: Berlin/Heidelberg, Germany, 2014; pp. 185–269.
111. Kirillov, A.A. Merits and demerits of the orbit method. *Bull. Am. Math. Soc.* **1999**, *36*, 433–488. [CrossRef]
112. Cartan, E. La structure des groupes de transformations continus et la théorie du trièdre mobile. *Bull. Sci. Math.* **1910**, *34*, 250–284. (In French)
113. Cartan, E. *Leçons sur les Invariants Intégraux*; Hermann: Paris, France, 1922. (In French)
114. Cartan, E. Les récentes généralisations de la notion d'espace. *Bull. Sci. Math.* **1924**, *48*, 294–320. (In French)
115. Cartan, E. *Le rôle de la Théorie des Groupes de Lie dans L'évolution de la Géométrie Modern*; C.R. Congrès International: Oslo, Norway, 1936; Volume 1, pp. 92–103. (In French)
116. Libermann, P. La géométrie différentielle d'Elie Cartan à Charles Ehresmann et André Lichnerowicz. In *Géométrie au XXe siècle, 1930-2000: Histoire et Horizons*; Hermann: Paris, France, 2005. (In French)
117. Koszul, J.L. Sur la forme hermitienne canonique des espaces homogènes complexes. *Can. J. Math.* **1955**, *7*, 562–576. (In French) [CrossRef]
118. Koszul, J.L. *Exposés sur les Espaces Homogènes Symétriques*; Publicação da Sociedade de Matematica de São Paulo: São Paulo, Brazil, 1959. (In French)
119. Koszul, J.L. Domaines bornées homogènes et orbites de groupes de transformations affines. *Bull. Soc. Math. Fr.* **1961**, *89*, 515–533. (In French)
120. Koszul, J.L. Ouverts convexes homogènes des espaces affines. *Math. Z.* **1962**, *79*, 254–259. (In French) [CrossRef]
121. Koszul, J.L. Variétés localement plates et convexité. *Osaka J. Math.* **1965**, *2*, 285–290. (In French)
122. Koszul, J.L. *Lectures on Groups of Transformations*; Tata Institute of Fundamental Research: Bombay, India, 1965.
123. Koszul, J.L. Déformations des variétés localement plates. *Ann. Inst. Fourier* **1968**, *18*, 103–114. (In French) [CrossRef]
124. Koszul, J.L. Trajectoires convexes de groupes affines unimodulaires. In *Essays on Topology and Related Topics*; Springer: Berlin, Germany, 1970; pp. 105–110.

125. Vinberg, E.B. The theory of homogeneous convex cones. *Trudy Moskovskogo Matematicheskogo Obshchestva* **1963**, *12*, 303–358.
126. Vinberg, E.B. Structure of the group of automorphisms of a homogeneous convex cone. *Trudy Moskovskogo Matematicheskogo Obshchestva* **1965**, *13*, 56–83. (In Russian)
127. Byande, P.M.; Ngakeu, F.; Boyom, M.N.; Wolak, R. KV-cohomology and differential geometry of affinely flat manifolds. Information geometry. *Afr. Diaspora J. Math.* **2012**, *14*, 197–226.
128. Byande, P.M. *Des Structures Affines à la Géométrie de L'information*; Omniscriptum: Saarbrücken, France, 2012.
129. Nguiffo Boyom, M. Sur les structures affines homotopes à zéro des groupes de Lie. *J. Differ. Geom.* **1990**, *31*, 859–911. (In French)
130. Nguiffo Boyom, M. Structures localement plates dans certaines variétés symplectiques. *Math. Scand.* **1995**, *76*, 61–84. (In French) [CrossRef]
131. Nguiffo Boyom, M. The cohomology of Koszul-Vinberg algebras. *Pac. J. Math.* **2006**, *225*, 119–153. [CrossRef]
132. Nguiffo Boyom, M. *Some Lagrangian Invariants of Symplectic Manifolds, Geometry and Topology of Manifolds*; Banach Center Institute of Mathematics, Polish Academy of Sciences: Warsaw, Poland, 2007; Volume 76, pp. 515–525.
133. Nguiffo Boyom, M. Métriques kählériennes affinement plates de certaines variétés symplectiques. I. *Proc. Lond. Math. Soc.* **1993**, *2*, 358–380. (In French) [CrossRef]
134. Nguiffo Boyom, M.; Byande, P.M. *KV Cohomology in Information Geometry Matrix Information Geometry*; Springer: Heidelberg, Germany, 2013; pp. 69–92.
135. Nguiffo Boyom, M. Transversally Hessian foliations and information geometry I. *Am. Inst. Phys. Proc.* **2014**, *1641*, 82–89.
136. Nguiffo Boyom, M.; Wolak, R. Transverse Hessian metrics information geometry MaxEnt 2014. *AIP. Conf. Proc. Am. Inst. Phys.* **2015**. [CrossRef]
137. Vey, J. *Sur une Notion D'hyperbolicité des Variables Localement Plates. Thèse de Troisième Cycle de Mathématiques Pures*; Faculté des Sciences de l'université de Grenoble: Grenoble, France, 1969. (In French)
138. Vey, J. Sur les Automorphismes affines des ouverts convexes saillants. Annali della scuola normale superiore di pisa. *Classe Sci.* **1970**, *24*, 641–665. (In French)
139. Barbaresco, F. Koszul information geometry and Souriau Lie group thermodynamics. In AIP Conference Proceedings, Proceedings of MaxEnt'14 Conference, Amboise, France, 21–26 September 2014.
140. Lesne, A. Shannon entropy: A rigorous notion at the crossroads between probability, information theory, dynamical systems and statistical physics. *Math. Struct. Comput. Sci.* **2014**, *24*, e240311. [CrossRef]
141. Fréchet, M.R. Sur l'extension de certaines évaluations statistiques au cas de petits échantillons. *Rev. Inst. Int. Stat.* **1943**, *11*, 182–205. (In French) [CrossRef]
142. Fréchet, M.R. Les espaces abstraits topologiquement affines. *Acta Math.* **1925**, *47*, 25–52. [CrossRef]
143. Fréchet, M.R. Les éléments aléatoires de nature quelconque dans un espace distancié. *Ann. Inst. Henri Poincaré* **1948**, *10*, 215–310.
144. Fréchet, M.R. Généralisations de la loi de probabilité de Laplace. *Ann. Inst. Henri Poincaré* **1951**, *12*, 1–29. (In French)
145. Shima, H. *The Geometry of Hessian Structures*; World Scientific: Singapore, 2007.
146. Shima, H. Geometry of Hessian Structures. In *Springer Lecture Notes in Computer Science*; Nielsen, F., Frederic, B., Eds.; Springer: Berlin/Heidelberg, Germany, 2013; Volume 8085, pp. 37–55.
147. Crouzeix, J.P. A relationship between the second derivatives of a convex function and of its conjugate. *Math. Program.* **1977**, *3*, 364–365. [CrossRef]
148. Hiriart-Urruty, J.B. A new set-valued second-order derivative for convex functions. In *Mathematics for Optimization*; Elsevier: Amsterdam, The Netherlands, 1986.
149. Bakhvalov, N.S. Memorial: Nikolai Nikolaevitch Chentsov. *Theory Probab. Appl.* **1994**, *38*, 506–515. [CrossRef]
150. Chentsov, N.N. *Statistical Decision Rules and Optimal Inference*; American Mathematical Society: Providence, RI, USA, 1982.
151. Berezin, F. Quantization in complex symmetric spaces. *Izv. Akad. Nauk SSSR Ser. Math.* **1975**, *9*, 363–402. [CrossRef]
152. Bhatia, R. *Positive Definite Matrices*; Princeton University Press: Princeton, NJ, USA, 2007.
153. Bhatia, R. The bipolar decomposition. *Linear Algebra Appl.* **2013**, *439*, 3031–3037. [CrossRef]

154. Bini, D.A.; Garoni, C.; Iannazzo, B.; Capizzano, S.S.; Sesana, D. Asymptotic Behaviour and Computation of Geometric-Like Means of Toeplitz Matrices, SLA14 Conference, Kalamata, Greece, September 2014; Available online: http://noether.math.uoa.gr/conferences/sla2014/sites/default/files/Iannazzo.pdf (accessed on 8–12 September 2014).
155. Bini, D.A.; Garoni, C.; Iannazzo, B.; Capizzano, S.S. Geometric means of toeplitz matrices by positive parametrizations. **2016**, in press.
156. Calvo, M.; Oller, J.M. An explicit solution of information geodesic equations for the multivariate normal model. *Stat. Decis.* **1991**, *9*, 119–138. [CrossRef]
157. Calvo, M.; Oller, J.M. A distance between multivariate normal distributions based in an embedding into the Siegel group. *J. Multivar. Anal. Arch.* **1990**, *35*, 223–242. [CrossRef]
158. Calvo, M.; Oller, J.M. A distance between elliptical distributions based in an embedding into the Siegel group. *J. Comput. Appl. Math.* **2002**, *145*, 319–334. [CrossRef]
159. Chevallier, E.; Barbaresco, F.; Angulo, J. Probability density estimation on the hyperbolic space applied to radar processing. In *Geometric Science of Information Proceedings*; Lecture Notes in Computer Science; Springer: Berlin/Heidelberg, Germany, 2015; Volume 9389, pp. 753–761.
160. Chevallier, E.; Forget, T.; Barbaresco, F.; Angulo, J. Kernel Density Estimation on the Siegel Space Applied to Radar Processing. Available online: https://hal-ensmp.archives-ouvertes.fr/hal-01344910/document (accessed on 24 October 2016).
161. Costa, S.I.R.; Santosa, S.A.; Strapasson, J.E. Fisher information distance: A geometrical reading. *Discret. Appl. Math.* **2015**, *197*, 59–69. [CrossRef]
162. Jeuris, B.; Vandebril, R.; Vandereycken, B. A survey and comparison of contemporary algorithms for computing the matrix geometric mean. *Electron. Trans. Numer. Anal.* **2012**, *39*, 379–402.
163. Jeuris, B. Riemannian Optimization for Averaging Positive Definite Matrices. Ph.D. Thesis, Katholieke Universiteit Leuven, Leuven, Belgium, 2015.
164. Jeuris, B.; Vandebril, R. *The Kähler Mean of Block-Toeplitz Matrices with Toeplitz Structured Blocks*; Department of Computer Science, KU Leuven: Leuven, Belgium, 2015.
165. Maliavin, P. Invariant or quasi-invariant probability measures for infinite dimensional groups, Part II: Unitarizing measures or Berezinian measures. *Jpn. J. Math.* **2008**, *3*, 19–47. [CrossRef]
166. Strapasson, J.E.; Porto, J.P.S.; Costa, S.I.R. On bounds for the Fisher-Rao distance between multivariate normal distributions. *AIP Conf. Proc.* **2015**, *1641*, 313–320.
167. Hua, L.K. *Harmonic Analysis of Functions of Several Complex Variables in the Classical Domains*; American Mathematical Society: Providence, RI, USA, 1963.
168. Siegel, C.L. Symplectic geometry. *Am. J. Math.* **1943**, *65*, 1–86. [CrossRef]
169. Yoshizawa, S.; Tanabe, K. Dual differential geometry associated with the Kullback-Leibler information on the Gaussian distributions and its 2-parameters deformations. *SUT J. Math.* **1999**, *35*, 113–137.
170. Skovgaard, L.T. *A Riemannian Geometry of the Multivariate Normal Model*; Technical Report for Stanford University: Stanford, CA, USA, April 1981.
171. Deza, M.M.; Deza, E. *Encyclopedia of Distances*, 3rd ed.; Springer: Berlin/Heidelberg, Germany, 2013; p. 242.
172. Casalis, M. Familles exponentielles naturelles invariantes par un groupe de translations. *C. R. Acad. Sci. Ser. I Math.* **1988**, *307*, 621–623. (In French)
173. Casalis, M. Familles Exponentielles Naturelles Invariantes par un Groupe. Ph.D. Thesis, Thèse de l'Université Paul Sabatier, Toulouse, France, 1990. (In French)
174. Casalis, M. Familles exponentielles naturelles sur rd invariantes par un groupe. *Int. Stat. Rev.* **1991**, *59*, 241–262. (In French) [CrossRef]
175. Casalis, M. Les familles exponentielles à variance quadratique homogène sont des lois de Wishart sur un cône symétrique. *C. R. Acad. Sci. Ser. I Math.* **1991**, *312*, 537–540. (In French)
176. Casalis, M.; Letac, G. Characterization of the Jørgensen set in generalized linear models. *Test* **1994**, *3*, 145–162. [CrossRef]
177. Casalis, M.; Letac, G. The Lukacs-Olkin-Rubin characterization of the Wishart distributions on symmetric cone. *Ann. Stat.* **1996**, *24*, 763–786. [CrossRef]
178. Casalis, M. The 2d + 4 simple quadratic natural exponential families on Rd. *Ann. Stat.* **1996**, *24*, 1828–1854.
179. Letac, G. A characterization of the Wishart exponential families by an invariance property. *J. Theor. Probab.* **1989**, *2*, 71–86. [CrossRef]

180. Letac, G. *Lectures on Natural Exponential Families and Their Variance Functions, Volume 50 of Monografias de Matematica (Mathematical Monographs)*; Instituto de Matematica Pura e Aplicada (IMPA): Rio de Janeiro, Brazil, 1992.
181. Letac, G. Les familles exponentielles statistiques invariantes par les groupes du Cône et du paraboloïde de revolution. In *Journal of Applied Probability, Volume 31, Studies in Applied Probability*; Takacs, L., Galambos, J., Gani, J., Eds.; Applied Probability Trust: Sheffield, UK, 1994; pp. 71–95.
182. Barndorff-Nielsen, O.E. Differential geometry and statistics: Some mathematical aspects. *Indian J. Math.* **1987**, *29*, 335–350.
183. Barndorff-Nielsen, O.E.; Jupp, P.E. Yokes and symplectic structures. *J. Stat. Plan Inference* **1997**, *63*, 133–146. [CrossRef]
184. Barndorff-Nielsen, O.E.; Jupp, P.E. Statistics, yokes and symplectic geometry. *Annales de la Faculté des sciences de Toulouse: Mathématiques* **1997**, *6*, 389–427. [CrossRef]
185. Barndorff-Nielsen, O.E. *Information and Exponential Families in Stattistical Theory*; Wiley: New York, NY, USA, 2014.
186. Jespersen, N.C.B. On the structure of transformation models. *Ann. Stat.* **1999**, *17*, 195–208.
187. Skovgaard, L.T. A Riemannian geometry of the multivariate normal model. *Scand. J. Stat.* **1984**, *11*, 211–223.
188. Han, M.; Park, F.C. DTI segmentation and fiber tracking using metrics on multivariate normal distributions. *J. Math. Imaging Vis.* **2014**, *49*, 317–334. [CrossRef]
189. Imai, T.; Takaesu, A.; Wakayama, M. Remarks on geodesics for multivariate normal models. *J. Math. Ind.* **2011**, *3*, 125–130.
190. Inoue, H. Group theoretical study on geodesics for the elliptical models. In *Geometric Science of Information Proceedings*; Lecture Notes in Computer Science; Springer: Berlin/Heidelberg, Germany, 2015; Volume 9389, pp. 605–614.
191. Pilté, M.; Barbaresco, F. Tracking quality monitoring based on information geometry and geodesic shooting. In Proceedings of the 17th International Radar Symposium (IRS), Krakow, Poland, 10–12 May 2016; pp. 1–6.
192. Eriksen, P.S. *(k, 1)* Exponential transformation models. *Scand. J. Stat.* **1984**, *11*, 129–145.
193. Eriksen, P. *Geodesics Connected with the Fisher Metric on the Multivariate Normal Manifold*; Technical Report 86-13; Institute of Electronic Systems, Aalborg University: Aalborg, Denmark, 1986.
194. Eriksen, P.S. Geodesics connected with the Fisher metric on the multivariate normal manifold. In Proceedings of the GST Workshop, Lancaster, UK, 28–31 October 1987.
195. Feragen, A.; Lauze, F.; Hauberg, S. Geodesic exponential kernels: When curvature and linearity conflict. In Proceedings of the IEEE Conference on Computer Vision and Pattern Recognition (CVPR), 8–10 June 2015; pp. 3032–3042.
196. Besse, A.L. *Einstein Manifolds, Ergebnisse der Mathematik und ihre Grenzgebiete*; Springer: Berlin/Heidelberg, Germany, 1986.
197. Tumpach, A.B. Infinite-dimensional hyperkähler manifolds associated with Hermitian-symmetric affine coadjoint orbits. *Ann. Inst. Fourier* **2009**, *59*, 167–197. [CrossRef]
198. Tumpach, A.B. Classification of infinite-dimensional Hermitian-symmetric affine coadjoint orbits. *Forum Math.* **2009**, *21*, 375–393. [CrossRef]
199. Tumpach, A.B. Variétés Kählériennes et Hyperkählériennes de Dimension Infinie. Ph.D. Thesis, Ecole Polytechnique, Paris, France, 26 July 2005.
200. Neeb, K.-H. Infinite-dimensional groups and their representations. In *Lie Theory*; Birkhäuser: Basel, Switzerland, 2004.
201. Gauduchon, P. Calabi's Extremal Kähler Metrics: An Elementary Introduction. Available online: germanio.math.unifi.it/wp-content/uploads/2015/03/dercalabi.pdf (accessed on 27 October 2016).
202. Biquard, O.; Gauduchon, P. Hyperkähler Metrics on Cotangent Bundles of Hermitian Symmetric Spaces. Available online: https://www.math.ens.fr/~biquard/aarhus96.pdf (accessed on 27 October 2016).
203. Biquard, O.; Gauduchon, P. La métrique hyperkählérienne des orbites coadjointes de type symétrique d'un groupe de Lie complexe semi-simple. *Comptes Rendus de l'Académie des Sciences* **1996**, *323*, 1259–1264. (In French)
204. Biquard, O.; Gauduchon, P. Géométrie hyperkählérienne des espaces hermitiens symétriques complexifiés. *Séminaire de Théorie Spectrale et Géométrie* **1998**, *16*, 127–173. [CrossRef]

205. Chaperon, M. *Jets, Transversalité, Singularités: Petite Introduction aux Grandes Idées de René Thom*; Kouneiher, J., Flament, D., Nabonnand, P., Szczeciniarz, J.-J., Eds.; Géométrie au Vingtième Siècle, Histoire et Horizons: Hermann, Paris, 2005; pp. 246–256.
206. Chaperon, M. Generating maps, invariant manifolds, conjugacy. *J. Geom. Phys.* **2015**, *87*, 76–85. [CrossRef]
207. Viterbo, C. Symplectic topology as the geometry of generating functions. *Math. Ann.* **1992**, *292*, 685–710. [CrossRef]
208. Viterbo, C. Generating functions, symplectic geometry and applications. In Proceedings of the International Congress of Mathematics, Zürich, Switzerland, 3–11 August 1994.
209. Dazord, P.; Weinstein, A. *Symplectic, Groupoids, and Integrable Systems*; Springer: Berlin/Heidelberg, Germany, 1991; pp. 99–128.
210. Drinfeld, V.G. Hamiltonian structures on Lie groups. *Sov. Math. Dokl.* **1983**, *27*, 68–71.
211. Thom, R. Une théorie dynamique de la Morphogenèse. In *Towards a Theoretical Biology I*; Waddington, C.H., Ed.; University of Edinburgh Press: Edinburgh, UK, 1966; pp. 52–166.
212. Thom, R. *Stabilité Structurelle et Morphogénèse*, 2nd ed.; Inter Editions: Paris, France, 1977.
213. Ingarden, R.S.; Nakagomi, T. The second order extension of the Gibbs state. *Open Syst. Inf. Dyn.* **1992**, *1*, 243–258. [CrossRef]
214. Ingarden, R.S.; Meller, J. Temperatures in Linguistics as a Model of Thermodynamics. *Open Syst. Inf. Dyn.* **1994**, *2*, 211–230. [CrossRef]
215. Nencka, H.; Streater, R.F. Information Geometry for some Lie algebras. *Infin. Dimens. Anal. Quantum Probab. Relat. Top.* **1999**, *2*, 441–460. [CrossRef]
216. Burdet, G.; Perrin, M.; Perroud, M. Generating functions for the affine symplectic group. *Comm. Math. Phys.* **1978**, *3*, 241–254. [CrossRef]
217. Berthoz, A. *Le Sens du Movement*; Odile Jacob Edirot: Paris, France, 1997. (In French)
218. Afgoustidis, A. Invariant Harmonic Analysis and Geometry in the Workings of the Brain. Available online: https://hal-univ-diderot.archives-ouvertes.fr/tel-01343703 (accessed on 17 October 2016).
219. Souriau, J.M. Innovaxiom—Interview of Jean-Marie Souriau. Available online: https://www.youtube.com/watch?v=Lb_TWYqBUS4 (accessed on 27 October 2016).
220. Souriau, J.M. Quantique ? Alors c'est Géométrique. Available online: http://www.ahm.msh-paris.fr/Video.aspx?domain=84fa1a68-95c0-4c74-aed7-06055edaca16&language=fr&metaDescriptionId=dd3bd275-8372-4130-976b-847c36156a83&mediatype=VideoWithShots (accessed on 27 October 2016).
221. Masseau, D. *Les marges des Lumières Françaises (1750–1789)*; Dix-huitième Siècle Année: Paris, France, 2005; Volume 37, pp. 638–639. (In French)
222. Cioran, E. *Précis de Décomposition Poche*; Gallimard: Paris, France, 1977.
223. Rao, C.R. Information and the accuracy attainable in the estimation of statistical parameters. *Bull. Calcutta Math. Soc.* **1945**, *37*, 81–91.
224. Burbea, J.; Rao, C.R. Entropy differential metric, distance and divergence measures in probability spaces: A unified approach. *J. Multivar. Anal.* **1982**, *12*, 575–596. [CrossRef]
225. Legendre, A.M. *Mémoire Sur L'intégration de Quelques Equations aux Différences Partielles*; Mémoires de l'Académie des Sciences: Paris, France, 1787; pp. 309–351. (In French)
226. Darboux, G. *Leçons sur la Théorie Générale des Surfaces et les Applications Géométriques du Calcul Infinitésimal: Premiere Partie (Généralités, Coordonnées Curvilignes, Surface Minima)*; Gauthier-Villars: Paris, France, 1887. (In French)
227. Balian, R.; Alhassid, Y.; Reinhardt, H. Dissipation in many-body systems: A geometric approach based on information theory. *Phys. Rep.* **1986**, *131*, 1–146. [CrossRef]
228. Balian, R.; Balazs, N. Equiprobability, inference and entropy in quantum theory. *Ann. Phys.* **1987**, *179*, 97–144. [CrossRef]
229. Balian, R. On the principles of quantum mechanics. *Am. J. Phys.* **1989**, *57*, 1019–1027. [CrossRef]
230. Balian, R. *From Microphysics to Macrophysics: Methods and Applications of Statistical Physics*; Springer: Heidelberg, Germany, 1991 & 1992; Volumes I and II.
231. Balian, R. Incomplete descriptions and relevant entropies. *Am. J. Phys.* **1999**, *67*, 1078–1090. [CrossRef]

232. Balian, R. Entropy, a protean concept. In *Poincaré Seminar 2003*; Dalibard, J., Duplantier, B., Rivasseau, V., Eds.; Birkhauser: Basel, Switzerland, 2004; pp. 119–144.
233. Balian, R. Information in statistical physics. In *Studies in History and Philosophy of Modern Physics, Part B*; Elsevier: Amsterdam, The Netherlands, 2005.
234. Balian, R. The entropy-based quantum metric. *Entropy* **2014**, *16*, 3878–3888. [CrossRef]
235. Balian, R. *François Massieu et les Potentiels Thermodynamiques, Évolution des Disciplines et Histoire des Découvertes*; Académie des Sciences: Avril, France, 2015.

7

Syntactic Parameters and a Coding Theory Perspective on Entropy and Complexity of Language Families

Matilde Marcolli

Department of Mathematics, California Institute of Technology, Pasadena, CA 91125, USA; matilde@caltech.edu;

Academic Editors: Frédéric Barbaresco, Frank Nielsen and Kevin H. Knuth

Abstract: We present a simple computational approach to assigning a measure of complexity and information/entropy to families of natural languages, based on syntactic parameters and the theory of error correcting codes. We associate to each language a binary string of syntactic parameters and to a language family a binary code, with code words the binary string associated to each language. We then evaluate the code parameters (rate and relative minimum distance) and the position of the parameters with respect to the asymptotic bound of error correcting codes and the Gilbert–Varshamov bound. These bounds are, respectively, related to the Kolmogorov complexity and the Shannon entropy of the code and this gives us a computationally simple way to obtain estimates on the complexity and information, not of individual languages but of language families. This notion of complexity is related, from the linguistic point of view to the degree of variability of syntactic parameter across languages belonging to the same (historical) family.

Keywords: syntax; principles and parameters; error-correcting codes; asymptotic bound; Kolmogorov complexity; Gilbert–Varshamov bound; Shannon entropy

---

1. Introduction

We propose an approach, based on Longobardi's parametric comparison method (PCM) and the theory of error-correcting codes, to a quantitative evaluation of the "complexity" of a language family. One associates to a collection of languages to be analyzed with the PCM a binary (or ternary) code with one code word for each language in the family and each word consisting of the binary values of the syntactic parameters of that language. The ternary case allows for an additional parameter state that takes into account certain phenomena of entailment of parameters. We then consider a different kind of parameters: the code parameters of the resulting code, which in coding theory account for the efficiency of the coding and decoding procedures. These can be compared with some classical bounds of coding theory: the asymptotic bound, the Gilbert–Varshamov (GV) bound, *etc.* The position of the code parameters with respect to some of these bounds provides quantitative information on the variability of syntactic parameters within and across historical-linguistic families. While computations carried out for languages belonging to the same historical family yield codes below the GV curve, comparisons across different historical families can give examples of isolated codes lying above the asymptotic bound.

1.1. *Principles and Parameters*

The generative approach to linguistics relies on the notion of a Universal Grammar (UG) and a related universal list of syntactic parameters. In the Principles and Parameters model, developed since [1], these are thought of as binary valued parameters or "switches" that set the grammatical structure of a given language. Their universality makes it possible to obtain comparisons, at the syntactic level, between arbitrary pairs of natural languages.

A PCM was introduced in [2] as a quantitative method in historical linguistics, for comparison of languages within and across historical families at the syntactic instead of the lexical level. Evidence was given in [3,4] that the PCM gives reliable information on the phylogenetic tree of the family of Indo-European languages.

The PCM relies essentially on constructing a metric on a family of languages based on the relative Hamming distance between the sets of parameters as a measure of relatedness. The phylogenetic tree is then constructed on the basis of this datum of relative distances, see [3].

More work on syntactic phylogenetic reconstructions, involving a larger set of languages and parameters is ongoing, [5]. Syntactic parameters of world languages have also been used recently for investigations on the topology and geometry of syntactic structures and for statistical physics models of language evolution, [6–8].

Publicly available data of syntactic parameters of world languages can be obtained from databases such as Syntactic Structures of World Languages (SSWL) [9] or TerraLing [10] or World Atlas of Language Structures (WALS) [11]. The data of syntactic parameters used in the present paper are taken from Table A of [3].

1.2. Syntactic Parameters, Codes and Code Parameters

Our purpose in this paper is to connect the PCM approach to the mathematical theory of error-correcting codes. We associate a code to any group of languages one wishes to analyze via the PCM, which has one code word for each language. If one uses a number n of syntactic parameters, then the code C sits in the space $\mathbb{F}_2^n$, where the elements of $\mathbb{F}_2 = \{0,1\}$ correspond to the two $\mp$ possible values of each parameter, and the code word of a language is the string of values of its n parameters. We also consider a version with codes on an alphabet $\mathbb{F}_3$ of three letters which allows for the possibility that some of the parameters may be made irrelevant by entailment from other parameters. In this case we use the letter $0 \in \mathbb{F}_3$ for the irrelevant parameters and the nonzero values ± 1 for the parameters that are set in the language.

In the theory of error-correcting codes, see [12], one assigns to a code $C \subset \mathbb{F}_q^n$ two code parameters: $R = \log_q(\#C)/n$, the transmission rate of the code, and $\delta = d/n$ the relative minimum distance of the code, where d is the miminum Hamming distance between pairs of distinct code words. It is well known in coding theory that "good codes" are those that maximize both parameters, compatibly with several constraints relating R and δ. Consider the function $f : \mathcal{C}_q \to [0,1]^2$ from the space $\mathcal{C}_q$ of q-ary codes to the unit square, that assigns to a code C its code parameters, $f(C) = (\delta(C), R(C))$. A point (δ, R) in the range of f has finite (respectively, infinite) multiplicity if the preimage $f^{-1}(\delta, R)$ is a finite set (respectively, an infinite set). It was proved in [13] that there is a curve $R = \alpha_q(\delta)$ in the space of code parameters, the asymptotic bound, that separates code points that fill a dense region and that have infinite multiplicity from isolated code points that only have finite multiplicity. These better but more elusive codes are typically obtained through algebro-geometric constructions, see [13–15]. The asymptotic bound was related to Kolmogorov complexity in [16].

1.3. Position with Respect to the Asymptotic Bound

Given a collection of languages one wants to compare through their syntactic parameters, one can ask natural questions about the position of the resulting code in the space of code parameters and with respect to the asymptotic bound. The theory of error correcting codes tells us that codes above the asymptotic bound are very rare. Indeed, we considered various sets of languages, and for each choice of a set of languages we considered an associated code, with a code word for each language in the set, given by its list of syntactic parameters. When computing the code parameters of the resulting code, one finds that, in a range of cases we looked at, when the languages in the chosen set belong to the same historical-linguistic family the resulting code lies below the asymptotic bound (and in fact below the Gilbert–Varshamov curve). This provides a precise quantitative bound to the possible spread of syntactic parameters compared to the size of the family, in terms of the number of different languages belonging to the same historico-linguistic group.

However, we also show that, if one considers sets of languages that do not belong to the same historical-linguistic family, then one can obtain codes that lie above the asymptotic bound, a fact that reflects, in code theoretic terms, the much greater variability of syntactic parameters. The result is in itself not surprising, but the point we wish to make is that the theory of error-correcting codes provides a natural setting where quantitative statements of this sort can be made using methods already developed for the different purposes of coding theory. We conclude by listing some new linguistic questions that arise by considering the parametric comparison method under this coding theory perspective.

1.4. Complexity of Languages and Language Families

The study of natural languages from the point of view of complexity theory has been of significant interest to linguists in recent years. The approaches typically followed focus on assigning a reasonable measure of complexity to individual languages and comparing complexities across different languages. For example, a notion of morphological complexity was studied in [17]. An approach to defining Kolmogorov complexity of languages on the basis of syntactic parameters was developed in [18]. A notion of language complexity based on the production rules of a generative grammar was considered in [19], in the setting of (finite) formal languages. For a more general computational perspective on the complexity of natural languages, see [20]. The idea of distinguishing languages by complexity is not without controversy in Linguistics. A very interesting general discussion of the problem and its evolution in the field can be found in [21].

In the present paper, we argue in favor of a somewhat different perspective, where we assign an estimate of complexity not to individual languages but to groups of languages, and in particular (historical) language families. Our version of complexity is measuring how "spread out" the syntactic parameters can be, across the languages that belong to the same family. As we outlined in the previous subsections, this is measured by assigning to the language family a code, whose code words record the syntactic parameters of the individual languages in the family, then computing its code parameters and evaluating the position of the resulting code points with respect to curves like the asymptotic bound or the Gilbert–Varshamov line. The reason why this position carries complexity information lies in the subtle relation between the asymptotic bound and Kolmogorov complexity, recently derived by Manin and the author in [16], which we will review briefly in this paper.

2. Language Families as Codes

The Principles and Parameters model of Linguistics assigns to every natural language L a set of binary values parameters that describe properties of the syntactic structure of the language.

Let F be a *language family*, by which we mean a finite collection $F = \{L_1, \ldots, L_m\}$ of languages. This may coincide with a family in the historical sense, such as the Indo-European family, or a smaller subset of languages related by historic origin and development (e.g., the Indo-Iranian, or Balto–Svalic languages), or simply any collection of languages one is interested in comparing at the parametric level, even if they are spread across different historical families.

We denote by n be the number of parameters used in the parametric comparison method. We do not fix, a priori, a value for n, and we consider it a variable of the model. We will discuss below how one views, in our perspective, the issue of the independence of parameters.

After fixing an enumeration of the parameters, that is, a bijection between the set of parameters and the set $\{1, \ldots, n\}$, we associate to a language family F a code $C = C(F)$ in $\mathbb{F}_2^n$, with one code word for each language $L \in F$, with the code word $w = w(L)$ given by the list of parameters $w = (x_1, \ldots, x_n)$, $x_i \in \mathbb{F}_2$ of the language. For simplicity of notation, we just write L for the word $w(L)$ in the following.

In this model, we only consider binary parameters with values ± 1 (here identified with letters 0 or 1 in $\mathbb{F}_2$) and we ignore parameters in a neutralized state following implications across parameters, as in the datasets of [3,4]. The entailment of parameters, that is, the phenomenon by which a particular value of one parameter (but not the complementary value) renders another parameter irrelevant, was addressed in greater detail in [22]. We first discuss a version of our coding theory model that does not incorporate entailment. We will then comment in Section 2.7 below on how the model can be modified to incorporate this phenomenon.

The idea that natural languages can be described, at the level of their core grammatical structures, in terms of a string of binary characters (code words) was already used extensively in [23].

2.1. Code Parameters

In the theory of error-correcting codes, one assigns two main parameters to a code C, the *transmission rate* and the *relative minimum distance*. More precisely, a binary code $C \subset \mathbb{F}_2^n$ is an $[n,k,d]_2$-code if the number of code words is $\#C = 2^k$, that is,

$$k = \log_2 \#C, \tag{1}$$

where k need not be an integer, and the minimal Hamming distance between code words is

$$d = \min_{L_1 \neq L_2 \in C} d_H(L_1, L_2), \tag{2}$$

where the Hamming distance is given by

$$d_H(L_1, L_2) = \sum_{i=1}^{n} |x_i - y_i|,$$

for $L_1 = (x_i)_{i=1}^n$ and $L_2 = (y_i)_{i=1}^n$ in C. The transmission rate of the code C is given by

$$R = \frac{k}{n}. \tag{3}$$

One denotes by $\delta_H(L_1, L_2)$ the relative Hamming distance

$$\delta_H(L_1, L_2) = \frac{1}{n} \sum_{i=1}^{n} |x_i - y_i|,$$

and one defines the relative minimum distance of the code C as

$$\delta = \frac{d}{n} = \min_{L_1 \neq L_2 \in C} \delta_H(L_1, L_2). \tag{4}$$

In coding theory, one would like to construct codes that simultaneously optimize both parameters (δ, R): a larger value of R represents a faster transmission rate (better encoding), and a larger value of δ represents the fact that code words are sufficiently sparse in the ambient space $\mathbb{F}_2^n$ (better decoding, with better error-correcting capability). Constraints on this optimization problem are expressed in the form of bounds in the space of (δ, R) parameters, see [12,13].

In our setting, the R parameter measures the ratio between the logarithmic size of the number of languages encompassing the given family and the total number of parameters, or equivalently how densely the given language family is in the ambient configuration space $\mathbb{F}_2^n$ of parameter possibilities. The parameter δ is the minimum, over all pairs of languages in the given family, of the relative Hamming distance used in the PCM method of [3,4].

2.2. Parameter Spoiling

In the theory of error-correcting codes, one considers *spoiling operations* on the code parameters. Applied to an $[n,k,d]_2$-code C, these produce, respectively, new codes with the following description (see Section 1.1.1 of [24]):

- A code $C_1 = C \star_i f$ in $\mathbb{F}_2^{n+1}$, for a map $f : C \to \mathbb{F}_2$, whose code words are of the form $(x_1,\dots,x_{i-1},f(x_1,\dots,x_n),x_i,\dots,x_n)$ for $w = (x_1,\dots,x_n) \in C$. If f is a constant function, C_1 is an $[n+1,k,d]_2$-code. If all pairs $w,w' \in C$ with $d_H(w,w') = d$ have $f(w) \neq f(w')$, then C_1 is an $[n+1,k,d+1]_2$-code.
- A code $C_2 = C\star_i$ in $\mathbb{F}_2^{n-1}$, whose code words are given by the projections

$$(x_1,\dots,x_{i-1},x_{i+1},\dots,x_n)$$

of code words $(x_1,\dots,x_{i-1},x_i,x_{i+1},\dots,x_n)$ in C. This is an $[n-1,k,d-1]_2$-code, except when all pairs $w,w' \in C$ with $d_H(w,w') = d$ have the same letter x_i, in which case it is an $[n-1,k,d]_2$-code.
- A code $C_3 = C(a,i) \subset C \subset \mathbb{F}_2^n$, given by the level set $C(a,i) = \{w = (x_k)_{k=1}^n \in C \mid x_i = a\}$. Taking $C(a,i)\star_i$ gives an $[n-1,k',d']_2$-code with $k-1 \leq k' < k$, and $d' \geq d$.

The same spoiling operations hold for q-ary codes $C \subset \mathbb{F}_q^n$, for any fixed q.

In our setting, where C is the code obtained from a family of languages, according to the procedure described above, the first spoiling operation can be seen as the effect of considering one more syntactic parameter, which is dependent on the other parameters, hence describing a function $F : \mathbb{F}_2^n \to \mathbb{F}_2$, whose restriction to C gives the function $f : C \to \mathbb{F}_2$. In particular, the case where f is constant on C represents the situation in which the new parameter adds no useful comparison information for the selected family of languages. The second spoiling operation consists in forgetting one of the parameters, and the third corresponds to forming subfamilies of the given family of languages, by grouping together those languages with a set value of one of the syntactic parameters. Thus, all these spoiling operations have a clear meaning from the point of view of the linguistic PCM.

2.3. Examples

We consider the same list of 63 parameters used in [3] (see Section 5.3.1 and Table A). This choice of parameters follows the *modularized global parameterization* method of [2], for the Determiner Phrase module. They encompass parameters dealing with person, number, and gender (1–6 on their list), parameters of definiteness (7–16 in their list), of countability (17–24), genitive structure (25–31), adjectival and relative modification (32–14), position and movement of the head noun (42–50), demonstratives and other determiners (51–50 and 60–63), possessive pronouns (56–59); see Section 5.3.1 and Section 5.3.2 of [3] for more details.

Our very simple examples here are just meant to clarify our notation: they consist of some collections of languages selected from the list of 28, mostly Indo-European, languages considered in [3]. In each group we consider we eliminate the parameters that are entailed from others, and we focus on a shorter list, among the remaining parameters, that will suffice to illustrate our viewpoint.

Example 1. Consider a code C formed out of the languages $\ell_1 =$ Italian, $\ell_2 =$ Spanish, and $\ell_3 =$ French, and let us consider only the first six syntactic parameters of Table A of [3], so that $C \subset \mathbb{F}_2^n$ with $n = 6$. The code words for the three languages are

ℓ_1	1	1	1	0	1	1
ℓ_2	1	1	1	1	1	1
ℓ_3	1	1	1	0	1	0

This has code parameters $(R = \log_2(3)/6 = 0.2642, \delta = 1/6)$, which satisfy $R < 1 - H_2(\delta)$, hence they lie below the GV curve (see Equation (8) below). We use this code to illustrate the three spoiling operations mentioned above.

- Throughout the entire set of 28 languages considered in [3], the first two parameters are set to the same value 1, hence for the purpose of comparative analysis within this family, we can regard a code like the above as a twice spoiled code $C = C' \star_1 f_1 = (C'' \star_2 f_2) \star_1 f_1$ where both f_1 and f_2 are constant equal to 1 and $C'' \subset \mathbb{F}_2^4$ is the code obtained from the above by canceling the first two letters in each code word.
- Conversely, we have $C'' = C'\star_2$ and $C' = C\star_1$, in terms of the second spoiling operation described above.
- To illustrate the third spoiling operation, one can see, for instance, that $C(0,4) = \{\ell_1, \ell_3\}$, while $C(1,6) = \{\ell_2, \ell_3\}$.

2.4. The Asymptotic Bound

The spoiling operations on codes were used in [13] to prove the existence of an *asymptotic bound* in the space of code parameters (δ, R), see also [16,24,25] for more detailed properties of the asymptotic bound.

Let $\mathcal{V}_q \subset [0,1]^2 \cap \mathbb{Q}^2$ denote the space of code parameters (δ, R) of codes $C \subset \mathbb{F}_q^n$ and let $\mathcal{U}_q$ be the set of all limit points of $\mathcal{V}_q$. The set $\mathcal{U}_q$ is characterized in [13] as

$$\mathcal{U}_q = \{(\delta, R) \in [0,1]^2 \mid R \leq \alpha_q(\delta)\}$$

for a continuous, monotonically decreasing function $\alpha_q(\delta)$ (the asymptotic bound). Moreover, code parameters lying in $\mathcal{U}_q$ are realized with infinite multiplicity, while code points in $\mathcal{V}_q \setminus (\mathcal{V}_q \cap \mathcal{U}_q)$ have finite multiplicity and correspond to the *isolated codes*, see [13,16].

Codes lying above the asymptotic bound are codes which have extremely good transmission rate and relative minimum distance, hence very desirable from the coding theory perspective. The fact that the corresponding code parameters are not limit points of other code parameters and only have finite multiplicity reflect the fact that such codes are very difficult to reach or approximate. Isolated codes are known to arise from algebro-geometric constructions, [14,15].

Relatively little is known about the asymptotic bound: the question of the computability of the function $\alpha_q(\delta)$ was recently addressed in [25] and the relation to Kolmogorov complexity was investigated in [16]. There are explicit upper and lower bounds for the function $\alpha_q(\delta)$, see [12], including the Plotkin bound

$$\alpha_q(\delta) = 0, \quad \text{for} \quad \delta \geq \frac{q-1}{q}; \tag{5}$$

the singleton bound, which implies that $R = \alpha_q(\delta)$ lies below the line $R + \delta = 1$; the Hamming bound

$$\alpha_q(\delta) \leq 1 - H_q(\frac{\delta}{2}), \tag{6}$$

where $H_q(x)$ is the q-ary Shannon entropy

$$x \log_q(q-1) - x \log_q(x) - (1-x)\log_q(1-x)$$

which is the usual Shannon entropy for $q = 2$,

$$H_2(x) = -x\log_2(x) - (1-x)\log_2(1-x). \tag{7}$$

One also has a lower bound given by the Gilbert–Varshamov bound

$$\alpha_q(\delta) \geq 1 - H_q(\delta) \tag{8}$$

The Gilbert–Varshamov curve can be characterized in terms of the behavior of sufficiently random codes, in the sense of the Shannon Random Code Ensemble, see [26,27], while the asymptotic bound can be characterized in terms of Kolmogorov complexity, see [16].

2.5. Code Parameters of Language Families

From the coding theory viewpoint, it is natural to ask whether there are codes C, formed out of a choice of a collection of natural languages and their syntactic parameters, whose code parameters lie above the asymptotic bound curve $R = \alpha_2(\delta)$.

For instance, a code C whose code parameters violate the Plotkin bound (5) must be an isolated code above the asymptotic bound. This means constructing a code C with $\delta \geq 1/2$, that is, such that any pair of code words $w \neq w' \in C$ differ by at least half of the parameters. A direct examination of the list of parameters in Table A of [3] and Figure 7 of [4] shows that it is very difficult to find, within the same historical linguistic family (e.g., the Indo-European family) pairs of languages L_1, L_2 with $\delta_H(L_1, L_2) \geq 1/2$. For example, among the syntactic relative distances listed in Figure 7 of [4] one finds only the pair (Farsi, Romanian) with a relative distance of 0.5. Other pairs come close to this value, for example Farsi and French have a relative distance of 0.483, but French and Romanian only differ by 0.162.

One has better chances of obtaining codes above the asymptotic bound if one compares languages that are not so closely related at the historical level.

Example 2. Consider the set $C = \{L_1, L_2, L_3\}$ with languages L_1 = Arabic, L_2 = Wolof, and L_3 = Basque. We exclude from the list of Table A of [3] all those parameters that are entailed and made irrelevant by some other parameter in at least one of these three chosen languages. This gives us a list of 25 remaining parameters, which are those numbered as 1–5, 7, 10, 20–21, 25, 27–29, 31–32, 34, 37, 42, 50–53, 55–57 in [3], and the following three code words:

L_1	1	1	1	1	1	1	0	1	0	1	0	1	0	1	1	1	1	1	1	0	1	0	0	0	0
L_2	1	1	1	0	0	1	1	0	1	0	1	0	0	1	0	1	1	0	0	1	1	1	1	1	1
L_3	1	1	0	1	0	0	1	0	0	0	1	1	1	0	1	1	0	1	1	1	1	1	1	0	0

This example, although very simple and quite artificial in the choice of languages, already suffices to produce a code C that lies above the asymptotic bound. In fact, we have $d_H(L_1, L_2) = 16$, $d_H(L_2, L_3) = 13$ and $d_H(L_1, L_3) = 13$, so that $\delta = 0.52$. Since $R > 0$, the code point (δ, R) violates the Plotkin bound, hence it lies above the asymptotic bound.

It would be more interesting to find a code C consisting of languages belonging to the same historical-linguistic family (outside of the Indo-European group), that lies above the asymptotic bound. Such examples would correspond to linguistic families that exhibit a very strong variability of the syntactic parameters, in a way that is quantifiable through the properties of C as a code.

If one considers the 22 Indo-European languages in [3] with their parameters, one obtains a code C that is below the Gilbert–Varshamov line, hence below the asymptotic bound by Equation (8). A few other examples, taken from other non Indo-European historical-linguistic families, computed using those parameters reported in the SSWL database (for example the set of Malayo–Polynesian languages currently recorded in SSWL) also give codes whose code parameters lie below the Gilbert–Varshamov curve. One can conjecture that any code C constructed out of natural languages belonging to the same historical-linguistic family will be below the asymptotic bound (or perhaps below the GV bound), which would provide a quantitative bound on the possible spread of syntactic parameters within a historical family, given the size of the family. Examples like the simple one constructed above, using languages not belonging to the same historical family show that, to the contrary, across different historical families one encounters a greater variability of syntactic parameters. To our knowledge, no systematic study of parameter variability from this coding theory perspective has been implemented so far.

Ongoing work of the author is considering a systematic analysis of language families, based on the SSWL database of syntactic parameters, using this coding theory technique. This will include an analysis of how much the conclusions about the spreading of syntactic parameters across language families obtained with this technique depends on data pre-processing like the removal of spoiling features and what can be retained as an objective property of a set of languages. Moreover, a further purpose of this ongoing study is to combine the coding theory approach and the measures of complexity for groups of languages described in the present paper with the spin glass dynamical models of language change considered in [8], which was aimed at studying dynamically the spreading of syntactic parameters across groups of languages. The aim is to introduce complexity measures based on coding theory as part of the energy landscape of the spin glass model, following the suggestion of [28], on analogies between the roles of complexity in the theory of computation and energy in physical theories. These results, along with a more detailed analysis of the codes and code parameters of various language families, will appear in forthcoming work.

2.6. Comparison with Other Bounds

Another possible question one can consider in this setting is how the codes obtained from syntactic parameters of a given set of natural languages compare with other known families of error correcting codes and with other bounds in the space of code parameters.

For instance, it is known that an important improvement over the behavior of typical random codes can be obtained by considering codes determined by algebro-geometric curves defined over a finite field $\mathbb{F}_q$. Let $N_q(X) = \#X(\mathbb{F}_q)$ be the number of points over $\mathbb{F}_q$ of the curve X, and let $N_q(g) = \max N_q(X)$, with the maximum taken over all genus g curves X over $\mathbb{F}_q$. As shown in Theorem 2.3.22 of [12], asymptotically the $N_q(g)$ satisfy the Drinfeld–Vladut bound

$$A(q) := \limsup_{q\to\infty} \frac{N_q(g)}{g} \leq \sqrt{q} - 1,$$

and as shown in Section 3.4.1 of [12], this determines an algebro-geometric bound

$$\alpha_q(\delta) \geq R_{AG}(\delta) = 1 - \frac{1}{A(q)} - \delta$$

and the asymptotic Tsfasman–Vladut–Zink bound

$$\alpha_q(\delta) \geq R_{TVZ}(\delta) = 1 - (\sqrt{q} - 1)^{-1} - \delta.$$

The Tsfasman–Vladut–Zink line $R_{TVZ}(\delta) = 1 - (\sqrt{q} - 1)^{-1} - \delta$ lies entirely below the GV line for $q < 49$ (Theorem 3.4.4 of [12]).

A probabilistic argument given in Section 3.4.2 of [12] shows that highly non-random codes coming from algebraic curves can be asymptotically better than random codes (for sufficiently large q) as they cluster around the TVZ line. However, for $q = 2$ or $q = 3$, as in the case of codes from syntactic parameters of groups of languages that we consider here, the TVZ line lies below the GV line, hence any example that lies above the GV bound also behaves better than the the algebro-geometric bound. Such examples, like the one given above, for the three languages Arabic, Wolof, Basque, are very rare among codes obtained from syntactic parameters of languages, as they require the choice of a group of languages that are all very far from each other syntactically, with very large relative Hamming distances between syntactic parameters.

On the other hand, even for cases of groups of languages for which the resulting code parameters are below the GV line, it is still possible to get some additional information by comparing the position of the code parameters to other curves obtained from other bounds, such as the Blokh–Zyablow bound or the Katsman–Tsfasman–Vladut bound, see Appendix A.2.1 of [12] for a summary of all these different bounds.

For example, the first example given above, with the three languages Italian, Spanish, French and a string of six syntactic parameters, gives a code with code parameters that are below the GV line, but above both the Blokh–Zyablow and the Katsman–Tsfasman–Vladut, according to the table of asymptotic bounds given in Appendix A.2.4 of [12].

2.7. Entailment and Dependency of Parameters

In the discussion above we did not incorporate in our model the fact that certain syntactic parameters can entail other parameters in such a way that one particular value of one of the parameters renders another parameter irrelevant or not defined, see the discussion in Section 5.3.2 of [3].

One possible way to alter the previous construction to account for these phenomena is to consider the codes C associated to families of languages as codes in $\mathbb{F}_3^n$, where n is the number of parameters, as before, and the set of values is now given by $\{-1, 0, +1\} = \mathbb{F}_3$, with ± 1 corresponding to the binary values of the parameters that are set for a given language and value 0 assigned to those parameters that are made irrelevant for the given language, by entailment from other parameters, or are not defined. This allows us to consider the full range of parameters used in [3,4]. We revisit Example 2 considered above.

Example 3. Let $C = \{L_1, L_2, L_3\}$ be the code obtained from the languages L_1 = Arabic, L_2 = Wolof, and L_3 = Basque, as a code in $\mathbb{F}_3^n$ with $n = 63$, using the entire list of parameters in [3]. The code parameters $(R = 0.0252, \delta = 0.4643)$ of this code no longer violate the Plotkin bound. In fact, the parameters satisfy $R < 1 - H_3(\delta)$ so the code C now also lies below the GV bound.

Thus, the effect of including the entailed syntactic parameters in the comparison spoils the code parameters enough that they fall in the area below the GV bound.

Notice that what we propose here is different from the counting used in [3], where the relative distances $\delta_H(L_1, L_2)$ are normalized with respect to the number of non-zero parameters (which therefore varies with the choice of the pair (L_1, L_2)) rather than the total number n of parameters. While this has the desired effect of getting rid of insignificant parameters that spoil the code, it has the undesirable property of producing codes with code words of varying lengths, while counting only those parameters that have no zero-values over the entire family of languages, as in Example 2 avoids this problem. Adapting the coding theory results about the asymptotic bound to codes with words of variable length may be desirable for other reasons as well, but it will require an investigation beyond the scope of the present paper.

More generally, there are various kinds of dependencies among syntactic parameters. Some sets of hierarchical relations are discussed, for instance, in [29].

By the spoiling operations $C \star_i f$ of codes described above, we know that if some of the syntactic parameters considered are functions of other parameters, the resulting code parameters of $C \star_i f$ are worse than the parameters of the code C where only independent parameters were considered.

Part of the reason why code parameters of groups of languages in the family analyzed in [3] end up in the region below the asymptotic and the GV bound may be an artifact of the presence of dependences among the chosen 63 syntactic parameters. From the coding theory perspective, the parametric comparison method works best on a smaller set of independent parameters than on a larger set that includes several dependencies.

Entailment relations between syntactic parameters play an important role in the dynamical models of language evolutions constructed in [8], based on spin glass models in statistical physics.

Notice that the type of entailment relations we consider here are only of a rather special form, where a parameter is made undefined by effect of the value of another parameter (hence the use of the value 0 for the undetermined parameter). There are more general forms of entailment that we do not consider here, but which will be discussed in more detail in upcoming work. For example, one can have a situation with two languages in which a parameter is entailed by the values of two other parameters, but entailed to two different values in the two languages. In this case, the proposal above need to be modified, because this entailed parameter should contribute to the Hamming distance between the two languages. In such a situation the entailed parameter should increase, rather than spoil, the efficiency of the code. Keeping entailed parameters can be used for error-correcting purposes, as contributing to error detection. The role of entailment of parameters was considered in [8], in the use of spin glass models for language change, where the entailment relations appear as couplings at the vertices (interaction terms) between different Ising/Potts models on the same underlying graph of language interactions. In upcoming work, now in preparation, we will discuss how treating different forms of entailment of parameters in the coding theory setting described here related to the treatment of entailment relations in the spin glass model of [8].

3. Entropy and Complexity for Language Families

3.1. Why the Asymptotic Bound?

In the examples discussed above we compared the position of the code point associated to a given set of languages to certain curves in the space of code parameters. In particular, we focused on the asymptotic bound curve and the Gilbert–Varshamov curve. It should be pointed out that these two curves have a very different nature.

The asymptotic bound is the only curve that separates regions in the space of parameters that correspond to code points with entirely different behavior. As shown in [13,24], code points in the area below the asymptotic bound are realized with infinite multiplicity and fill densely the region, while code points that lie above the asymptotic bound are isolated and realized with finite multiplicity.

The Gilbert–Varshamov curve, by contrast, is related to the statistical behavior of sufficiently random codes (as we recall in Section 3.2 below), but does not separate two regions with significantly different behavior in the space of code points. Thus, in this respect, the asymptotic bound is a more natural curve to consider than the Gilbert–Varshamov curve.

Thus, a heuristic interpretation of the position of codes obtained from groups of languages, with respect to the asymptotic bound can be understood as follows. The position of a code point above or below the asymptotic bound reflects a very different behavior of the corresponding code with respect to how easily "deformable" it is. The sporadic codes that lie above the asymptotic bound are rigid objects, in contrast to the deformable objects below the asymptotic bound. In terms of properties of the distribution of syntactic parameters within a set of languages, this different nature of the associated code can be seen as a measure of the degree of "deformability" of the parameter distribution: in languages that belong to the same historical linguistic families, the parameter distribution has evolved historically along with the development of the family's phylogenetic tree, and one expects that correspondingly the code parameters will indicate a higher degree of "deformability" of the corresponding code. If a group of languages is chosen that belong to very different historical families, on the contrary, one expects that the distribution of syntactic parameters will not necessarily lead any longer to a code that has the same kind of deformability property: code points above the asymptotic bound may be realizable by this type of language groups.

There is no similar interpretation for the position of the code point with respect to the Gilbert–Varshamov line. An interpretation of that position can be sought in terms of Shannon entropy, as we discuss below. Summarizing: the main conceptual distinction between the Gilbert–Varshamov line and the asymptotic bound is that the GV line represents only a statistical phenomenon, as we review below, while the asymptotic bound represents a true separation between two classes of structurally different codes, in the sense explained above.

3.2. Entropy and Statistics of the Gilbert–Varshamov Line

The Gilbert–Varshamov line $R = 1 - H_q(\delta)$ can be characterized statisticallly. Such a statistical description can be obtained by considering the Shannon Random Code Ensemble (SRCE). These are random codes obtained by choosing code words as independent random variables with respect to a uniform Bernoulli measure, so that a code is described by a randomly chosen set of different words of length n occurring with probability q^{-n}, see [26,27]. There is no a priori reason why the type of codes we consider here, with code words formed using the syntactic parameters of natural languages, would be linear. Thus, we consider the general setting of unstructured codes, as in Section V of [27].

The Hamming volume $Vol_q(n, d = n\delta)$, that is, the number of words of length n at Hamming distance at most d from a given one, can be estimated in terms of the q-ary Shannon entropy

$$H_q(\delta) = \delta \log_q(q-1) - \delta \log_q \delta - (1-\delta) \log_q(1-\delta)$$

in the form

$$q^{(H_q(\delta)-o(1))n} \leq Vol_q(n, d = n\delta) = \sum_{j=0}^{d} \binom{n}{j} (q-1)^j \leq q^{H_q(\delta)n}.$$

The expectation value for the random variable counting the number of unordered pairs of distinct code words with Hamming distance at most d is then estimated as

$$\mathbb{E} \sim \binom{q^k}{2} Vol_q(n,d) q^{-n} \sim q^{n(H_q(\delta)-1+2R)+o(n)}.$$

This estimate is then used (see [26,27]) to show that the probability to have codes in the SRCE with $H_q(\delta) \geq \max\{1-2R, 0\} + \epsilon$ is bounded by $q^{-\epsilon n(1+o(1))}$. By a similar argument (see Section V of [27] and Proposition 2.2 of [16]) it is shown that the probability that $H_q(\delta) \geq 1 - R + \epsilon$ is bounded by $q^{-n\epsilon(1+o(1))}$.

While, by this type of argument, one can see the Gilbert–Varshamov line as representing the typical behavior of sufficiently random codes, the asymptotic bound does not have a similar statistical interpretation. It does have, however, a relation to Kolmogorov complexity, which is relevant to the point of view discussed in the present paper. The relation between asymptotic bound of error correcting codes and Kolmogorov complexity was described in [16]. We recall it in the rest of this section, along with its implications for the linguistic applications we are considering.

3.3. Kolmogorov Complexity

We refer the reader to [30] for an extensive treatment of Kolmogorov complexity and its properties. We recall here some basic facts we need in the following.

Let $T_{\mathcal{U}}$ be a universal Turing machine, that is, a Turing machine that can simulate any other arbitrary Turing machine, by reading on tape both the input and the description of the Turing machine it should simulate. A prefix Turing machine is a Turing machine with unidirectional input and output tapes and bidirectional work tapes. The set of programs P on which a prefix Turing machine halts forms a prefix code.

Given a string w in an alphabet $\mathfrak{A}$, the prefix Kolmogorov complexity is given by minimal length of a program for which the universal prefix Turing machine $T_{\mathcal{U}}$ outputs w,

$$\mathcal{K}_{T_{\mathcal{U}}}(w) = \min_{P: T_{\mathcal{U}}(P)=w} \ell(P).$$

There is a universality property. Namely, given any other prefix Turing machine T, one has

$$\mathcal{K}_T(w) \leq \mathcal{K}_{T_{\mathcal{U}}}(w) + c_T,$$

where the shift is by a bounded constant, independent of w. The constant c_T is the Kolmogorov complexity of the program needed to describe T so that $T_{\mathcal{U}}$ can simulate it.

A variant of the notion of Kolmogorov complexity described above is given by conditional Kolmogorov complexity,

$$\mathcal{K}_{T_{\mathcal{U}}}(w \mid \ell(w)) = \min_{P:T_{\mathcal{U}}(P,\ell(w))=w} \ell(P),$$

where the length $\ell(w)$ is given, and made available to the machine $T_{\mathcal{U}}$. One then has

$$\mathcal{K}(w \mid \ell(w)) \leq \ell(w) + c,$$

because if $\ell(w)$ is known, then a possible program is just to write out w. This means that then $\ell(w) + c$ is just number of bits needed for the transmission of w plus the print instructions.

An upper bound is given by

$$\mathcal{K}_{T_{\mathcal{U}}}(w) \leq \mathcal{K}_{T_{\mathcal{U}}}(w \mid \ell(w)) + 2\log \ell(w) + c.$$

If one does not know a priori $\ell(w)$, one needs to signal the end of the description of w. For this it suffices to have a "punctuation method", and one can see that this has the effect of adds the term $2\log \ell(w)$ in the above estimate. In particular, any program that produces a description of w is an upper bound on Kolmogorov complexity $\mathcal{K}_{T_{\mathcal{U}}}(w)$.

One can think of Kolmogorov complexity in terms of data compression: the shortest description of w is also its most compressed form. Upper bounds for Kolmogorov complexity are therefore provided easily by data compression algorithms. However, while providing upper bounds for complexity is straightforward, the situation with lower bounds is entirely different: constructing a lower bound runs into a fundamental obstacle caused by the fact that the halting problem is unsolvable. As a consequence, Kolmogorov complexity is not a computable function. Indeed, suppose one would list programs P_k (with increasing lengths) and run them through the machine $T_{\mathcal{U}}$. If the machine halts on P_k with output w, then $\ell(P_k)$ is an approximation to $\mathcal{K}_{T_{\mathcal{U}}}(w)$. However, there may be an earlier P_j in the list such that $T_{\mathcal{U}}$ has not yet halted on P_j. If $T_{\mathcal{U}}$ eventually halts also on P_j and outputs w, then $\ell(P_j)$ will be a better approximation to $\mathcal{K}_{T_{\mathcal{U}}}(w)$. So one would be able to compute $\mathcal{K}_{T_{\mathcal{U}}}(w)$ if one could tell exactly on which programs P_k the machine $T_{\mathcal{U}}$ halts, but that is indeed the unsolvable halting problem.

Kolmogorov complexity and Shannon entropy are related: one can view Shannon entropy as an averaged version of Kolmogorov complexity in the following sense (see Section 2.3 of [31]). Suppose given independent random variables X_k, distributed according to Bernoulli measure $\mathbb{P} = \{p_a\}_{a\in\mathfrak{A}}$ with $p_a = \mathbb{P}(X = a)$. The Shannon entropy is given by

$$S(X) = -\sum_{a\in\mathfrak{A}} \mathbb{P}(X = a) \log \mathbb{P}(X = a).$$

There exists a $c > 0$, such that, for all $n \in \mathbb{N}$,

$$S(X) \leq \frac{1}{n} \sum_{w\in\mathcal{W}^n} \mathbb{P}(w)\, \mathcal{K}(w \mid \ell(w)) \leq S(X) + \frac{\#\mathfrak{A} \log n}{n} + \frac{c}{n}.$$

The expectation value

$$\lim_{n\to\infty} \mathbb{E}(\frac{1}{n}\mathcal{K}(X_1 \cdots X_n \mid n)) = S(X)$$

shows that the average expected Kolmogorov complexity for length n descriptions approaches the Shannon entropy in the limit when $n \to \infty$.

3.4. Kolmogorov Complexity and the Asymptotic Bound

We recall here briefly the result of [16] linking the asymptotic bound of error correcting codes to Kolmogorov complexity.

As we discussed above, only the asymptotic bound marks a significant change of behavior of codes across the curve (isolated code points with finite multiplicity versus accumulation points with infinite multiplicity). In this sense this curve is very different from all the other bounds in the space of code parameters. However, there is no explicit expression for the curve $R = \alpha_q(\delta)$ that gives the asymptotic bound. Indeed, even the question of the computability of the function $R = \alpha_q(\delta)$ is a priori unclear. This question was formulated explicitly in [25].

It is proved in [16] that the asymptotic bound $R = \alpha_q(\delta)$ becomes computable given an oracle that can list codes by increasing Kolmogorov complexity. Given such an oracle, one can provide an explicit iterative (algorithmic) procedure for constructing the asymptotic bound. This implies that the asymptotic bound is "at worst as non-computable as Kolmogorov complexity".

Consider the set $X = \mathcal{C}_q$ of (unstructured) q-ary codes and the set $Y \subset [0,1]^2$ of code points and the computable function $f : X \to Y$ that assigns to a code $C \in X$ its code parameters $(R(C), \delta(C)) \in Y$. Let Y_{fin} and Y_∞ be, respectively, the subsets of the space of code points that correspond to code points realized with finite and with infinite multiplicity. The algorithm iteratively produces two sets A_m and B_m that approximate, respectively, Y_∞ and Y_{fin} by $Y_{fin} = \cup_{m\geq 1} B_m$ and $Y_\infty = \cup_{m\geq 1}(\cap_{n\geq 0} A_{m+n})$. The inductive construction starts by choosing an increasing sequence of positive integers N_m and setting $B_1 = \emptyset$ and taking A_1 to be the set of code points y with $\nu_Y^{-1}(y) \leq N_1$, where $\nu_Y : \mathbb{N} \to Y$ is a fixed enumeration of the set of rational points $[0,1]^2 \cap \mathbb{Q}^2$ where code points belong.

General estimates on the behavior of (exponential) Kolmogorov complexity under composition of total recursive functions (see [30], Section VI.9 of [32]) show that, for a composition $F = f_0(f_1(t_1, \ldots, t_m), \cdots, f_n(t_1, \ldots, t_m), t_{m+1}, \ldots, t_\ell)$ of recursive functions the Kolmogorov complexity satisfies

$$\mathcal{K}(F) \leq c \cdot \prod_{i=1}^{n} \mathcal{K}(f_i) \cdot \left(\log \prod_{i=1}^{n} \mathcal{K}(f_i) \right)^{n-1},$$

for a fixed f_0 and varying f_i, $i \geq 1$.

Consider the total recursive function $F(x) = (f(x), n(x))$ with

$$n(x) = \#\{x' \mid \nu_X^{-1}(x') \leq \nu_X^{-1}(x), \, f(x') = f(x)\}$$

where $\nu_X : \mathbb{N} \to X$ is an enumeration of the space of codes. Consider the enumerable sets $X_m := \{x \in X \mid n(x) = m\}$ and $Y_m := f(X_m) \subset Y$, with $Y_\infty = \cap_m f(X_m)$ and $Y_{fin} = f(X) \setminus Y_\infty$. For $\varphi : f(X) \to X_1$, defined as f^{-1} on $f(X_1) = f(X)$, applying the composition rule for exponential Kolmogorov complexity, it is shown in Proposition 3.1 of [16] that, for $x \in X_1$ and $y = f(x)$, one has $\mathcal{K}(x) = \mathcal{K}(\varphi(y)) \leq c_\varphi \cdot \mathcal{K}(y) \leq c\nu_Y^{-1}(y)$, hence

$$\mathcal{K}_{T_\mathcal{U}}(x) \leq c \cdot \nu_Y^{-1}(y).$$

Using the same type of estimate of Kolmogorov complexity for composition of recursive functions, it is then shown in Proposition 3.2 [16] that, for $y \in Y_\infty$ and $m \geq 1$, and for a unique $x_m \in X$, with $y = f(x_m)$, $n(x_m) = m$ and $c = c(f, u, v, \nu_X, \nu_Y) > 0$, one finds

$$\mathcal{K}_{T_\mathcal{U}}(x_m) \leq c \cdot \nu_Y^{-1}(y) \, m \, \log(\nu_Y^{-1}(y) m).$$

To construct inductively A_{m+1} and B_{m+1}, given A_m and B_m, one takes A_{m+1} to consist of the elements in the list

$$\mathcal{L}_{m+1} = \{y \in f(X) \,:\, \nu_Y^{-1}(y) \leq N_{m+1},\ \exists x \in X, \text{ with } y = f(x) \text{ and } n(x) = m+1\}.$$

Here one invokes the oracle, which ensures that, if such x exists, then it must be contained in a finite list of points $x \in X$ with bounded complexity

$$\mathcal{K}_{T_{\mathcal{U}}}(x_m) \leq c \cdot \nu_Y^{-1}(y)\, m\, \log(\nu_Y^{-1}(y) m).$$

One then takes B_{m+1} to consist of the remaining elements in the list $\mathcal{L}_{m+1}$. We refer the reader to [16] for a more detailed formulation.

More generally, the argument of [16] recalled above shows that, for a recursive function $f : \mathbb{Z}_+ \to \mathbb{Q}$, determining which values have infinite multiplicities is computable given an oracle that enumerate integers in order of Kolmogorov complexity.

As discussed in [16,24], the asymptotic bound can also be seen as the phase transition curve for a quantum statistical mechanical system constructed out of the space of codes, where the partition function of the system weights codes according to their Kolmogorov complexity. This is as close to a "statistical description" of the asymptotic bound that one can achieve.

In comparison with the behavior of random codes (codes whose complexity is comparable to their size), which concentrate in the region bounded by the Gilbert–Varshamov line, when ordering codes by complexity, non-random codes of lower complexity populate the region above, with code points accumulating in the intermediate region bounded by the asymptotic bound. That intermediate region thus, in a sense, reflects the difference between Shannon entropy and complexity.

3.5. Entropy and Complexity Estimates for Language Families

On the basis of the considerations of the previous sections and of the results of [16,24] recalled above, we propose a way to assign a quantitative estimate of entropy and complexity to a given set of natural languages.

As before let $C = \{L_1, \ldots, L_k\}$ be a binary (or ternary) code where the code words L_i are the binary (ternary) strings of syntactic parameters of a set of languages L_i. We define the *entropy* of the language family $\{L_1, \ldots, L_k\}$ as the q-ary Shannon entropy $H_q(\delta(C))$, where q is either 2 or 3 for binary or ternary codes, and $\delta(C)$ is the relative minimum distance parameter of the code C. We also define the *entropy gap* of the language family $\{L_1, \ldots, L_k\}$ as the value of $H_q(\delta(C)) - 1 + R(C)$, which measures the distance of the code point $(R(C), \delta(C))$ from the Gilbert–Varshamov line, that is, from the behavior of a typical random code.

As a source of estimates of complexity of a language family $\{L_1, \ldots, L_k\}$ one can consider any upper bound on Kolmogorov complexity of the code C. A possible approach, which contains more linguistic input, would be to provide estimates of complexity for each individual language in the family and then compare these. Estimates of complexity for individual languages have been considered in the literature, some of them based on the description of languages in terms of their syntactic parameters. For instance, following [18], for a syntactic parameter Π with possible values $v \in \{\pm 1\}$, the Kolmogorov complexity of Π set to value v is given by

$$\mathcal{K}(\Pi = v) = \min_{\tau \text{ expressing } \Pi} \mathcal{K}_{T_{\mathcal{U}}}(\tau),$$

with the minimum taken over the complexities of all the parse trees that express the syntactic parameter Π and require $\Pi = v$ to be grammatical in the language. Notice that, in this approach, the syntactic parameters are not just regarded as binary or ternary values, but one needs to consider actual parse trees of sentences in the language that express the parameter. Thus, such an approach to complexity has the advantage that it is very rich in linguistic information. However, it is at the same time computationally very difficult to realize.

What we are proposing here is a much simpler way to obtain an estimate of complexity for a language family $\{L_1, \ldots, L_k\}$, which is not based on estimating complexity of the individual languages in the family, but which is aimed at detecting how spread out and diversified the syntactic parameters are across the family, by estimating the position of the code point $(R(C), \delta(C))$ of the associated code C with respect to the asymptotic bound $R = \alpha_q(\delta)$. This can be estimated in terms of the distance to other curves in the space of code parameters (R, δ) that constrain the asymptotic bound from above and below, such as the Plotkin bound, Hamming bound, and Gilbert–Varshamov bound, as in the examples discussed in the previous sections.

4. Conclusions

We proposed an approach to estimating entropy and complexity of groups of natural languages (language families), based on the linguistic parametric comparison method (PCM) of [2,22] via the mathematical theory of error-correcting codes, by assigning a code to a family of languages to be analyzed with the PCM, and investigating its position in the space of code parameters, with respect to the asymptotic bound and the GV bound. We have shown that there are examples of languages not belonging to the same historical-linguistic family that yield isolated codes above the asymptotic bound, while languages belonging to the same historical-linguistic family appear to give rise to codes below the bound, though a more systematic analysis would be needed to map code parameters of different language groups. We have also shown that, from these coding theory perspective, it is preferable to exclude from the PCM all those parameters that are entailed and made irrelevant by other parameters, as those spoil the properties of the resulting code and produce code parameters that are artificially low with respect to the asymptotic bound and the GV bound.

The approach proposed here, based on the PCM and the theory of error-correcting codes, suggests a few new linguistic questions that may be suitable for treatment with coding theory methods:

1. Do languages belonging to the same historical-linguistic family always yield codes below the asymptotic bound or the GV bound? How often does the same happen across different linguistic families? How much can code parameters be improved by eliminating spoiling effects caused by dependencies and entailment of syntactic parameters?
2. Codes near the GV curve are typically coming from the Shannon Random Code Ensemble, where code words and letters of code words behave like independent random variables, see [26,27]. Are there families of languages whose associated codes are located near the GV bound? Do their syntactic parameters mimic the uniform Poisson distribution of random codes?
3. The asymptotic bound for error-correcting codes was related in [16] to Kolmogorov complexity, and the measure of complexity for language families that we proposed here is estimated in terms of the position of the code point with respect to the asymptotic bound. There are other notions of complexity, most notably the type of organized complexities discussed in [33–35]. Can these be related to loci in the space of code parameters? What do these represent when applied to codes obtained from syntactic parameters of a set of natural languages?
4. Is there a more direct linguistic complexity measure associated to a family of natural languages that would relate to the position of the resulting code above or below the asymptotic bound?
5. Codes and the asymptotic bound in the space of code parameters were recently studied using methods from quantum statistical mechanics, operator algebra and fractal geometry, [24,36]. Can some of these mathematical methods be employed in the linguistic parametric comparison method?

The observational results reported here are still preliminary. The following topics should be consolidated:

- How much the conclusions obtained for a given family of languages will depend on data pre-processing (removal of "spoiling" features, *etc.*)
- To what extent the proposed criterion (above or below the asymptotic bound) is really an objective property of a set of languages.

This will be addressed more thoroughly in future work. The concern about the effect of data pre-processing in paticular requires more analysis, that will be developed in further ongoing work, as outlined at the end of Section 2.5.

Acknowledgments: The author's research is supported by NSF grants DMS-1201512 and PHY-1205440, and by the Perimeter Institute for Theoretical Physics. The author thanks the referees for their useful comments.

References

1. Chomsky, N. *Lectures on Government and Binding*; Foris: Dordrecht, The Netherlands, 1981.
2. Longobardi, G. Methods in parametric linguistics and cognitive history. *Linguist. Var. Yearb.* **2003**, *3*, 101–138.
3. Longobardi, G.; Guardiano, C. Evidence for syntax as a signal of historical relatedness. *Lingua* **2009**, *119*, 1679–1706.
4. Longobardi, G.; Guardiano, C.; Silvestri, G.; Boattini, A.; Ceolin, A. Toward a syntactic phylogeny of modern Indo-European languages. *J. Hist. Linguist.* **2013**, *3*, 122–152.
5. Aziz, S.; Huynh, V.L.; Warrick, D.; Marcolli, M. Syntactic Phylogenetic Trees. 2016, In Preparation.
6. Park, J.J.; Boettcher, R.; Zhao, A.; Mun, A.; Yuh, K.; Kumar, V.; Marcolli, M. Prevalence and recoverability of syntactic parameters in sparse distributed memories. 2015, arXiv:1510.06342.
7. Port, A.; Gheorghita, I.; Guth, D.; Clark, J.M.; Liang, C.; Dasu, S.; Marcolli, M. Persistent Topology of Syntax. 2015, arXiv:1507.05134.
8. Siva, K.; Tao, J.; Marcolli, M. Spin Glass Models of Syntax and Language Evolution. 2015, arXiv:1508.00504.
9. Syntactic Structures of the World's Languages (SSWL) Database of Syntactic Parameters. Available online: http://sswl.railsplayground.net (accessed on 18 March 2016).
10. TerraLing. Available online: http://www.terraling.com (accessed on 18 March 2016).
11. Haspelmath, M.; Dryer, M.S.; Gil, D.; Comrie, B. *The World Atlas of Language Structures*; Oxford University Press: Oxford, UK, 2005.
12. Tsfasman, M.A.; Vladut, S.G. Algebraic-Geometric Codes. In *Mathematics and Its Applications (Soviet Series)*; Springer: Amsterdam, the Netherlands, 1991; Volume 58.
13. Manin, Y.I. What is the maximum number of points on a curve over $\mathbb{F}_2$? *J. Fac. Sci. Univ. Tokyo Sect. 1A Math.* **1982**, *28*, 715–720.
14. Tsfasman, M.A.; Vladut, S.G.; Zink, T. Modular curves, Shimura curves, and Goppa codes, better than Varshamov–Gilbert bound. *Math. Nachr.* **1982**, *109*, 21–28.
15. Vladut, S.G.; Drinfel'd, V.G. Number of points of an algebraic curve. *Funct. Anal. Appl.* **1983**, *17*, 68–69.
16. Manin, Y.I.; Marcolli, M. Kolmogorov complexity and the asymptotic bound for error-correcting codes. *J. Differ. Geom.* **2014**, *97*, 91–108.
17. Bane, M. Quantifying and measuring morphological complexity. In Proceedings of the 26th West Coast Conference on Formal Linguistics, Berkeley, CA, USA, 27–29 April 2007.
18. Clark, R. *Kolmogorov Complexity and the Information Content of Parameters*; Institute for Research in Cognitive Science: Philadelphia, PA, USA, 1994.
19. Tuza, Z. On the context-free production complexity of finite languages. *Discret. Appl. Math.* **1987**, *18*, 293–304.
20. Barton, G.E.; Berwick, R.C.; Ristad, E.S. *Computational Complexity and Natural Language*; MIT Press: Cambrige, MA, USA, 1987.
21. Sampson, G.; Gil, D.; Trudgill, P. (Eds.) *Language Complexity as an Evolving Variable*; Oxford University Press: Oxford, UK, 2009.
22. Longobardi, G. A minimalist program for parametric linguistics? In *Organizing Grammar: Linguistic Studies in Honor of Henk van Riemsdijk*; Broekhuis, H.; Corver, N.; Huybregts, M.; Kleinhenz, U.; Koster, J., Eds.; Mouton de Gruyter: Berlin, Germany, 2005; pp. 407–414.
23. Clark, R.; Roberts, I. A computational model of language learnability and language change. *Linguist. Inq.* **1993**, *24*, 299–345.

24. Manin, Y.I.; Marcolli, M. Error-correcting codes and phase transitions. *Math. Comput. Sci.* **2001**, *5*, 133–170.
25. Manin, Y.I. A computability challenge: Asymptotic bounds and isolated error-correcting codes. 2011, arXiv:1107.4246.
26. Barg, A.; Forney, G.D. Random codes: minimum distances and error exponents. *IEEE Trans. Inf. Theory* **2002**, *48*, 2568–2573.
27. Coffey, J.T.; Goodman, R.M. Any code of which we cannot think is good. *IEEE Trans. Inf. Theory* **1990**, *36*, 1453–1461.
28. Manin, Y.I. Complexity vs Energy: Theory of Computation and Theoretical Physics. 2014, arXiv:1302.6695.
29. Baker, M.C. *The Atoms of Language: The Mind's Hidden Rules of Grammar*; Basic Books: New York, NY, USA, 2001.
30. Li, M.; Vitányi, P. *An Introduction to Kolmogorov Complexity and Its Applications*; Springer: New York, NY, USA, 2008.
31. Grünwald, P.; Vitányi, P. Shannon Information and Kolmogorov Complexity. 2004, arXiv:cs/0410002.
32. Manin, Y.I. *A Course in Mathematical Logic for Mathematicians*, 2nd ed; Springer: New York, NY, USA, 2010.
33. Bennett, C.; Gacs, P.; Li, M.; Vitanyi, P.; Zurek, W. Information distance. *IEEE Trans. Inf. Theory* **1998**, *44*, 1407–1423.
34. Delahaye, J.P. *Complexité Aléatoire et Complexité Organisée*; Éditions Quæ: Paris, France, 2009. (In French)
35. Gell-Mann, M.; Lloyd, S. Information measures, effective complexity, and total information. *Complexity* **1996**, *2*, 44–52.
36. Marcolli, M.; Perez, C. Codes as fractals and noncommutative spaces. *Math. Comput. Sci.* **2012**, *6*, 199–215.

8

Guaranteed Bounds on Information-Theoretic Measures of Univariate Mixtures using Piecewise Log-Sum-Exp Inequalities

Frank Nielsen [1,2,*] **and Ke Sun** [3]

1 Computer Science Department LIX, École Polytechnique, 91128 Palaiseau Cedex, France
2 Sony Computer Science Laboratories Inc, Tokyo 141-0022, Japan
3 King Abdullah University of Science and Technology, Thuwal 23955, Saudi Arabia; sunk.edu@gmail.com
* Correspondence: Frank.Nielsen@acm.org.

Academic Editor: Antonio M. Scarfone

Abstract: Information-theoretic measures, such as the entropy, the cross-entropy and the Kullback–Leibler divergence between two mixture models, are core primitives in many signal processing tasks. Since the Kullback–Leibler divergence of mixtures provably does not admit a closed-form formula, it is in practice either estimated using costly Monte Carlo stochastic integration, approximated or bounded using various techniques. We present a fast and generic method that builds algorithmically closed-form lower and upper bounds on the entropy, the cross-entropy, the Kullback–Leibler and the α-divergences of mixtures. We illustrate the versatile method by reporting our experiments for approximating the Kullback–Leibler and the α-divergences between univariate exponential mixtures, Gaussian mixtures, Rayleigh mixtures and Gamma mixtures.

Keywords: information geometry; mixture models; α-divergences; log-sum-exp bounds

---

1. Introduction

Mixture models are commonly used in signal processing. A typical scenario is to use mixture models [1–3] to smoothly model histograms. For example, Gaussian Mixture Models (GMMs) can be used to convert grey-valued images into binary images by building a GMM fitting the image intensity histogram and then choosing the binarization threshold as the average of the Gaussian means [1]. Similarly, Rayleigh Mixture Models (RMMs) are often used in ultrasound imagery [2] to model histograms, and perform segmentation by classification. When using mixtures, a fundamental primitive is to define a proper statistical distance between them. The Kullback–Leibler (KL) divergence [4], also called relative entropy or information discrimination, is the most commonly used distance. Hence the main target of this paper is to faithfully measure the KL divergence. Let $m(x) = \sum_{i=1}^{k} w_i p_i(x)$ and $m'(x) = \sum_{i=1}^{k'} w'_i p'_i(x)$ be two finite statistical density mixtures of k and k' components, respectively. Notice that the Cumulative Density Function (CDF) of a mixture is like its density also a convex combinations of the component CDFs. However, beware that a mixture is not a sum of random variables (RVs). Indeed, sums of RVs have convolutional densities. In statistics, the mixture components $p_i(x)$ are often parametric: $p_i(x) = p(x;\theta_i)$, where θ_i is a vector of parameters. For example, a mixture of Gaussians (MoG also used as a shortcut instead of GMM) has each component distribution parameterized by its mean μ_i and its covariance matrix Σ_i (so that the parameter vector is $\theta_i = (\mu_i, \Sigma_i)$). Let $\mathcal{X} = \{x \in \mathbb{R} : p(x;\theta) > 0\}$ be the support of the component distributions. Denote by $H_\times(m,m') = -\int_{\mathcal{X}} m(x) \log m'(x) \mathrm{d}x$ the cross-entropy [4] between two continuous mixtures of densities m and m', and denote by $H(m) = H_\times(m,m) = \int_\mathcal{X} m(x)\log\frac{1}{m(x)}\mathrm{d}x = -\int_\mathcal{X} m(x)\log m(x)\mathrm{d}x$ the Shannon entropy [4]. Then the Kullback–Leibler divergence between m and m' is given by:

$$\mathrm{KL}(m : m') = H_\times(m,m') - H(m) = \int_\mathcal{X} m(x)\log\frac{m(x)}{m'(x)}\mathrm{d}x \geq 0. \tag{1}$$

The notation ":" is used instead of the usual comma "," notation to emphasize that the distance is not a metric distance since neither is it symmetric ($\mathrm{KL}(m : m') \neq \mathrm{KL}(m' : m)$), nor does it satisfy the triangular inequality [4] of metric distances ($\mathrm{KL}(m : m') + \mathrm{KL}(m' : m'') \not\geq \mathrm{KL}(m : m'')$). When the natural base of the logarithm is chosen, we get a differential entropy measure expressed in nat units. Alternatively, we can also use the base-2 logarithm ($\log_2 x = \frac{\log x}{\log 2}$) and get the entropy expressed in bit units. Although the KL divergence is available in closed-form for many distributions (in particular as equivalent Bregman divergences for exponential families [5], see Appendix C), it was proven that the Kullback–Leibler divergence between two (univariate) GMMs is not analytic [6] (see also the particular case of a GMM of two components with the same variance that was analyzed in [7]). See Appendix A for an analysis. Note that the differential entropy may be negative. For example, the differential entropy of a univariate Gaussian distribution is $\log(\sigma\sqrt{2\pi e})$, and is therefore negative when the standard variance $\sigma < \frac{1}{\sqrt{2\pi e}} \approx 0.242$. We consider continuous distributions with entropies well-defined (entropy may be undefined for singular distributions like Cantor's distribution [8]).

1.1. Prior Work

Many approximation techniques have been designed to beat the computationally intensive Monte Carlo (MC) stochastic estimation: $\widehat{\mathrm{KL}}_s(m : m') = \frac{1}{s}\sum_{i=1}^{s}\log\frac{m(x_i)}{m'(x_i)}$ with $x_1,\ldots,x_s \sim m(x)$ (s independently and identically distributed (i.i.d.) samples $x_1,\ldots,x_s$). The MC estimator is asymptotically consistent, $\lim_{s\to\infty}\widehat{\mathrm{KL}}_s(m : m') = \mathrm{KL}(m : m')$, so that the "true value" of the KL of mixtures is estimated in practice by taking a very large sample (say, $s = 10^9$). However, we point out that the MC estimator gives as output a stochastic *approximation*, and therefore does not guarantee deterministic bounds (confidence intervals may be used). Deterministic lower and upper bounds of the integral can be obtained by various numerical integration techniques using quadrature rules. We refer to [9–12] for the current state-of-the-art approximation techniques and bounds on the KL of GMMs. The latest work for computing the entropy of GMMs is [13]. It considers arbitrary finely tuned bounds of the entropy of isotropic Gaussian mixtures (a case encountered when dealing with KDEs, kernel density estimators). However, there is a catch in the technique of [13]: It relies on solving the unique roots of some log-sum-exp equations (See Theorem 1 of [13], p. 3342) that do not admit a closed-form solution. Thus it is a hybrid method that contrasts with our combinatorial approach. Bounds of the KL divergence between mixture models can be generalized to bounds of the likelihood function of mixture models [14], because log-likelihood is just the KL between the empirical distribution and the mixture model up to a constant shift.

In information geometry [15], a mixture family of linearly independent probability distributions $p_1(x),\ldots,p_k(x)$ is defined by the convex combination of those non-parametric component distributions: $m(x;\eta) = \sum_{i=1}^{k}\eta_i p_i(x)$ with $\eta_i > 0$ and $\sum_{i=1}^{k}\eta_i = 1$. A mixture family induces a dually flat space where the Kullback–Leibler divergence is equivalent to a Bregman divergence [5,15] defined on the η-parameters. However, in that case, the Bregman convex generator $F(\eta) = \int m(x;\eta)\log m(x;\eta)\mathrm{d}x$ (the Shannon information) is not available in closed-form. Except for the family of multinomial distributions that is both a mixture family (with closed-form $\mathrm{KL}(m : m') = \sum_{i=1}^{k} m_i\log\frac{m_i}{m'_i}$, the discrete KL [4]) and an exponential family [15].

1.2. Contributions

In this work, we present a simple and efficient method that builds algorithmically a closed-form formula that guarantees both deterministic lower and upper bounds on the KL divergence within an additive factor of $\log k + \log k'$. We then further refine our technique to get improved adaptive bounds. For univariate GMMs, we get the non-adaptive bounds in $O(k \log k + k' \log k')$ time, and the adaptive bounds in $O(k^2 + k'^2)$ time. To illustrate our generic technique, we demonstrate it based on Exponential Mixture Models (EMMs), Gamma mixtures, RMMs and GMMs. We extend our preliminary results on KL divergence [16] to other information theoretical measures such as the differential entropy and α-divergences.

1.3. Paper Outline

The paper is organized as follows. Section 2 describes the algorithmic construction of the formula using piecewise log-sum-exp inequalities for the cross-entropy and the Kullback–Leibler divergence. Section 3 instantiates this algorithmic principle to the entropy and discusses related works. Section 4 extends the proposed bounds to the family of alpha divergences. Section 5 discusses an extension of the lower bound to f-divergences. Section 6 reports our experimental results on several mixture families. Finally, Section 7 concludes this work by discussing extensions to other statistical distances. Appendix A proves that the Kullback–Leibler divergence of mixture models is not analytic [6]. Appendix B reports the closed-form formula for the KL divergence between scaled and truncated distributions of the same exponential family [17] (that include Rayleigh, Gaussian and Gamma distributions among others). Appendix C shows that the KL divergence between two mixtures can be approximated by a Bregman divergence.

2. A Generic Combinatorial Bounding Algorithm Based on Density Envelopes

Let us bound the cross-entropy $H_\times(m : m')$ by deterministic lower and upper bounds, $L_\times(m : m') \leq H_\times(m : m') \leq U_\times(m : m')$, so that the bounds on the Kullback–Leibler divergence $\mathrm{KL}(m : m') = H_\times(m : m') - H_\times(m : m)$ follows as:

$$L_\times(m : m') - U_\times(m : m) \leq \mathrm{KL}(m : m') \leq U_\times(m : m') - L_\times(m : m). \tag{2}$$

Since the cross-entropy of two mixtures $\sum_{i=1}^{k} w_i p_i(x)$ and $\sum_{j=1}^{k'} w'_j p'_j(x)$:

$$H_\times(m : m') = -\int_{\mathcal{X}} \left(\sum_{i=1}^{k} w_i p_i(x) \right) \log \left(\sum_{j=1}^{k'} w'_j p'_j(x) \right) \mathrm{d}x \tag{3}$$

has a log-sum term of positive arguments, we shall use bounds on the log-sum-exp (lse) function [18,19]:

$$\mathrm{lse}\left(\{x_i\}_{i=1}^{l} \right) = \log \left(\sum_{i=1}^{l} e^{x_i} \right).$$

We have the following basic inequalities:

$$\max\{x_i\}_{i=1}^{l} < \mathrm{lse}\left(\{x_i\}_{i=1}^{l} \right) \leq \log l + \max\{x_i\}_{i=1}^{l}. \tag{4}$$

The left-hand-side (LHS) strict inequality holds because $\sum_{i=1}^{l} e^{x_i} > \max\{e^{x_i}\}_{i=1}^{l} = \exp\left(\max\{x_i\}_{i=1}^{l}\right)$ since $e^x > 0, \forall x \in \mathbb{R}$. The right-hand-side (RHS) inequality follows from the fact that $\sum_{i=1}^{l} e^{x_i} \leq l \max\{e^{x_i}\}_{i=1}^{l} = l \exp(\max\{x_i\}_{i=1}^{l})$, and equality holds if and only if $x_1 = \cdots = x_l$. The lse function is convex but not strictly convex, see exercise 7.9 [20]. It is known [21] that the conjugate of the lse function is the negative entropy restricted to the probability simplex. The lse function enjoys the following translation identity property: $\mathrm{lse}\left(\{x_i\}_{i=1}^{l} \right) = c + \mathrm{lse}\left(\{x_i - c\}_{i=1}^{l} \right), \forall c \in \mathbb{R}$. Similarly, we can also lower bound the lse function by $\log l + \min\{x_i\}_{i=1}^{l}$. We write equivalently that for l positive numbers $x_1, \ldots, x_l$,

$$\max\left\{\log \max\{x_i\}_{i=1}^{l}, \log l + \log \min\{x_i\}_{i=1}^{l}\right\} \leq \log \sum_{i=1}^{l} x_i \leq \log l + \log \max\{x_i\}_{i=1}^{l}. \quad (5)$$

In practice, we seek matching lower and upper bounds that minimize the bound gap. The gap of that ham-sandwich inequality in Equation (5) is $\min\{\log \frac{\max_i x_i}{\min_i x_i}, \log l\}$, which is upper bounded by $\log l$.

A mixture model $\sum_{j=1}^{k'} w'_j p'_j(x)$ must satisfy

$$\begin{aligned} &\max\left\{\max\{\log w'_j p'_j(x)\}_{j=1}^{k'},\ \log k' + \min\{\log w'_j p'_j(x)\}_{j=1}^{k'}\right\} \\ &\leq \log\left(\sum_{j=1}^{k'} w'_j p'_j(x)\right) \leq \log k' + \max\{\log w'_j p'_j(x)\}_{j=1}^{k'} \end{aligned} \quad (6)$$

point-wisely for any $x \in \mathcal{X}$. Therefore we shall bound the integral term $\int_{\mathcal{X}} m(x) \log\left(\sum_{j=1}^{k'} w'_j p'_j(x)\right) \mathrm{d}x$ in Equation (3) using piecewise lse inequalities where the min and max are kept unchanged. We get

$$L_\times(m : m') = -\int_{\mathcal{X}} m(x) \max\{\log w'_j p'_j(x)\}_{j=1}^{k'} \mathrm{d}x - \log k', \quad (7)$$

$$U_\times(m : m') = -\int_{\mathcal{X}} m(x) \max\left\{\min\{\log w'_j p'_j(x)\}_{j=1}^{k'} + \log k',\ \max\{\log w'_j p'_j(x)\}_{j=1}^{k'}\right\} \mathrm{d}x. \quad (8)$$

In order to calculate $L_\times(m : m')$ and $U_\times(m : m')$ efficiently using closed-form formula, let us compute the upper and lower envelopes of the k' real-valued functions $\{w'_j p'_j(x)\}_{j=1}^{k'}$ defined on the support $\mathcal{X}$, that is, $\mathcal{E}_{\mathrm{U}}(x) = \max\{w'_j p'_j(x)\}_{j=1}^{k'}$ and $\mathcal{E}_{\mathrm{L}}(x) = \min\{w'_j p'_j(x)\}_{j=1}^{k'}$. These envelopes can be computed exactly using techniques of computational geometry [22,23] provided that we can calculate the roots of the equation $w'_r p'_r(x) = w'_s p'_s(x)$, where $w'_r p'_r(x)$ and $w'_s p'_s(x)$ are a pair of weighted components. (Although this amounts to solve quadratic equations for Gaussian or Rayleigh distributions, the roots may not always be available in closed form, e.g. in the case of Weibull distributions.)

Let the envelopes be combinatorially described by ℓ elementary interval pieces in the form $I_r = (a_r, a_{r+1})$ partitioning the support $\mathcal{X} = \uplus_{r=1}^{\ell} I_r$ (with $a_1 = \min \mathcal{X}$ and $a_{\ell+1} = \max \mathcal{X}$). Observe that on each interval I_r, the maximum of the functions $\{w'_j p'_j(x)\}_{j=1}^{k'}$ is given by $w'_{\delta(r)} p'_{\delta(r)}(x)$, where $\delta(r)$ indicates the weighted component dominating all the others, i.e., the arg max of $\{w'_j p'_j(x)\}_{j=1}^{k'}$ for any $x \in I_r$, and the minimum of $\{w'_j p'_j(x)\}_{j=1}^{k'}$ is given by $w'_{\epsilon(r)} p'_{\epsilon(r)}(x)$.

To fix ideas, when mixture components are univariate Gaussians, the upper envelope $\mathcal{E}_{\mathrm{U}}(x)$ amounts to find equivalently the lower envelope of k' parabolas (see Figure 1) which has linear complexity, and can be computed in $O(k' \log k')$-time [24], or in output-sensitive time $O(k' \log \ell)$ [25], where ℓ denotes the number of parabola segments in the envelope. When the Gaussian mixture components have all the same weight and variance (e.g., kernel density estimators), the upper envelope amounts to find a lower envelope of cones: $\min_j |x - \mu'_j|$ (a Voronoi diagram in arbitrary dimension).

Figure 1. Lower envelope of parabolas corresponding to the upper envelope of weighted components of a Gaussian mixture with $k' = 3$ components.

To proceed once the envelopes have been built, we need to calculate two types of definite integrals on those elementary intervals: (i) the probability mass in an interval $\int_a^b p(x)\mathrm{d}x = \Phi(b) - \Phi(a)$ where Φ denotes the Cumulative Distribution Function (CDF); and (ii) the partial cross-entropy $-\int_a^b p(x)\log p'(x)\mathrm{d}x$ [26]. Thus let us define these two quantities:

$$\begin{aligned} C_{i,j}(a,b) &= -\int_a^b w_i p_i(x)\log(w'_j p'_j(x))\mathrm{d}x, &(9)\\ M_i(a,b) &= -\int_a^b w_i p_i(x)\mathrm{d}x. &(10)\end{aligned}$$

By Equations (7) and (8), we get the bounds of $H_\times(m:m')$ as

$$\begin{aligned} L_\times(m:m') &= \sum_{r=1}^{\ell}\sum_{s=1}^{k} C_{s,\delta(r)}(a_r,a_{r+1}) - \log k',\\ U_\times(m:m') &= \sum_{r=1}^{\ell}\sum_{s=1}^{k} \min\left\{C_{s,\delta(r)}(a_r,a_{r+1}),\ C_{s,\epsilon(r)}(a_r,a_{r+1}) - M_s(a_r,a_{r+1})\log k'\right\}. \end{aligned} \tag{11}$$

The size of the lower/upper bound formula depends on the envelope complexity ℓ, the number k of mixture components, and the closed-form expressions of the integral terms $C_{i,j}(a,b)$ and $M_i(a,b)$. In general, when a pair of weighted component densities intersect in at most p points, the envelope complexity is related to the Davenport–Schinzel sequences [27]. It is quasi-linear for bounded $p = O(1)$, see [27].

Note that in symbolic computing, the Risch semi-algorithm [28] solves the problem of computing indefinite integration in terms of elementary functions provided that there exists an oracle (hence the term "semi-algorithm") for checking whether an expression is equivalent to zero or not (however it is unknown whether there exists an algorithm implementing the oracle or not).

We presented the technique by bounding the cross-entropy (and entropy) to deliver lower/upper bounds on the KL divergence. When only the KL divergence needs to be bounded, we rather consider the ratio term $\frac{m(x)}{m'(x)}$. This requires to partition the support $\mathcal{X}$ into elementary intervals by overlaying the critical points of both the lower and upper envelopes of $m(x)$ and $m'(x)$, which can be done in linear time. In a given elementary interval, since $\max\{k\min_i\{w_ip_i(x)\}, \max_i\{w_ip_i(x)\}\} \le m(x) \le k\max_i\{w_ip_i(x)\}$, we then consider the inequalities:

$$\frac{\max\{k\min_i\{w_ip_i(x)\}, \max_i\{w_ip_i(x)\}\}}{k'\max_j\{w'_jp'_j(x)\}} \le \frac{m(x)}{m'(x)} \le \frac{k\max_i\{w_ip_i(x)\}}{\max\{k'\min_j\{w'_jp'_j(x)\}, \max_j\{w'_jp'_j(x)\}\}}. \tag{12}$$

We now need to compute definite integrals of the form $\int_a^b w_1p(x;\theta_1)\log\frac{w_2p(x;\theta_2)}{w_3p(x;\theta_3)}\mathrm{d}x$ (see Appendix B for explicit formulas when considering scaled and truncated exponential families [17]). (Thus for exponential families, the ratio of densities removes the auxiliary carrier measure term.)

We call these bounds CELB and CEUB for Combinatorial Envelope Lower and Upper Bounds, respectively.

2.1. Tighter Adaptive Bounds

We shall now consider shape-dependent bounds improving over the additive $\log k + \log k'$ non-adaptive bounds. This is made possible by a decomposition of the lse function explained as follows. Let $t_i(x_1, \ldots, x_k) = \log\left(\sum_{j=1}^{k} e^{x_j - x_i}\right)$. By translation identity of the lse function,

$$\operatorname{lse}(x_1, \ldots, x_k) = x_i + t_i(x_1, \ldots, x_k) \tag{13}$$

for all $i \in [k]$. Since $e^{x_j - x_i} = 1$ if $j = i$, and $e^{x_j - x_i} > 0$, we have necessarily $t_i(x_1, \ldots, x_k) > 0$ for any $i \in [k]$. Since Equation (13) is an identity for all $i \in [k]$, we minimize the residual $t_i(x_1, \ldots, x_k)$ by maximizing x_i. Denoting by $x_{(1)}, \ldots, x_{(k)}$ the sequence of numbers sorted in non-decreasing order, the decomposition

$$\operatorname{lse}(x_1, \ldots, x_k) = x_{(k)} + t_{(k)}(x_1, \ldots, x_k) \tag{14}$$

yields the smallest residual. Since $x_{(j)} - x_{(k)} \le 0$ for all $j \in [k]$, we have

$$t_{(k)}\,(x_1, \ldots, x_k) = \log\left(1 + \sum_{j=1}^{k-1} e^{x_{(j)} - x_{(k)}}\right) \le \log k.$$

This shows the bounds introduced earlier can indeed be improved by a more accurate computation of the residual term $t_{(k)}\,(x_1, \ldots, x_k)$.

When considering 1D GMMs, let us now bound $t_{(k)}(x_1, \ldots, x_k)$ in a combinatorial range $I_r = (a_r, a_{r+1})$. Let $\delta = \delta(r)$ denote the index of the dominating weighted component in this range. Then,

$$\forall x \in I_r, \forall i, \quad \exp\left(-\log\sigma_i - \frac{(x-\mu_i)^2}{2\sigma_i^2} + \log w_i\right) \le \exp\left(-\log\sigma_\delta - \frac{(x-\mu_\delta)^2}{2\sigma_\delta^2} + \log w_\delta\right).$$

Thus we have:

$$\log m(x) = \log\frac{w_\delta}{\sigma_\delta\sqrt{2\pi}} - \frac{(x-\mu_\delta)^2}{2\sigma_\delta^2} + \log\left(1 + \sum_{i \ne \delta} \exp\left(-\frac{(x-\mu_i)^2}{2\sigma_i^2} + \log\frac{w_i}{\sigma_i} + \frac{(x-\mu_\delta)^2}{2\sigma_\delta^2} - \log\frac{w_\delta}{\sigma_\delta}\right)\right).$$

Now consider the ratio term:

$$\rho_{i,\delta}(x) = \exp\left(-\frac{(x-\mu_i)^2}{2\sigma_i^2} + \log\frac{w_i\sigma_\delta}{w_\delta\sigma_i} + \frac{(x-\mu_\delta)^2}{2\sigma_\delta^2}\right).$$

It is maximized in $I_r = (a_r, a_{r+1})$ by maximizing equivalently the following quadratic equation:

$$l_{i,\delta}(x) = -\frac{(x-\mu_i)^2}{2\sigma_i^2} + \log\frac{w_i\sigma_\delta}{w_\delta\sigma_i} + \frac{(x-\mu_\delta)^2}{2\sigma_\delta^2}.$$

Setting the derivative to zero ($l'_{i,\delta}(x) = 0$), we get the root (when $\sigma_i \ne \sigma_\delta$)

$$x_{i,\delta} = \left(\frac{\mu_\delta}{\sigma_\delta^2} - \frac{\mu_i}{\sigma_i^2}\right) \Big/ \left(\frac{1}{\sigma_\delta^2} - \frac{1}{\sigma_i^2}\right).$$

If $x_{i,\delta} \in I_r$, the ratio $\rho_{i,\delta}(x)$ can be bounded in the slab I_r by considering the extreme values of the three element set $\{\rho_{i,\delta}(a_r), \rho_{i,\delta}(x_{i,\delta}), \rho_{i,\delta}(a_{r+1})\}$. Otherwise $\rho_{i,\delta}(x)$ is monotonic in I_r, its bounds in I_r

are given by $\{\rho_{i,\delta}(a_r), \rho_{i,\delta}(a_{r+1})\}$. In any case, let $\rho_{i,\delta}^{\min}(r)$ and $\rho_{i,\delta}^{\max}(r)$ represent the resulting lower and upper bounds of $\rho_{i,\delta}(x)$ in I_r. Then t_δ is bounded in the range I_r by:

$$0 < \log\left(1 + \sum_{i \neq \delta} \rho_{i,\delta}^{\min}(r)\right) \leq t_\delta \leq \log\left(1 + \sum_{i \neq \delta} \rho_{i,\delta}^{\max}(r)\right) \leq \log k.$$

In practice, we always get better bounds using the shape-dependent technique at the expense of computing overall $O(k^2)$ intersection points of the pairwise densities. We call those bounds CEALB and CEAUB for Combinatorial Envelope Adaptive Lower Bound and Combinatorial Envelope Adaptive Upper Bound.

Let us illustrate one scenario where this adaptive technique yields very good approximations. Consider a GMM with all variance σ^2 tending to zero (a mixture of k Diracs). Then in a combinatorial slab I_r, we have $\rho_{i,\delta}^{\max}(r) \to 0$ for all $i \neq \delta$, and therefore we get tight bounds.

As a related technique, we could also upper bound $\int_{a_r}^{a_{r+1}} \log m(x) \mathrm{d}x$ by $(a_{r+1} - a_r) \log m(a_r, a_{r+1})$ where $m(x, x')$ denotes the maximal value of the mixture density in the range (x, x'). This maximal value is either found at the slab extremities, or is a mode of the GMM. It then requires to find the modes of a GMM [29,30], for which no analytical solution is known in general.

2.2. Another Derivation Using the Arithmetic-Geometric Mean Inequality

Let us start by considering the inequality of arithmetic and geometric weighted means (AGI, Arithmetic-Geometric Inequality) applied to the mixture component distributions:

$$m(x) = \sum_{i=1}^{k} w_i p(x; \theta_i) \geq \prod_{i=1}^{k} p(x; \theta_i)^{w_i}$$

with equality holds iff. $\theta_1 = \ldots = \theta_k$.

To get a tractable formula with a positive remainder of the log-sum term $\log m(x)$, we need to have the log argument greater or equal to 1, and thus we shall write the positive remainder:

$$R(x) = \log\left(\frac{m(x)}{\prod_{i=1}^{k} p(x; \theta_i)^{w_i}}\right) \geq 0.$$

Therefore, we can decompose the log-sum into a tractable part and a remainder as:

$$\log m(x) = \sum_{i=1}^{k} w_i \log p(x; \theta_i) + \log\left(\frac{m(x)}{\prod_{i=1}^{k} p(x; \theta_i)^{w_i}}\right). \tag{15}$$

For exponential families, the first term can be integrated accurately. For the second term, we notice that $\prod_{i=1}^{k} p(x; \theta_i)^{w_i}$ is a distribution in the same exponential family. We denote $p(x; \theta_0) = \prod_{i=1}^{k} p(x; \theta_i)^{w_i}$. Then

$$R(x) = \log\left(\sum_{i=1}^{k} w_i \frac{p(x; \theta_i)}{p(x; \theta_0)}\right)$$

As the ratio $p(x; \theta_i)/p(x; \theta_0)$ can be bounded above and below using techniques in Section 2.1, $R(x)$ can be correspondingly bounded. Notice the similarity between Equations (14) and (15). The key difference with the adaptive bounds is that, here we choose $p(x; \theta_0)$ instead of the dominating component in $m(x)$ as the "reference distribution" in the decomposition. This subtle difference is not presented in detail in our experimental studies but discussed here for completeness. Essentially, the gap of the bounds is up to the difference between the geometric average and the arithmetic average. In the extreme case that all mixture components are identical, this gap will reach zero. Therefore we expect good quality bounds with a small gap when the mixture components are similar as measured by KL divergence.

2.3. Case Studies

In the following, we instantiate the proposed method for several prominent cases on the mixture of exponential family distributions.

2.3.1. The Case of Exponential Mixture Models

An exponential distribution has density $p(x;\lambda) = \lambda\exp(-\lambda x)$ defined on $\mathcal{X} = [0,\infty)$ for $\lambda > 0$. Its CDF is $\Phi(x;\lambda) = 1 - \exp(-\lambda x)$. Any two components $w_1 p(x;\lambda_1)$ and $w_2 p(x;\lambda_2)$ (with $\lambda_1 \neq \lambda_2$) have a unique intersection point

$$x^\star = \frac{\log(w_1\lambda_1) - \log(w_2\lambda_2)}{\lambda_1 - \lambda_2} \tag{16}$$

if $x^\star \geq 0$; otherwise they do not intersect. The basic formulas to evaluate the bounds are

$$C_{i,j}(a,b) = \log\left(\lambda'_j w'_j\right) M_i(a,b) + w_i\lambda'_j \left[\left(a + \frac{1}{\lambda_i}\right)e^{-\lambda_i a} - \left(b + \frac{1}{\lambda_i}\right)e^{-\lambda_i b}\right], \tag{17}$$

$$M_i(a,b) = -\, w_i\left(e^{-\lambda_i a} - e^{-\lambda_i b}\right). \tag{18}$$

2.3.2. The Case of Rayleigh Mixture Models

A Rayleigh distribution has density $p(x;\sigma) = \frac{x}{\sigma^2}\exp\left(-\frac{x^2}{2\sigma^2}\right)$, defined on $\mathcal{X} = [0,\infty)$ for $\sigma > 0$. Its CDF is $\Phi(x;\sigma) = 1 - \exp\left(-\frac{x^2}{2\sigma^2}\right)$. Any two components $w_1 p(x;\sigma_1)$ and $w_2 p(x;\sigma_2)$ (with $\sigma_1 \neq \sigma_2$) must intersect at $x_0 = 0$ and can have at most one other intersection point

$$x^\star = \sqrt{\log\frac{w_1\sigma_2^2}{w_2\sigma_1^2} \Big/ \left(\frac{1}{2\sigma_1^2} - \frac{1}{2\sigma_2^2}\right)} \tag{19}$$

if the square root is well defined and $x^\star > 0$. We have

$$\begin{aligned} C_{i,j}(a,b) &= \log\frac{w'_j}{(\sigma'_j)^2} M_i(a,b) + \frac{w_i}{2(\sigma'_j)^2}\left[(a^2 + 2\sigma_i^2)e^{-\frac{a^2}{2\sigma_i^2}} - (b^2 + 2\sigma_i^2)e^{-\frac{b^2}{2\sigma_i^2}}\right] \\ &\quad - w_i\int_a^b \frac{x}{\sigma_i^2}\exp\left(-\frac{x^2}{2\sigma_i^2}\right)\log x\,dx, \end{aligned} \tag{20}$$

$$M_i(a,b) = -\, w_i\left(e^{-\frac{a^2}{2\sigma_i^2}} - e^{-\frac{b^2}{2\sigma_i^2}}\right). \tag{21}$$

The last term in Equation (20) does not have a simple closed form (it requires the exponential integral, Ei). One need a numerical integrator to compute it.

2.3.3. The Case of Gaussian Mixture Models

The Gaussian density $p(x;\mu,\sigma) = \frac{1}{\sqrt{2\pi}\sigma}e^{-(x-\mu)^2/(2\sigma^2)}$ has support $\mathcal{X} = \mathbb{R}$ and parameters $\mu \in \mathbb{R}$ and $\sigma > 0$. Its CDF is $\Phi(x;\mu,\sigma) = \frac{1}{2}\left[1 + \mathrm{erf}(\frac{x-\mu}{\sqrt{2}\sigma})\right]$, where erf is the Gauss error function. The intersection point $x^\star$ of two components $w_1 p(x;\mu_1,\sigma_1)$ and $w_2 p(x;\mu_2,\sigma_2)$ can be obtained by solving the quadratic equation $\log(w_1 p(x;\mu_1,\sigma_1)) = \log(w_2 p(x;\mu_2,\sigma_2))$, which gives at most two solutions. As shown in Figure 1, the upper envelope of Gaussian densities corresponds to the lower envelope of parabolas. We have

$$C_{i,j}(a,b) = M_i(a,b)\left(\log w'_j - \log \sigma'_j - \frac{1}{2}\log(2\pi) - \frac{1}{2(\sigma'_j)^2}\left((\mu'_j - \mu_i)^2 + \sigma_i^2\right)\right) + \frac{w_i\sigma_i}{2\sqrt{2\pi}(\sigma'_j)^2}\left[(a+\mu_i-2\mu'_j)e^{-\frac{(a-\mu_i)^2}{2\sigma_i^2}} - (b+\mu_i-2\mu'_j)e^{-\frac{(b-\mu_i)^2}{2\sigma_i^2}}\right], \tag{22}$$

$$M_i(a,b) = -\frac{w_i}{2}\left(\operatorname{erf}\left(\frac{b-\mu_i}{\sqrt{2}\sigma_i}\right) - \operatorname{erf}\left(\frac{a-\mu_i}{\sqrt{2}\sigma_i}\right)\right). \tag{23}$$

2.3.4. The Case of Gamma Distributions

For simplicity, we only consider gamma distributions with the shape parameter $k > 0$ fixed and the scale $\lambda > 0$ varying. The density is defined on $(0,\infty)$ as $p(x;k,\lambda) = \frac{x^{k-1}e^{-\frac{x}{\lambda}}}{\lambda^k\Gamma(k)}$, where $\Gamma(\cdot)$ is the gamma function. Its CDF is $\Phi(x;k,\lambda) = \gamma(k, x/\lambda)/\Gamma(k)$, where $\gamma(\cdot,\cdot)$ is the lower incomplete gamma function. Two weighted gamma densities $w_1p(x;k,\lambda_1)$ and $w_2p(x;k,\lambda_2)$ (with $\lambda_1 \neq \lambda_2$) intersect at a unique point

$$x^\star = \left(\log\frac{w_1}{\lambda_1^k} - \log\frac{w_2}{\lambda_2^k}\right) / \left(\frac{1}{\lambda_1} - \frac{1}{\lambda_2}\right) \tag{24}$$

if $x^\star > 0$; otherwise they do not intersect. From straightforward derivations,

$$C_{i,j}(a,b) = \log\frac{w'_j}{(\lambda'_j)^k\Gamma(k)}M_i(a,b) + w_i\int_a^b \frac{x^{k-1}e^{-\frac{x}{\lambda_i}}}{\lambda_i^k\Gamma(k)}\left(\frac{x}{\lambda'_j} - (k-1)\log x\right)dx, \tag{25}$$

$$M_i(a,b) = -\frac{w_i}{\Gamma(k)}\left(\gamma\left(k,\frac{b}{\lambda_i}\right) - \gamma\left(k,\frac{a}{\lambda_i}\right)\right). \tag{26}$$

Similar to the case of Rayleigh mixtures, the last term in Equation (25) relies on numerical integration.

3. Upper-Bounding the Differential Entropy of a Mixture

First, consider a finite parametric mixture model $m(x) = \sum_{i=1}^k w_i p(x;\theta_i)$. Using the chain rule of the entropy, we end up with the well-known lemma:

Lemma 1. *The entropy of a d-variate mixture is upper bounded by the sum of the entropy of its marginal mixtures:* $H(m) \leq \sum_{i=1}^d H(m_i)$, *where* m_i *is the 1D marginal mixture with respect to variable* x_i.

Since the 1D marginals of a multivariate GMM are univariate GMMs, we thus get a loose upper bound. A generic sample-based probabilistic bound is reported for the entropies of distributions with given support [31]: The method builds probabilistic upper and lower piecewisely linear CDFs based on an i.i.d. finite sample set of size n and a given deviation probability threshold. It then builds algorithmically between those two bounds the maximum entropy distribution [31] with a so-called string-tightening algorithm.

Instead, we proceed as follows: Consider finite mixtures of component distributions defined on the full support $\mathbb{R}^d$ that have finite component means and variances (like exponential families). Then we shall use the fact that the maximum entropy distribution with prescribed mean and variance is a Gaussian distribution, and conclude the upper bound by plugging the mixture mean and variance in the differential entropy formula of the Gaussian distribution. In general, the maximum entropy with moment constraints yields as a solution an exponential family.

Without loss of generality, consider GMMs in the form $m(x) = \sum_{i=1}^k w_i p(x;\mu_i,\Sigma_i)$ ($\Sigma_i = \sigma_i^2$ for univariate Gaussians). The mean $\bar{\mu}$ of the mixture is $\bar{\mu} = \sum_{i=1}^k w_i\mu_i$ and the variance is $\bar{\sigma}^2 = E[m^2] - E[m]^2$. Since $E[m^2] = \sum_{i=1}^k w_i \int x^2 p(x;\mu_i,\Sigma_i)\mathrm{d}x = \sum_{i=1}^k w_i \left(\mu_i^2 + \sigma_i^2\right)$, we deduce that

$$\bar{\sigma}^2 = \sum_{i=1}^k w_i(\mu_i^2 + \sigma_i^2) - \left(\sum_{i=1}^k w_i\mu_i\right)^2 = \sum_{i=1}^k w_i \left[(\mu_i - \bar{\mu})^2 + \sigma_i^2\right].$$

The entropy of a random variable with a prescribed variance $\bar{\sigma}^2$ is maximal for the Gaussian distribution with the same variance $\bar{\sigma}^2$, see [4]. Since the differential entropy of a Gaussian is $\log(\bar{\sigma}\sqrt{2\pi e})$, we deduce that the entropy of the GMM is upper bounded by

$$H(m) \le \frac{1}{2}\log(2\pi e) + \frac{1}{2}\log\sum_{i=1}^k w_i \left[(\mu_i - \bar{\mu})^2 + \sigma_i^2\right].$$

This upper bound can be easily generalized to arbitrary dimensionality. We get the following lemma:

Lemma 2. *The entropy of a d-variate GMM* $m(x) = \sum_{i=1}^k w_i p(x;\mu_i,\Sigma_i)$ *is upper bounded by* $\frac{d}{2}\log(2\pi e) + \frac{1}{2}\log\det\Sigma$, *where* $\Sigma = \sum_{i=1}^k w_i(\mu_i\mu_i^\top + \Sigma_i) - \left(\sum_{i=1}^k w_i\mu_i\right)\left(\sum_{i=1}^k w_i\mu_i^\top\right)$.

In general, exponential families have finite moments of any order [17]: In particular, we have $E[t(X)] = \nabla F(\theta)$ and $V[t(X)] = \nabla^2 F(\theta)$. For Gaussian distribution, we have the sufficient statistics $t(x) = (x, x^2)$ so that $E[t(X)] = \nabla F(\theta)$ yields the mean and variance from the log-normalizer. It is easy to generalize Lemma 2 to mixtures of exponential family distributions.

Note that this bound (called the Maximum Entropy Upper Bound in [13], MEUB) is tight when the GMM approximates a single Gaussian. It is fast to compute compared to the bound reported in [9] that uses Taylor' s expansion of the log-sum of the mixture density.

A similar argument cannot be applied for a lower bound since a GMM with a given variance may have entropy tending to $-\infty$. For example, assume the 2-component mixture's mean is zero, and that the variance approximates 1 by taking $m(x) = \frac{1}{2}G(x;-1,\epsilon) + \frac{1}{2}G(x;1,\epsilon)$ where G denotes the Gaussian density. Letting $\epsilon \to 0$, we get the entropy tending to $-\infty$.

We remark that our log-sum-exp inequality technique yields a $\log 2$ additive approximation range in the case of a Gaussian mixture with two components. It thus generalizes the bounds reported in [7] to GMMs with arbitrary variances that are not necessarily equal.

To see the bound gap, we have

$$\begin{aligned} -\sum_r \int_{I_r} m(x)\left(\log k + \log\max_i w_i p_i(x)\right)\mathrm{d}x &\le H(m) \\ &\le -\sum_r \int_{I_r} m(x)\max\left\{\log\max_i w_i p_i(x),\ \log k + \log\min_i w_i p_i(x)\right\}\mathrm{d}x. \end{aligned} \tag{27}$$

Therefore the gap is at most

$$\begin{aligned} \Delta &= \min\left\{\sum_r \int_{I_r} m(x)\log\frac{\max_i w_i p_i(x)}{\min_i w_i p_i(x)}\mathrm{d}x,\ \log k\right\} \\ &= \min\left\{\sum_s\sum_r \int_{I_r} w_s p_s(x)\log\frac{\max_i w_i p_i(x)}{\min_i w_i p_i(x)}\mathrm{d}x,\ \log k\right\}. \end{aligned} \tag{28}$$

Thus to compute the gap error bound of the differential entropy, we need to integrate terms in the form

$$\int w_a p_a(x)\log\frac{w_b p_b(x)}{w_c p_c(x)}\mathrm{d}x.$$

See Appendix B for a closed-form formula when dealing with exponential family components.

4. Bounding the α-Divergence

The α-divergence [15,32–34] between $m(x) = \sum_{i=1}^{k} w_i p_i(x)$ and $m'(x) = \sum_{i=1}^{k'} w_i' p_i'(x)$ is defined as

$$D_\alpha\left(m : m'\right) = \frac{1}{\alpha(1-\alpha)}\left(1 - \int_{\mathcal{X}} m(x)^\alpha m'(x)^{1-\alpha} \mathrm{d}x\right), \tag{29}$$

which clearly satisfies $D_\alpha\left(m : m'\right) = D_{1-\alpha}\left(m' : m\right)$. The α-divergence is *a family* of information divergences parametrized by $\alpha \in \mathbb{R} \setminus \{0,1\}$. Let $\alpha \to 1$, we get the KL divergence (see [35] for a proof):

$$\lim_{\alpha\to 1} D_\alpha(m : m') = \mathrm{KL}(m : m') = \int_{\mathcal{X}} m(x) \log \frac{m(x)}{m'(x)} \mathrm{d}x, \tag{30}$$

and $\alpha \to 0$ gives the reverse KL divergence:

$$\lim_{\alpha\to 0} D_\alpha(m : m') = \mathrm{KL}(m' : m).$$

Other interesting values [33] include $\alpha = 1/2$ (squared Hellinger distance), $\alpha = 2$ (Pearson Chi-square distance), $\alpha = -1$ (Neyman Chi-square distance), etc. Notably, the Hellinger distance is a valid distance metric which satisfies non-negativity, symmetry, and the triangle inequality. In general, $D_\alpha(m : m')$ only satisfies non-negativity so that $D_\alpha\left(m : m'\right) \geq 0$ for any $m(x)$ and $m'(x)$. It is neither symmetric nor admitting the triangle inequality. Minimization of α-divergences allows one to choose a trade-off between mode fitting and support fitting of the minimizer [36]. The minimizer of α-divergences including MLE as a special case has interesting connections with transcendental number theory [37].

To compute $D_\alpha\left(m : m'\right)$ for given $m(x)$ and $m'(x)$ reduces to evaluate the Hellinger integral [38,39]:

$$H_\alpha(m : m') = \int_{\mathcal{X}} m(x)^\alpha m'(x)^{1-\alpha} \mathrm{d}x, \tag{31}$$

which in general does not have a closed form, as it was known that the α-divergence of mixture models is not analytic [6]. Moreover, $H_\alpha(m : m')$ may diverge making the α-divergence unbounded. Once $H_\alpha(m : m')$ can be solved, the Rényi and Tsallis divergences [35] and in general Sharma–Mittal divergences [40] can be easily computed. Therefore the results presented here directly extend to those divergence families.

Similar to the case of KL divergence, the Monte Carlo stochastic estimation of $H_\alpha(m : m')$ can be computed either as

$$\hat{H}_\alpha^n\left(m : m'\right) = \frac{1}{n}\sum_{i=1}^{n}\left(\frac{m'(x_i)}{m(x_i)}\right)^{1-\alpha},$$

where $x_1, \ldots, x_n \sim m(x)$ are i.i.d. samples, or as

$$\hat{H}_\alpha^n\left(m : m'\right) = \frac{1}{n}\sum_{i=1}^{n}\left(\frac{m(x_i)}{m'(x_i)}\right)^{\alpha},$$

where $x_1, \ldots, x_n \sim m'(x)$ are i.i.d. In either case, it is consistent so that $\lim_{n\to\infty} \hat{H}_\alpha^n\left(m : m'\right) = H_\alpha\left(m : m'\right)$. However, MC estimation requires a large sample and does not guarantee deterministic bounds. The techniques described in [41] work in practice for very close distributions, and do not apply between mixture models. We will therefore derive combinatorial bounds for $H_\alpha(m : m')$. The structure of this Section is parallel with Section 2 with necessary reformulations for a clear presentation.

4.1. Basic Bounds

For a pair of given $m(x)$ and $m'(x)$, we only need to derive bounds of $H_\alpha(m : m')$ in Equation (31) so that $L_\alpha(m : m') \le H_\alpha(m : m') \le U_\alpha(m : m')$. Then the α-divergence $D_\alpha(m : m')$ can be bounded by a linear transformation of the range $[L_\alpha(m : m'),\ U_\alpha(m : m')]$. In the following we always assume without loss of generality $\alpha \ge 1/2$. Otherwise we can bound $D_\alpha(m : m')$ by considering equivalently the bounds of $D_{1-\alpha}(m' : m)$.

Recall that in each elementary slab I_r, we have

$$\max\left\{kw_{\epsilon(r)}p_{\epsilon(r)}(x), w_{\delta(r)}p_{\delta(r)}(x)\right\} \le m(x) \le kw_{\delta(r)}p_{\delta(r)}(x). \tag{32}$$

Notice that $kw_{\epsilon(r)}p_{\epsilon(r)}(x)$, $w_{\delta(r)}p_{\delta(r)}(x)$, and $kw_{\delta(r)}p_{\delta(r)}(x)$ are all single component distributions up to a scaling coefficient. The general thinking is to bound the multi-component mixture $m(x)$ by single component distributions in each elementary interval, so that the integral in Equation (31) can be computed in a piecewise manner.

For the convenience of notation, we rewrite Equation (32) as

$$c_{\nu(r)}p_{\nu(r)}(x) \le m(x) \le c_{\delta(r)}p_{\delta(r)}(x), \tag{33}$$

where

$$c_{\nu(r)}p_{\nu(r)}(x) := kw_{\epsilon(r)}p_{\epsilon(r)}(x) \quad \text{or} \quad w_{\delta(r)}p_{\delta(r)}(x), \tag{34}$$

$$c_{\delta(r)}p_{\delta(r)}(x) := kw_{\delta(r)}p_{\delta(r)}(x). \tag{35}$$

If $1/2 \le \alpha < 1$, then both x^α and $x^{1-\alpha}$ are monotonically increasing on $\mathbb{R}^+$. Therefore we have

$$A^\alpha_{\nu(r),\nu'(r)}(I_r) \le \int_{I_r} m(x)^\alpha m'(x)^{1-\alpha}\mathrm{d}x \le A^\alpha_{\delta(r),\delta'(r)}(I_r), \tag{36}$$

where

$$A^\alpha_{i,j}(I) = \int_I (c_ip_i(x))^\alpha \left(c'_jp'_j(x)\right)^{1-\alpha} \mathrm{d}x, \tag{37}$$

and I denotes an interval $I = (a,b) \subset \mathbb{R}$. The other case $\alpha > 1$ is similar by noting that x^α and $x^{1-\alpha}$ are monotonically increasing and decreasing on $\mathbb{R}^+$, respectively. In conclusion, we obtain the following bounds of $H_\alpha(m : m')$:

$$\text{If } 1/2 \le \alpha < 1,\ L_\alpha(m : m') = \sum_{r=1}^{\ell} A^\alpha_{\nu(r),\nu'(r)}(I_r), \quad U_\alpha(m : m') = \sum_{r=1}^{\ell} A^\alpha_{\delta(r),\delta'(r)}(I_r); \tag{38}$$

$$\text{if } \alpha > 1,\ L_\alpha(m : m') = \sum_{r=1}^{\ell} A^\alpha_{\nu(r),\delta'(r)}(I_r), \quad U_\alpha(m : m') = \sum_{r=1}^{\ell} A^\alpha_{\delta(r),\nu'(r)}(I_r). \tag{39}$$

The remaining problem is to compute the definite integral $A^\alpha_{i,j}(I)$ in the above equations. Here we assume all mixture components are in the same exponential family so that $p_i(x) = p(x;\theta_i) = h(x)\exp\left(\theta_i^\top t(x) - F(\theta_i)\right)$, where $h(x)$ is a base measure, $t(x)$ is a vector of sufficient statistics, and the function F is known as the cumulant generating function. Then it is straightforward from Equation (37) that

$$A^\alpha_{i,j}(I) = c_i^\alpha (c'_j)^{1-\alpha} \int_I h(x)\exp\left(\left(\alpha\theta_i + (1-\alpha)\theta'_j\right)^\top t(x) - \alpha F(\theta_i) - (1-\alpha)F(\theta'_j)\right)\mathrm{d}x. \tag{40}$$

If $1/2 \leq \alpha < 1$, then $\bar{\theta} = \alpha\theta_i + (1-\alpha)\theta'_j$ belongs to the natural parameter space $\mathcal{M}_\theta$. Therefore $A^{\alpha}_{i,j}(I)$ is bounded and can be computed from the CDF of $p(x;\bar{\theta})$ as

$$A^{\alpha}_{i,j}(I) = c^{\alpha}_i (c'_j)^{1-\alpha} \exp(F(\bar{\theta}) - \alpha F(\theta_i) - (1-\alpha)F(\theta'_j)) \int_I p(x;\bar{\theta})\,\mathrm{d}x. \tag{41}$$

The other case $\alpha > 1$ is more difficult: if $\bar{\theta} = \alpha\theta_i + (1-\alpha)\theta'_j$ still lies in $\mathcal{M}_\theta$, then $A^{\alpha}_{i,j}(I)$ can be computed by Equation (41). Otherwise we try to solve it by a numerical integrator. This is not ideal as the integral may diverge, or our approximation may be too loose to conclude. We point the reader to [42] and Equations (61)–(69) in [35] for related analysis with more details. As computing $A^{\alpha}_{i,j}(I)$ only requires $O(1)$ time, the overall computational complexity (without considering the envelope computation) is $O(\ell)$.

4.2. Adaptive Bounds

This section derives the shape-dependent bounds which improve the basic bounds in Section 4.1. We can rewrite a mixture model $m(x)$ in a slab I_r as

$$m(x) = w_{\zeta(r)} p_{\zeta(r)}(x) \left(1 + \sum_{i \neq \zeta(r)} \frac{w_i p_i(x)}{w_{\zeta(r)} p_{\zeta(r)}(x)}\right), \tag{42}$$

where $w_{\zeta(r)} p_{\zeta(r)}(x)$ is a weighted component in $m(x)$ serving as a *reference*. We only discuss the case that the reference is chosen as the dominating component, i.e., $\zeta(r) = \delta(r)$. However it is worth to note that the proposed bounds do not depend on this particular choice. Therefore the ratio

$$\frac{w_i p_i(x)}{w_{\zeta(r)} p_{\zeta(r)}(x)} = \frac{w_i}{w_{\zeta(r)}} \exp\left(\left(\theta_i - \theta_{\zeta(r)}\right)^\top t(x) - F(\theta_i) + F(\theta_{\zeta(r)})\right) \tag{43}$$

can be bounded in a sub-range of $[0,1]$ by analyzing the extreme values of $t(x)$ in the slab I_r. This can be done because $t(x)$ usually consists of polynomial functions with finite critical points which can be solved easily. Correspondingly the function $\left(1 + \sum_{i \neq \zeta(r)} \frac{w_i p_i(x)}{w_{\zeta(r)} p_{\zeta(r)}(x)}\right)$ in I_r can be bounded in a subrange of $[1,k]$, denoted as $[\omega_{\zeta(r)}(I_r), \Omega_{\zeta(r)}(I_r)]$. Hence

$$\omega_{\zeta(r)}(I_r) w_{\zeta(r)} p_{\zeta(r)}(x) \leq m(x) \leq \Omega_{\zeta(r)}(I_r) w_{\zeta(r)} p_{\zeta(r)}(x). \tag{44}$$

This forms better bounds of $m(x)$ than Equation (32) because each component in the slab I_r is analyzed more accurately. Therefore, we refine the fundamental bounds of $m(x)$ by replacing the Equations (34) and (35) with

$$c_{\nu(r)} p_{\nu(r)}(x) := \omega_{\zeta(r)}(I_r) w_{\zeta(r)} p_{\zeta(r)}(x), \tag{45}$$

$$c_{\delta(r)} p_{\delta(r)}(x) := \Omega_{\zeta(r)}(I_r) w_{\zeta(r)} p_{\zeta(r)}(x). \tag{46}$$

Then, the improved bounds of H_α are given by Equations (38) and (39) according to the above replaced definition of $c_{\nu(r)} p_{\nu(r)}(x)$ and $c_{\delta(r)} p_{\delta(r)}(x)$.

To evaluate $\omega_{\zeta(r)}(I_r)$ and $\Omega_{\zeta(r)}(I_r)$ requires iterating through all components in each slab. Therefore the computational complexity is increased to $O(\ell(k+k'))$.

4.3. Variance-Reduced Bounds

This section further improves the proposed bounds based on variance reduction [43]. By assumption, $\alpha \geq 1/2$, then $m(x)^\alpha m'(x)^{1-\alpha}$ is more similar to $m(x)$ rather than $m'(x)$. The ratio

$m(x)^\alpha m'(x)^{1-\alpha}/m(x)$ is likely to have a small variance when x varies inside a slab I_r, especially when α is close to 1. We will therefore bound this ratio term in

$$\int_{I_r} m(x)^\alpha m'(x)^{1-\alpha}\mathrm{d}x = \int_{I_r} m(x)\left(\frac{m(x)^\alpha m'(x)^{1-\alpha}}{m(x)}\right)\mathrm{d}x = \sum_{i=1}^{k}\int_{I_r} w_i p_i(x)\left(\frac{m'(x)}{m(x)}\right)^{1-\alpha}\mathrm{d}x. \quad (47)$$

No matter $\alpha < 1$ or $\alpha > 1$, the function $x^{1-\alpha}$ must be monotonic on $\mathbb{R}^+$. In each slab I_r, $(m'(x)/m(x))^{1-\alpha}$ ranges between these two functions:

$$\left(\frac{c'_{\nu'(r)}p'_{\nu'(r)}(x)}{c_{\delta(r)}p_{\delta(r)}(x)}\right)^{1-\alpha} \quad \text{and} \quad \left(\frac{c'_{\delta'(r)}p'_{\delta'(r)}(x)}{c_{\nu(r)}p_{\nu(r)}(x)}\right)^{1-\alpha}, \quad (48)$$

where $c_{\nu(r)}p_{\nu(r)}(x)$, $c_{\delta(r)}p_{\delta(r)}(x)$, $c'_{\nu'(r)}p'_{\nu'(r)}(x)$ and $c'_{\delta'(r)}p'_{\delta'(r)}(x)$ are defined in Equations (45) and (46). Similar to the definition of $A^\alpha_{i,j}(I)$ in Equation (37), we define

$$B^\alpha_{i,j,l}(I) = \int_I w_i p_i(x)\left(\frac{c'_l p'_l(x)}{c_j p_j(x)}\right)^{1-\alpha}\mathrm{d}x. \quad (49)$$

Therefore we have,

$$L_\alpha(m:m') = \min\mathcal{S}, \quad U_\alpha(m:m') = \max\mathcal{S},$$
$$\mathcal{S} = \left\{\sum_{r=1}^{\ell}\sum_{i=1}^{k} B^\alpha_{i,\delta(r),\nu'(r)}(I_r),\ \sum_{r=1}^{\ell}\sum_{i=1}^{k} B^\alpha_{i,\nu(r),\delta'(r)}(I_r)\right\}. \quad (50)$$

The remaining problem is to evaluate $B^\alpha_{i,j,l}(I)$ in Equation (49). Similar to Section 4.1, assuming the components are in the same exponential family with respect to the natural parameters θ, we get

$$B^\alpha_{i,j,l}(I) = w_i\frac{c'^{1-\alpha}_l}{c^{1-\alpha}_j}\exp\left(F(\bar\theta) - F(\theta_i) - (1-\alpha)F(\theta'_l) + (1-\alpha)F(\theta_j)\right)\int_I p(x;\bar\theta)\mathrm{d}x. \quad (51)$$

If $\bar\theta = \theta_i + (1-\alpha)\theta'_l - (1-\alpha)\theta_j$ is in the natural parameter space, $B^\alpha_{i,j,l}(I)$ can be computed from the CDF of $p(x;\bar\theta)$; otherwise $B^\alpha_{i,j,l}(I)$ can be numerically integrated by its definition in Equation (49). The computational complexity is the same as the bounds in Section 4.2, i.e., $O(\ell\,(k+k'))$.

We have introduced three pairs of deterministic lower and upper bounds that enclose the true value of α-divergence between univariate mixture models. Thus the gap between the upper and lower bounds provides the additive approximation factor of the bounds. We conclude by emphasizing that the presented methodology can be easily generalized to other divergences [35,40] relying on Hellinger-type integrals $H_{\alpha,\beta}(p:q) = \int p(x)^\alpha q(x)^\beta \mathrm{d}x$ like the γ-divergence [44] as well as entropy measures [45].

5. Lower Bounds of the f-Divergence

The f-divergence between two distributions $m(x)$ and $m'(x)$ (not necessarily mixtures) is defined for a *convex generator* f by:

$$D_f(m:m') = \int m(x) f\left(\frac{m'(x)}{m(x)}\right)\mathrm{d}x.$$

If $f(x) = -\log x$, then $D_f(m:m') = \mathrm{KL}(m:m')$.

Let us partition the support $\mathcal{X} = \uplus_{r=1}^{\ell} I_r$ arbitrarily into elementary ranges, which *do not necessarily correspond to the envelopes*. Denote by M_I the probability mass of a mixture $m(x)$ in the range I: $M_I = \int_I m(x)dx$. Then

$$D_f(m : m') = \sum_{r=1}^{\ell} M_{I_r} \int_{I_r} \frac{m(x)}{M_{I_r}} f\left(\frac{m'(x)}{m(x)}\right) \mathrm{d}x.$$

Note that in range I_r, $\frac{m(x)}{M_{I_r}}$ is a unit weight distribution. Thus by Jensen's inequality $f(E[X]) \leq E[f(X)]$, we get

$$D_f(m : m') \geq \sum_{r=1}^{\ell} M_{I_r} f\left(\int_{I_r} \frac{m(x)}{M_{I_r}} \frac{m'(x)}{m(x)} \mathrm{d}x\right) = \sum_{r=1}^{\ell} M_{I_r} f\left(\frac{M'_{I_r}}{M_{I_r}}\right). \tag{52}$$

Notice that the RHS of Equation (52) is the f-divergence between $(M_{I_1}, \cdots, M_{I_\ell})$ and $(M'_{I_1}, \cdots, M'_{I_\ell})$, denoted by $D_f^{\mathcal{I}}(m : m')$. In the special case that $\ell = 1$ and $I_1 = \mathcal{X}$, the above Equation (52) turns out to be the usual Gibbs' inequality: $D_f(m : m') \geq f(1)$, and Csiszár generator is chosen so that $f(1) = 0$. In conclusion, for a fixed (coarse-grained) countable partition of $\mathcal{X}$, we recover the well-know information monotonicity [46] of the f-divergences:

$$D_f(m : m') \geq D_f^{\mathcal{I}}(m : m') \geq 0.$$

In practice, we get closed-form lower bounds when $M_I = \int_a^b m(x)\mathrm{d}x = \Phi(b) - \Phi(a)$ is available in closed-form, where $\Phi(\cdot)$ denote the CDF. In particular, if $m(x)$ is a mixture model, then its CDF can be computed by linearly combining the CDFs of its components.

To wrap up, we have proved that coarse-graining by making a finite partition of the support $\mathcal{X}$ yields a lower bound on the f-divergence by virtue of the information monotonicity. Therefore, instead of doing Monte Carlo stochastic integration:

$$\hat{D}_f^n(m : m') = \frac{1}{n} \sum_{i=1}^{n} f\left(\frac{m'(x_i)}{m(x_i)}\right),$$

with $x_1, \ldots, x_n \sim_{\text{i.i.d.}} m(x)$, it could be better to sort those n samples and consider the coarse-grained partition:

$$\mathcal{I} = (-\infty, x_{(1)}] \cup \left(\uplus_{i=1}^{n-1} (x_{(i)}, x_{(i+1)}]\right) \cup (x_{(n)}, +\infty)$$

to get a *guaranteed lower bound* on the f-divergence. We will call this bound CGQLB for Coarse Graining Quantization Lower Bound.

Given a budget of n splitting points on the range $\mathcal{X}$, it would be interesting to find the best n points that maximize the lower bound $D_f^{\mathcal{I}}(m : m')$. This is ongoing research.

6. Experiments

We perform an empirical study to verify our theoretical bounds. We simulate four pairs of mixture models $\{(\mathtt{EMM}_1, \mathtt{EMM}_2), (\mathtt{RMM}_1, \mathtt{RMM}_2), (\mathtt{GMM}_1, \mathtt{GMM}_2), (\mathtt{GaMM}_1, \mathtt{GaMM}_2)\}$ as the test subjects. The component type is implied by the model name, where GaMM stands for Gamma mixtures. The components of each mixture model are given as follows.

1. $\mathtt{EMM}_1$'s components, in the form (λ_i, w_i), are given by $(0.1, 1/3)$, $(0.5, 1/3)$, $(1, 1/3)$; $\mathtt{EMM}_2$'s components are $(2, 0.2)$, $(10, 0.4)$, $(20, 0.4)$.
2. $\mathtt{RMM}_1$'s components, in the form (σ_i, w_i), are given by $(0.5, 1/3)$, $(2, 1/3)$, $(10, 1/3)$; $\mathtt{RMM}_2$ consists of $(5, 0.25)$, $(60, 0.25)$, $(100, 0.5)$.

3. $\mathtt{GMM}_1$'s components, in the form (μ_i, σ_i, w_i), are $(-5, 1, 0.05)$, $(-2, 0.5, 0.1)$, $(5, 0.3, 0.2)$, $(10, 0.5, 0.2)$, $(15, 0.4, 0.05)$, $(25, 0.5, 0.3)$, $(30, 2, 0.1)$; $\mathtt{GMM}_2$ consists of $(-16, 0.5, 0.1)$, $(-12, 0.2, 0.1)$, $(-8, 0.5, 0.1)$, $(-4, 0.2, 0.1)$, $(0, 0.5, 0.2)$, $(4, 0.2, 0.1)$, $(8, 0.5, 0.1)$, $(12, 0.2, 0.1)$, $(16, 0.5, 0.1)$.
4. $\mathtt{GaMM}_1$'s components, in the form (k_i, λ_i, w_i), are $(2, 0.5, 1/3)$, $(2, 2, 1/3)$, $(2, 4, 1/3)$; $\mathtt{GaMM}_2$ consists of $(2, 5, 1/3)$, $(2, 8, 1/3)$, $(2, 10, 1/3)$.

We compare the proposed bounds with Monte Carlo estimation with different sample sizes in the range $\{10^2, 10^3, 10^4, 10^5\}$. For each sample size configuration, we report the 0.95 confidence interval by Monte Carlo estimation using the corresponding number of samples. Figure 2a–d shows the input signals as well as the estimation results, where the proposed bounds CELB, CEUB, CEALB, CEAUB, CGQLB are presented as horizontal lines, and the Monte Carlo estimations over different sample sizes are presented as error bars. We can loosely consider the average Monte Carlo output with the largest sample size (10^5) as the underlying truth, which is clearly inside our bounds. This serves as an empirical justification on the correctness of the bounds.

A key observation is that the bounds can be *very tight*, especially when the underlying KL divergence has a large magnitude, e.g., $\mathrm{KL}(\mathtt{RMM}_2 : \mathtt{RMM}_1)$. This is because the gap between the lower and upper bounds is always guaranteed to be within $\log k + \log k'$. Because KL is unbounded [4], in the general case two mixture models may have a large KL. Then our approximation gap is relatively very small. On the other hand, we also observed that the bounds in certain cases, e.g., $\mathrm{KL}(\mathtt{EMM}_2 : \mathtt{EMM}_1)$, are not as tight as the other cases. When the underlying KL is small, the bounds are not as informative as the general case.

Comparatively, there is a significant improvement of the shape-dependent bounds (CEALB and CEAUB) over the combinatorial bounds (CELB and CEUB). In all investigated cases, the adaptive bounds can roughly shrink the gap by half of its original size at the cost of additional computation.

Note that, the bounds are accurate and must contain the true value. Monte Carlo estimation gives no guarantee on where the true value is. For example, in estimating $\mathrm{KL}(\mathtt{GMM}_1 : \mathtt{GMM}_2)$, Monte Carlo estimation based on 10^4 samples can go beyond our bounds! It therefore suffers from a larger estimation error.

CGQLB as a simple-to-implement technique shows surprising good performance in several cases, e.g., $\mathrm{KL}(\mathtt{RMM}_1, \mathtt{RMM}_2)$. Although it requires a large number of samples, we can observe that increasing sample size has limited effect on improving this bound. Therefore, in practice, one may intersect the range defined by CEALB and CEAUB with the range defined by CGQLB with a small sample size (e.g., 100) to get better bounds.

We simulates a set of Gaussian mixture models besides the above $\mathtt{GMM}_1$ and $\mathtt{GMM}_2$. Figure 3 shows the GMM densities as well as their differential entropy. A detailed explanation of the components of each GMM model is omitted for brevity.

The key observation is that CEUB (CEAUB) is *very tight* in most of the investigated cases. This is because that the upper envelope that is used to compute CEUB (CEAUB) gives a very good estimation of the input signal.

Notice that MEUB only gives an upper bound of the differential entropy as discussed in Section 3. In general the proposed bounds are tighter than MEUB. However, this is not the case when the mixture components are merged together and approximate one single Gaussian (and therefore its entropy can be well approximated by the Gaussian entropy), as shown in the last line of Figure 3.

For α-divergence, the bounds introduced in Sections 4.1–4.3 are denoted as "Basic", "Adaptive" and "VR", respectively. Figure 4 visualizes these GMMs and plots the estimations of their α-divergences against α. The red lines mean the upper envelope. The dashed vertical lines mean the elementary intervals. The components of $\mathtt{GMM}_1$ and $\mathtt{GMM}_2$ are more separated than $\mathtt{GMM}_3$ and $\mathtt{GMM}_4$. Therefore these two pairs present different cases. For a clear presentation, only VR (which is expected to be better than Basic and Adaptive) is shown. We can see that, visually in the big scale, VR tightly surrounds the true value.

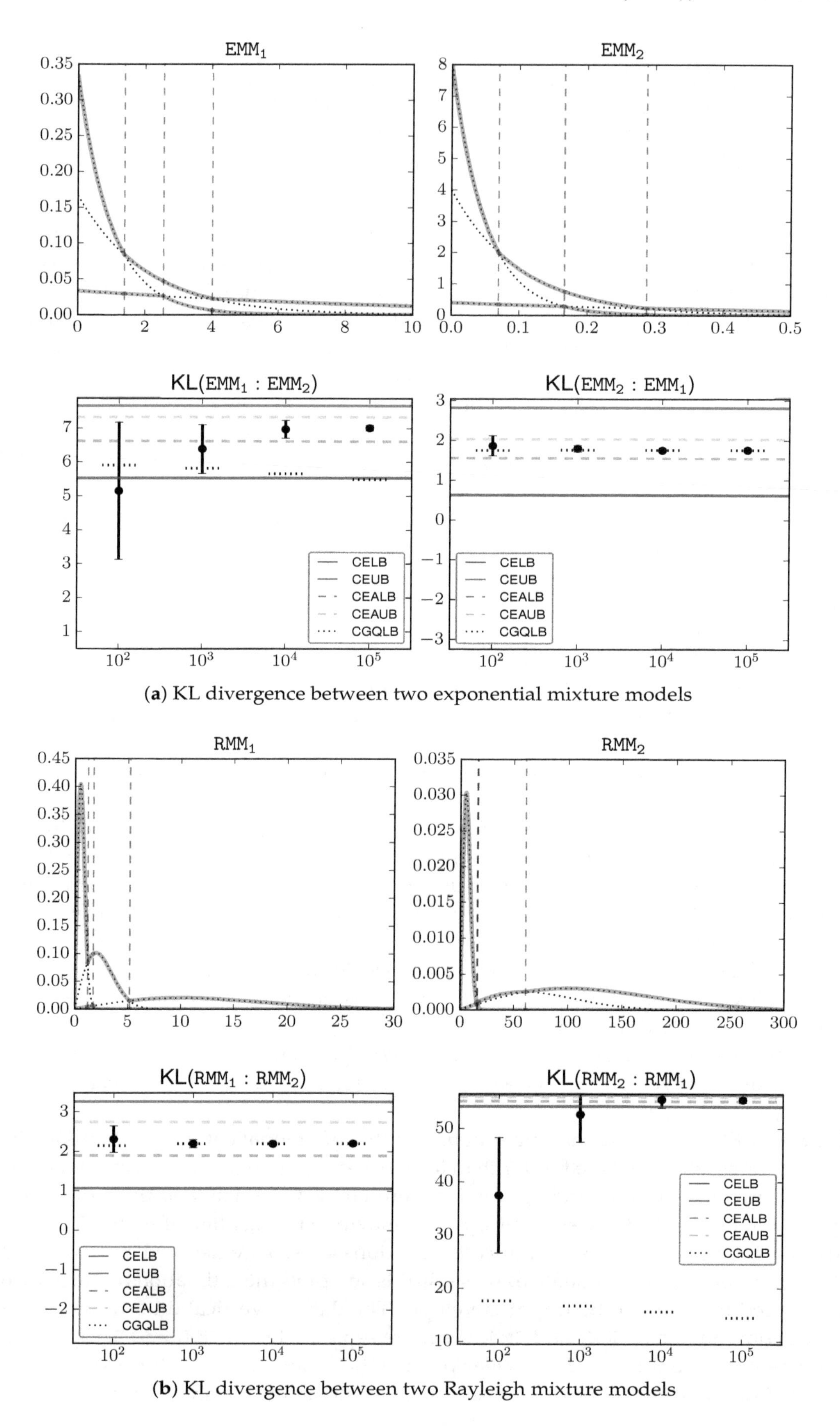

(**a**) KL divergence between two exponential mixture models

(**b**) KL divergence between two Rayleigh mixture models

Figure 2. *Cont.*

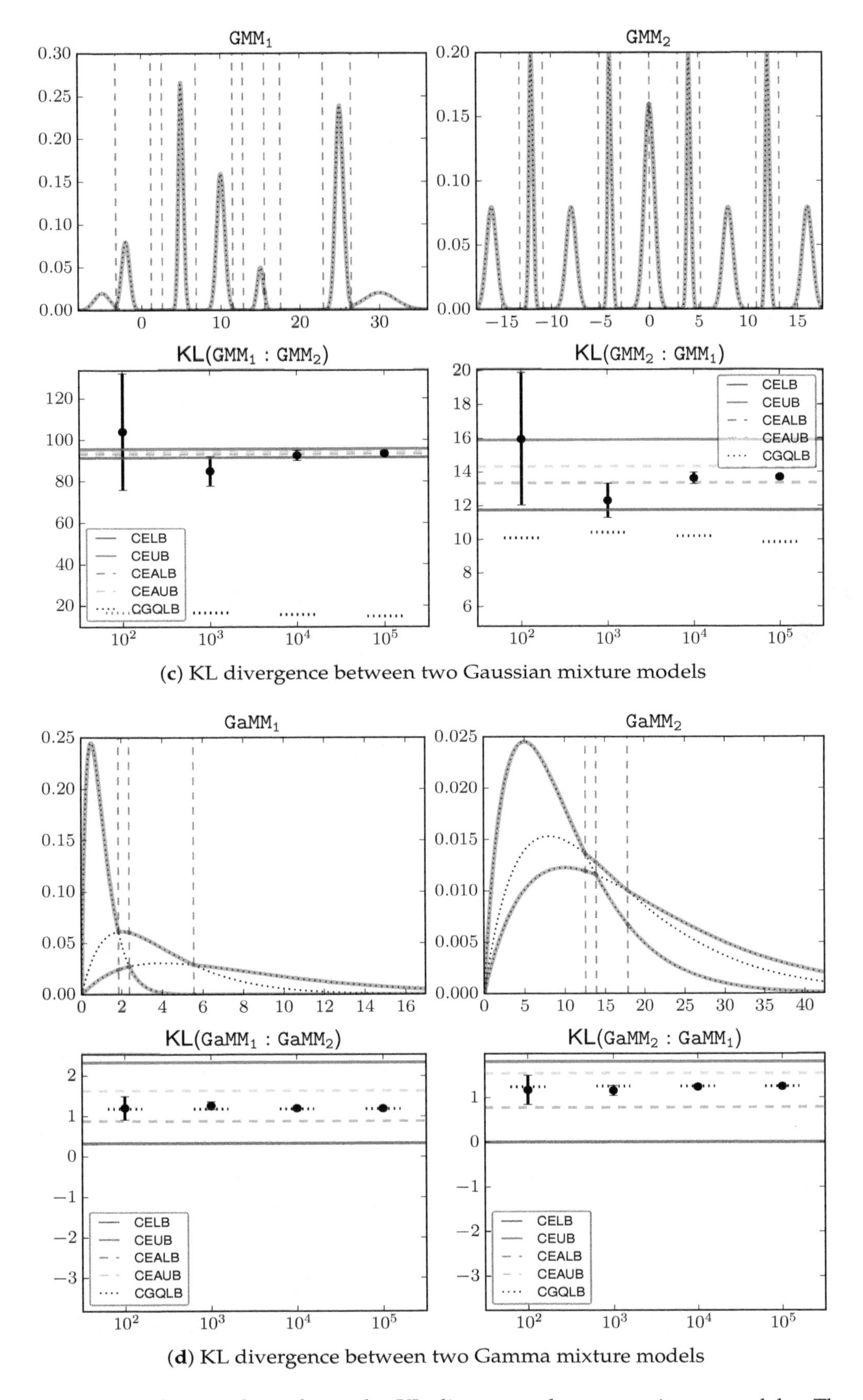

(c) KL divergence between two Gaussian mixture models

(d) KL divergence between two Gamma mixture models

Figure 2. Lower and upper bounds on the KL divergence between mixture models. The y-axis means KL divergence. Solid/dashed lines represent the combinatorial/adaptive bounds, respectively. The error-bars show the 0.95 confidence interval by Monte Carlo estimation using the corresponding sample size (x-axis). The narrow dotted bars show the CGQLB estimation w.r.t. the sample size.

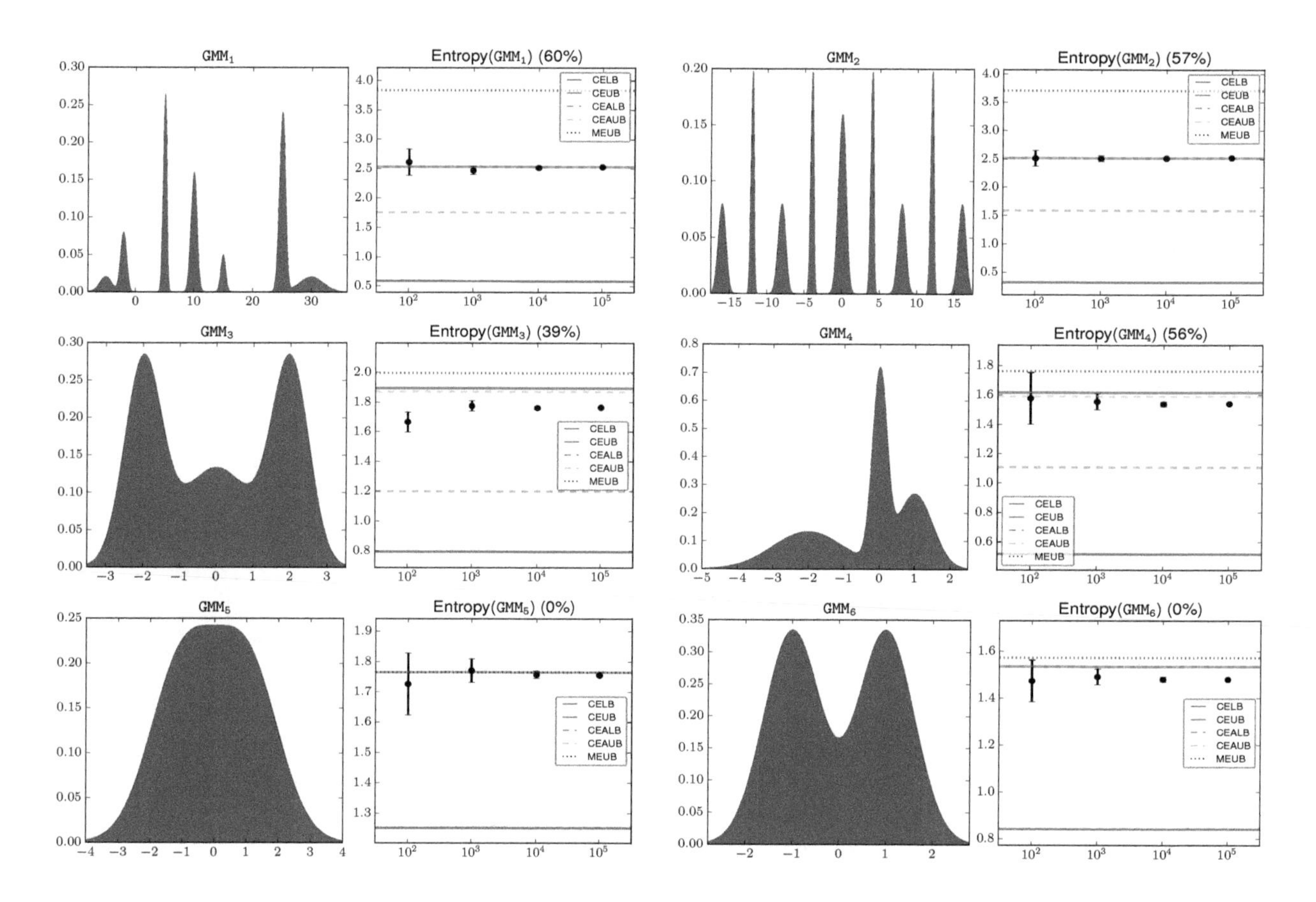

Figure 3. Lower and upper bounds on the differential entropy of Gaussian mixture models. On the left of each subfigure is the simulated GMM signal. On the right of each subfigure is the estimation of its differential entropy. Note that a subset of the bounds coincide with each other in several cases.

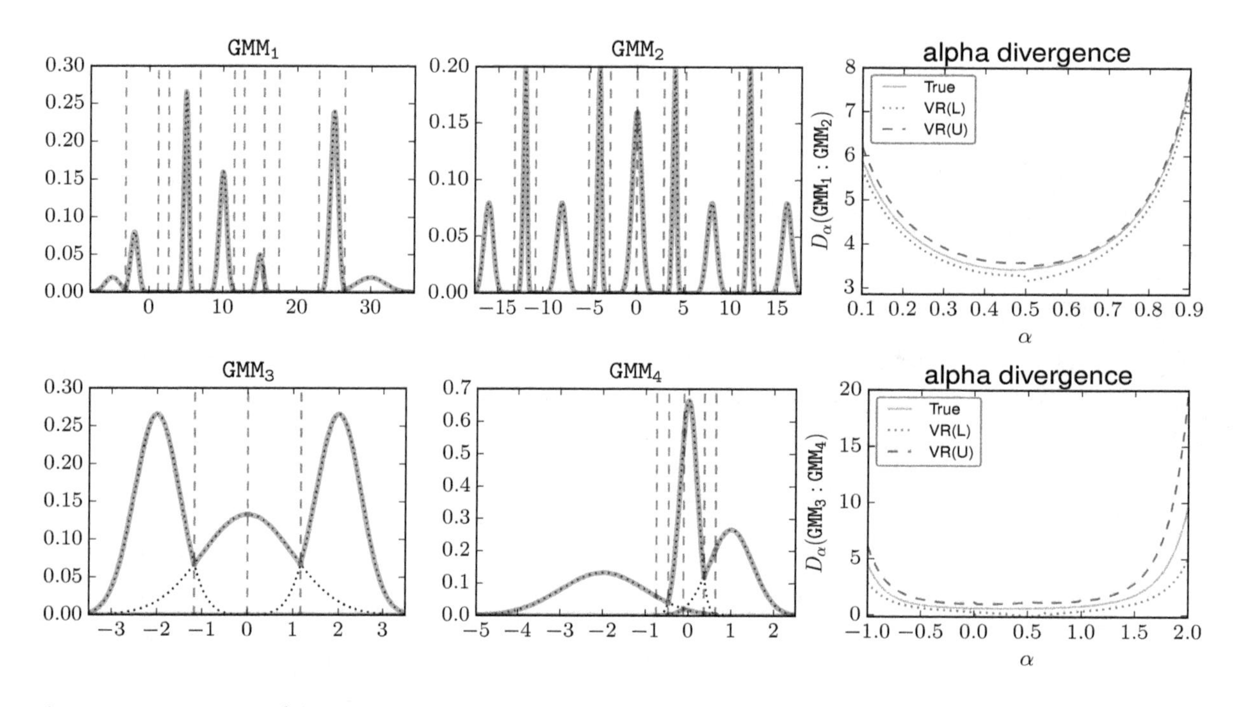

Figure 4. Two pairs of Gaussian Mixture Models and their α-divergences against different values of α. The "true" value of D_α is estimated by MC using 10^4 random samples. VR(L) and VR(U) denote the variation reduced lower and upper bounds, respectively. The range of α is selected for each pair for a clear visualization.

For a more quantitative comparison, Table 1 shows the estimated α-divergence by MC, Basic, Adaptive, and VR. As D_α is defined on $\mathbb{R} \setminus \{0, 1\}$, the KL bounds CE(A)LB and CE(A)UB are presented for $\alpha = 0$ or 1. Overall, we have the following order of gap size: Basic $>$ Adaptive $>$ VR, and VR is recommended in general for bounding α-divergences. There are certain cases that the upper VR bound is looser than Adaptive. In practice one can compute the intersection of these bounds as well as the trivial bound $D_\alpha(m : m') \geq 0$ to get the best estimation.

Table 1. The estimated D_α and its bounds. The 95% confidence interval is shown for MC.

	α	MC(10^2)	MC(10^3)	MC(10^4)	Basic		Adaptive		VR	
					L	U	L	U	L	U
GMM$_1$ & GMM$_2$	0	15.96 ± 3.9	12.30 ± 1.0	13.63 ± 0.3	11.75	15.89	12.96	14.63		
	0.01	13.36 ± 2.9	10.63 ± 0.8	11.66 ± 0.3	-700.50	11.73	-77.33	11.73	11.40	12.27
	0.5	3.57 ± 0.3	3.47 ± 0.1	3.47 ± 0.07	-0.60	3.42	3.01	3.42	3.17	3.51
	0.99	40.04 ± 7.7	37.22 ± 2.3	38.58 ± 0.8	-333.90	39.04	5.36	38.98	38.28	38.96
	1	104.01 ± 28	84.96 ± 7.2	92.57 ± 2.5	91.44	95.59	92.76	94.41		
GMM$_3$ & GMM$_4$	0	0.71 ± 0.2	0.63 ± 0.07	0.62 ± 0.02	0.00	1.76	0.00	1.16		
	0.01	0.71 ± 0.2	0.63 ± 0.07	0.62 ± 0.02	-179.13	7.63	-38.74	4.96	0.29	1.00
	0.5	0.82 ± 0.3	0.57 ± 0.1	0.62 ± 0.04	-5.23	0.93	-0.71	0.85	-0.18	1.19
	0.99	0.79 ± 0.3	0.76 ± 0.1	0.80 ± 0.03	-165.72	12.10	-59.76	9.11	0.37	1.28
	1	0.80 ± 0.3	0.77 ± 0.1	0.81 ± 0.03	0.00	1.82	0.31	1.40		

Note the similarity between KL in Equation (30) and the expression in Equation (47). We give without a formal analysis that: CEAL(U)B is equivalent to VR at the limit $\alpha \to 0$ or $\alpha \to 1$. Experimentally as we slowly set $\alpha \to 1$, we can see that VR is consistent with CEAL(U)B.

7. Concluding Remarks and Perspectives

We have presented a fast versatile method to compute bounds on the Kullback–Leibler divergence between mixtures by building algorithmic formulae. We reported on our experiments for various mixture models in the exponential family. For univariate GMMs, we get a guaranteed bound of the KL divergence of two mixtures m and m' with k and k' components within an additive approximation factor of $\log k + \log k'$ in $O\left((k + k') \log(k + k')\right)$-time. Therefore, the larger the KL divergence, the better the bound when considering a multiplicative $(1 + \alpha)$-approximation factor, since $\alpha = \frac{\log k + \log k'}{\mathrm{KL}(m:m')}$. The adaptive bounds are guaranteed to yield better bounds at the expense of computing potentially $O\left(k^2 + (k')^2\right)$ intersection points of pairwise weighted components.

Our technique also yields the bound for the Jeffreys divergence (the symmetrized KL divergence: $J(m, m') = \mathrm{KL}(m : m') + \mathrm{KL}(m' : m)$) and the Jensen–Shannon divergence [47] (JS):

$$\mathrm{JS}(m, m') = \frac{1}{2}\left(\mathrm{KL}\left(m : \frac{m + m'}{2}\right) + \mathrm{KL}\left(m' : \frac{m + m'}{2}\right)\right),$$

since $\frac{m+m'}{2}$ is a mixture model with $k + k'$ components. One advantage of this statistical distance is that it is symmetric, always bounded by $\log 2$, and its square root yields a metric distance [48]. The log-sum-exp inequalities may also be used to compute some Rényi divergences [35]:

$$R_\alpha(m, p) = \frac{1}{\alpha - 1} \log\left(\int m(x)^\alpha p(x)^{1-\alpha}\right) \mathrm{d}x,$$

when α is an integer, $m(x)$ a mixture and $p(x)$ a single (component) distribution. Getting fast guaranteed tight bounds on statistical distances between mixtures opens many avenues. For example, we may consider building hierarchical mixture models by merging iteratively two mixture components so that those pairs of components are chosen so that the KL distance between the full mixture and the simplified mixture is minimized.

In order to be useful, our technique is unfortunately limited to univariate mixtures: indeed, in higher dimensions, we can still compute the maximization diagram of weighted components (an additively weighted Bregman–Voronoi diagram [49,50] for components belonging to the same exponential family). However, it becomes more complex to compute in the elementary Voronoi cells V, the functions $C_{i,j}(V)$ and $M_i(V)$ (in 1D, the Voronoi cells are segments). We may obtain hybrid algorithms by approximating or estimating these functions. In 2D, it is thus possible to obtain lower and upper bounds on the Mutual Information [51] (MI) when the joint distribution $m(x,y)$ is a 2D mixture of Gaussians:

$$I(M;M') = \int m(x,y) \log \frac{m(x,y)}{m(x)m'(y)} \mathrm{d}x\mathrm{d}y.$$

Indeed, the marginal distributions $m(x)$ and $m'(y)$ are univariate Gaussian mixtures.

A Python code implementing those computational-geometric methods for reproducible research is available online [52].

Acknowledgments: The authors gratefully thank the referees for their comments. This work was carried out while Ke Sun was visiting Frank Nielsen at Ecole Polytechnique, Palaiseau, France.

Author Contributions: Frank Nielsen and Ke Sun contributed to the theoretical results as well as to the writing of the article. Ke Sun implemented the methods and performed the numerical experiments. All authors have read and approved the final manuscript.

Appendix A. The Kullback–Leibler Divergence of Mixture Models Is Not Analytic [6]

Ideally, we aim at getting a finite length closed-form formula to compute the KL divergence of finite mixture models. However, this is provably mathematically intractable [6] because of the log-sum term in the integral, as we shall prove below.

Analytic expressions encompass closed-form formula and may include special functions (e.g., Gamma function) but do not allow to use limits or integrals. An analytic function $f(x)$ is a C^∞ function (infinitely differentiable) such that around any point x_0 the k-order Taylor series $T_k(x) = \sum_{i=0}^{k} \frac{f^{(i)}(x_0)}{i!}(x-x_0)^i$ converges to $f(x)$: $\lim_{k\to\infty} T_k(x) = f(x)$ for x belonging to a neighborhood $N_r(x_0) = \{x : |x-x_0| \leq r\}$ of x_0, where r is called the radius of convergence. The analytic property of a function is equivalent to the condition that for each $k \in \mathbb{N}$, there exists a constant c such that $\left|\frac{\mathrm{d}^k f}{\mathrm{d}x^k}(x)\right| \leq c^{k+1} k!$.

To prove that the KL of mixtures is not analytic (hence does not admit a closed-form formula), we shall adapt the proof reported in [6] (in Japanese, we thank Professor Aoyagi for sending us his paper [6]). We shall prove that $\mathrm{KL}(p:q)$ is not analytic for $p(x) = G(x;0,1)$ and $q(x;w) = (1-w)G(x;0,1) + wG(x;1,1)$, where $w \in (0,1)$, and $G(x;\mu,\sigma) = \frac{1}{\sqrt{2\pi}\sigma}\exp(-\frac{(x-\mu)^2}{2\sigma^2})$ is the density of a univariate Gaussian of mean μ and standard deviation σ. Let $D(w) = \mathrm{KL}(p(x):q(x;w))$ denote the KL divergence between these two mixtures (p has a single component and q has two components).

We have

$$\log \frac{p(x)}{q(x;w)} = \log \frac{\exp\left(-\frac{x^2}{2}\right)}{(1-w)\exp\left(-\frac{x^2}{2}\right) + w\exp\left(-\frac{(x-1)^2}{2}\right)} = -\log(1 + w(e^{x-\frac{1}{2}} - 1)). \tag{A1}$$

Therefore

$$\frac{\mathrm{d}^k D}{\mathrm{d}w^k} = \frac{(-1)^k}{k} \int p(x)(e^{x-\frac{1}{2}} - 1)\mathrm{d}x.$$

Let x_0 be the root of the equation $e^{x-\frac{1}{2}} - 1 = e^{\frac{x}{2}}$ so that for $x \geq x_0$, we have $e^{x-\frac{1}{2}} - 1 \geq e^{\frac{x}{2}}$. It follows that:

$$\left|\frac{\mathrm{d}^k D}{\mathrm{d}w^k}\right| \geq \frac{1}{k}\int_{x_0}^{\infty} p(x) e^{\frac{kx}{2}} \mathrm{d}x = \frac{1}{k} e^{\frac{k^2}{8}} A_k$$

with $A_k = \int_{x_0}^{\infty} \frac{1}{\sqrt{2\pi}} \exp(-\frac{x-\frac{k}{2}}{2})\mathrm{d}x$. When $k \to \infty$, we have $A_k \to 1$. Consider $k_0 \in \mathbb{N}$ such that $A_{k_0} > 0.9$. Then the radius of convergence r is such that:

$$\frac{1}{r} \geq \lim_{k\to\infty} \left(\frac{1}{kk!} 0.9 \exp\left(\frac{k^2}{8}\right) \right)^{\frac{1}{k}} = \infty.$$

Thus the convergence radius is $r = 0$, and therefore the KL divergence is not an analytic function of the parameter w. The KL of mixtures is an example of a non-analytic smooth function. (Notice that the absolute value is not analytic at 0.)

Appendix B. Closed-Form Formula for the Kullback–Leibler Divergence between Scaled and Truncated Exponential Families

When computing approximation bounds for the KL divergence between two mixtures $m(x)$ and $m'(x)$, we end up with the task of computing $\int_{\mathcal{D}} w_a p_a(x) \log \frac{w_b' p_b'(x)}{w_c' p_c'(x)} \mathrm{d}x$ where $\mathcal{D} \subseteq \mathcal{X}$ is a subset of the full support $\mathcal{X}$. We report a generic formula for computing these formula when the mixture (scaled and truncated) components belong to the same exponential family [17]. An exponential family has canonical log-density written as $l(x;\theta) = \log p(x;\theta) = \theta^\top t(x) - F(\theta) + k(x)$, where $t(x)$ denotes the sufficient statistics, $F(\theta)$ the log-normalizer (also called cumulant function or partition function), and $k(x)$ an auxiliary carrier term.

Let $\mathrm{KL}(w_1 p_1 : w_2 p_2 : w_3 p_3) = \int_{\mathcal{X}} w_1 p_1(x) \log \frac{w_2 p_2(x)}{w_3 p_3(x)} \mathrm{d}x = H_\times(w_1 p_1 : w_3 p_3) - H_\times(w_1 p_1 : w_2 p_2)$. Since it is a difference of two cross-entropies, we get for three distributions belonging to the same exponential family [26] the following formula:

$$\mathrm{KL}(w_1 p_1 : w_2 p_2 : w_3 p_3) = w_1 \log \frac{w_2}{w_3} + w_1 (F(\theta_3) - F(\theta_2) - (\theta_3 - \theta_2)^\top \nabla F(\theta_1)).$$

Furthermore, when the support is restricted, say to support range $\mathcal{D} \subseteq \mathcal{X}$, let $m_{\mathcal{D}}(\theta) = \int_{\mathcal{D}} p(x;\theta)\mathrm{d}x$ denote the mass and $\tilde{p(x;\theta)} = \frac{p(x;\theta)}{m_{\mathcal{D}}(\theta)}$ the normalized distribution. Then we have:

$$\int_{\mathcal{D}} w_1 p_1(x) \log \frac{w_2 p_2(x)}{w_3 p_3(x)} \mathrm{d}x = m_{\mathcal{D}}(\theta_1)(\mathrm{KL}(w_1 \tilde{p}_1 : w_2 \tilde{p}_2 : w_3 \tilde{p}_3)) - \log \frac{w_2 m_{\mathcal{D}}(\theta_3)}{w_3 m_{\mathcal{D}}(\theta_2)}.$$

When $F_{\mathcal{D}}(\theta) = F(\theta) - \log m_{\mathcal{D}}(\theta)$ is strictly convex and differentiable then $\tilde{p(x;\theta)}$ is an exponential family and the closed-form formula follows straightforwardly. Otherwise, we still get a closed-form but need more derivations. For univariate distributions, we write $\mathcal{D} = (a,b)$ and $m_{\mathcal{D}}(\theta) = \int_a^b p(x;\theta)\mathrm{d}x = P_\theta(b) - P_\theta(a)$ where $P_\theta(a) = \int^a p(x;\theta)\mathrm{d}x$ denotes the cumulative distribution function.

The usual formula for truncated and scaled Kullback–Leibler divergence is:

$$\mathrm{KL}_{\mathcal{D}}(wp(x;\theta) : w'p(x;\theta')) = w m_{\mathcal{D}}(\theta) \left(\log \frac{w}{w'} + B_F(\theta' : \theta) \right) + w(\theta' - \theta)^\top \nabla m_{\mathcal{D}}(\theta), \tag{B1}$$

where $B_F(\theta' : \theta)$ is a Bregman divergence [5]:

$$B_F(\theta' : \theta) = F(\theta') - F(\theta) - (\theta' - \theta)^\top \nabla F(\theta).$$

This formula extends the classic formula [5] for full regular exponential families (by setting $w = w' = 1$ and $m_{\mathcal{D}}(\theta) = 1$ with $\nabla m_{\mathcal{D}}(\theta) = 0$).

Similar formulæ are available for the cross-entropy and entropy of exponential families [26].

Appendix C. On the Approximation of KL between Smooth Mixtures by a Bregman Divergence [5]

Clearly, since Bregman divergences are always finite while KL divergences may diverge, we need extra conditions to assert that the KL between mixtures can be approximated by Bregman divergences.

We require that the Jeffreys divergence between mixtures be finite in order to approximate the KL between mixtures by a Bregman divergence. We loosely derive this observation (Careful derivations will be reported elsewhere) using two different approaches:

- First, continuous mixture distributions have smooth densities that can be arbitrarily closely approximated using a single distribution (potentially multi-modal) belonging to the Polynomial Exponential Families [53,54] (PEFs). A polynomial exponential family of order D has log-likelihood $l(x;\theta) \propto \sum_{i=1}^{D} \theta_i x^i$: Therefore, a PEF is an exponential family with polynomial sufficient statistics $t(x) = (x, x^2, \ldots, x^D)$. However, the log-normalizer $F_D(\theta) = \log \int \exp(\theta^\top t(x)) \mathrm{d}x$ of a D-order PEF is not available in closed-form: It is computationally intractable. Nevertheless, the KL between two mixtures $m(x)$ and $m'(x)$ can be theoretically approximated closely by a Bregman divergence between the two corresponding PEFs: $\mathrm{KL}(m(x) : m'(x)) \simeq \mathrm{KL}(p(x;\theta) : p(x;\theta')) = B_{F_D}(\theta' : \theta)$, where θ and θ' are the natural parameters of the PEF family $\{p(x;\theta)\}$ approximating $m(x)$ and $m'(x)$, respectively (i.e., $m(x) \simeq p(x;\theta)$ and $m'(x) \simeq p(x;\theta')$). Notice that the Bregman divergence of PEFs has necessarily finite value but the KL of two smooth mixtures can potentially diverge (infinite value), hence the conditions on Jeffreys divergence to be finite.

- Second, consider two finite mixtures $m(x) = \sum_{i=1}^{k} w_i p_i(x)$ and $m'(x) = \sum_{j=1}^{k'} w'_j p'_j(x)$ of k and k' components (possibly with heterogeneous components $p_i(x)$'s and $p'_j(x)$'s), respectively. In information geometry, a mixture family is the set of convex combination of fixed component densities. Thus in statistics, a mixture is understood as a convex combination of parametric components while in information geometry a mixture family is the set of convex combination of fixed components. Let us consider the mixture families $\{g(x;(w,w'))\}$ generated by the $D = k + k'$ fixed components $p_1(x), \ldots, p_k(x), p'_1(x), \ldots, p'_{k'}(x)$:

$$\left\{ g(x;(w,w')) = \sum_{i=1}^{k} w_i p_i(x) + \sum_{j=1}^{k'} w'_j p'_j(x) \ : \ \sum_{i=1}^{k} w_i + \sum_{j=1}^{k'} w'_j = 1 \right\}$$

  We can approximate arbitrarily finely (with respect to total variation) mixture $m(x)$ for any $\epsilon > 0$ by $g(x;\alpha) \simeq (1-\epsilon)m(x) + \epsilon m'(x)$ with $\alpha = ((1-\epsilon)w, \epsilon w')$ (so that $\sum_{i=1}^{k+k'} \alpha_i = 1$) and $m'(x) \simeq g(x;\alpha') = \epsilon m(x) + (1-\epsilon)m'(x)$ with $\alpha' = (\epsilon w, (1-\epsilon)w')$ (and $\sum_{i=1}^{k+k'} \alpha'_i = 1$). Therefore $\mathrm{KL}(m(x) : m'(x)) \simeq \mathrm{KL}(g(x;\alpha) : g(x;\alpha')) = B_{F^*}(\alpha : \alpha')$, where $F^*(\alpha) = \int g(x;\alpha) \log g(x;\alpha) \mathrm{d}x$ is the Shannon information (negative Shannon entropy) for the composite mixture family. Again, the Bregman divergence $B_{F^*}(\alpha : \alpha')$ is necessarily finite but $\mathrm{KL}(m(x) : m'(x))$ between mixtures may be potentially infinite when the KL integral diverges (hence, the condition on Jeffreys divergence finiteness). Interestingly, this Shannon information can be arbitrarily closely approximated when considering isotropic Gaussians [13]. Notice that the convex conjugate $F(\theta)$ of the continuous Shannon neg-entropy $F^*(\eta)$ is the log-sum-exp function on the inverse soft map.

References

1. Huang, Z.K.; Chau, K.W. A new image thresholding method based on Gaussian mixture model. *Appl. Math. Comput.* **2008**, *205*, 899–907.
2. Seabra, J.; Ciompi, F.; Pujol, O.; Mauri, J.; Radeva, P.; Sanches, J. Rayleigh mixture model for plaque characterization in intravascular ultrasound. *IEEE Trans. Biomed. Eng.* **2011**, *58*, 1314–1324.
3. Julier, S.J.; Bailey, T.; Uhlmann, J.K. Using Exponential Mixture Models for Suboptimal Distributed Data Fusion. In Proceedings of the 2006 IEEE Nonlinear Statistical Signal Processing Workshop, Cambridge, UK, 13–15 September 2006; IEEE: New York, NY, USA, 2006; pp. 160–163.
4. Cover, T.M.; Thomas, J.A. *Elements of Information Theory*; John Wiley & Sons: Hoboken, NJ, USA, 2012.
5. Banerjee, A.; Merugu, S.; Dhillon, I.S.; Ghosh, J. Clustering with Bregman divergences. *J. Mach. Learn. Res.* **2005**, *6*, 1705–1749.

6. Watanabe, S.; Yamazaki, K.; Aoyagi, M. *Kullback Information of Normal Mixture is Not an Analytic Function*; Technical Report of IEICE Neurocomputing; The Institute of Electronics, Information and Communication Engineers, Tokyo, Japan, 2004; pp. 41–46. (In Japanese)
7. Michalowicz, J.V.; Nichols, J.M.; Bucholtz, F. Calculation of differential entropy for a mixed Gaussian distribution. *Entropy* **2008**, *10*, 200–206.
8. Pichler, G.; Koliander, G.; Riegler, E.; Hlawatsch, F. Entropy for singular distributions. In Proceedings of the IEEE International Symposium on Information Theory (ISIT), Honolulu, HI, USA, 29 June–4 July 2014; pp. 2484–2488.
9. Huber, M.F.; Bailey, T.; Durrant-Whyte, H.; Hanebeck, U.D. On entropy approximation for Gaussian mixture random vectors. In Proceedings of the IEEE International Conference on Multisensor Fusion and Integration for Intelligent Systems, Seoul, Korea, 20–22 August 2008; IEEE: New York, NY, USA, 2008; pp. 181–188.
10. Yamada, M.; Sugiyama, M. Direct importance estimation with Gaussian mixture models. *IEICE Trans. Inf. Syst.* **2009**, *92*, 2159–2162.
11. Durrieu, J.L.; Thiran, J.P.; Kelly, F. Lower and upper bounds for approximation of the Kullback-Leibler divergence between Gaussian Mixture Models. In Proceedings of the IEEE International Conference on Acoustics, Speech and Signal Processing (ICASSP), Kyoto, Japan, 25–30 March 2012; IEEE: New York, NY, USA, 2012; pp. 4833–4836.
12. Schwander, O.; Marchand-Maillet, S.; Nielsen, F. Comix: Joint estimation and lightspeed comparison of mixture models. In Proceedings of the 2016 IEEE International Conference on Acoustics, Speech and Signal Processing, ICASSP 2016, Shanghai, China, 20–25 March 2016; pp. 2449–2453.
13. Moshksar, K.; Khandani, A.K. Arbitrarily Tight Bounds on Differential Entropy of Gaussian Mixtures. *IEEE Trans. Inf. Theory* **2016**, *62*, 3340–3354.
14. Mezuman, E.; Weiss, Y. A Tight Convex Upper Bound on the Likelihood of a Finite Mixture. *arXiv* **2016**, arXiv:1608.05275.
15. Amari, S.-I. *Information Geometry and Its Applications*; Springer: Tokyo, Japan, 2016; Volume 194.
16. Nielsen, F.; Sun, K. Guaranteed Bounds on the Kullback–Leibler Divergence of Univariate Mixtures. *IEEE Signal Process. Lett.* **2016**, *23*, 1543–1546.
17. Nielsen, F.; Garcia, V. Statistical exponential families: A digest with flash cards. *arXiv* **2009**, arXiv:0911.4863.
18. Calafiore, G.C.; El Ghaoui, L. *Optimization Models*; Cambridge University Press: Cambridge, UK, 2014.
19. Shen, C.; Li, H. On the dual formulation of boosting algorithms. *IEEE Trans. Pattern Anal. Mach. Intell.* **2010**, *32*, 2216–2231.
20. Beck, A. *Introduction to Nonlinear Optimization: Theory, Algorithms, and Applications with MATLAB*; Society for Industrial and Applied Mathematics: Philadelphia, PA, USA, 2014.
21. Boyd, S.; Vandenberghe, L. *Convex Optimization*; Cambridge University Press: Cambridge, UK, 2004.
22. De Berg, M.; van Kreveld, M.; Overmars, M.; Schwarzkopf, O.C. *Computational Geometry*; Springer: Heidelberg, Germany, 2000.
23. Setter, O.; Sharir, M.; Halperin, D. *Constructing Two-Dimensional Voronoi Diagrams via Divide-and-Conquer of Envelopes in Space*; Springer: Heidelberg, Germany, 2010.
24. Devillers, O.; Golin, M.J. Incremental algorithms for finding the convex hulls of circles and the lower envelopes of parabolas. *Inf. Process. Lett.* **1995**, *56*, 157–164.
25. Nielsen, F.; Yvinec, M. An output-sensitive convex hull algorithm for planar objects. *Int. J. Comput. Geom. Appl.* **1998**, *8*, 39–65.
26. Nielsen, F.; Nock, R. Entropies and cross-entropies of exponential families. In Proceedings of the 17th IEEE International Conference on Image Processing (ICIP), Hong Kong, China, 26–29 September 2010; IEEE: New York, NY, USA, 2010; pp. 3621–3624.
27. Sharir, M.; Agarwal, P.K. *Davenport-Schinzel Sequences and Their Geometric Applications*; Cambridge University Press: Cambridge, UK, 1995.
28. Bronstein, M. Algorithms and computation in mathematics. In *Symbolic Integration. I. Transcendental Functions*; Springer: Berlin, Germany, 2005.
29. Carreira-Perpinan, M.A. Mode-finding for mixtures of Gaussian distributions. *IEEE Trans. Pattern Anal. Mach. Intell.* **2000**, *22*, 1318–1323.

30. Aprausheva, N.N.; Sorokin, S.V. Exact equation of the boundary of unimodal and bimodal domains of a two-component Gaussian mixture. *Pattern Recognit. Image Anal.* **2013**, *23*, 341–347.
31. Learned-Miller, E.; DeStefano, J. A probabilistic upper bound on differential entropy. *IEEE Trans. Inf. Theory* **2008**, *54*, 5223–5230.
32. Amari, S.-I. α-Divergence Is Unique, Belonging to Both f-Divergence and Bregman Divergence Classes. *IEEE Trans. Inf. Theory* **2009**, *55*, 4925–4931.
33. Cichocki, A.; Amari, S.I. Families of Alpha- Beta- and Gamma-Divergences: Flexible and Robust Measures of Similarities. *Entropy* **2010**, *12*, 1532–1568.
34. Póczos, B.; Schneider, J. On the Estimation of α-Divergences. In Proceedings of the 14th International Conference on Artificial Intelligence and Statistics, Ft. Lauderdale, FL, USA, 11–13 April 2011; pp. 609–617.
35. Nielsen, F.; Nock, R. On Rényi and Tsallis entropies and divergences for exponential families. *arXiv* **2011**, arXiv:1105.3259.
36. Minka, T. *Divergence Measures and Message Passing*; Technical Report MSR-TR-2005-173; Microsoft Research: Cambridge, UK, 2005.
37. Améndola, C.; Drton, M.; Sturmfels, B. Maximum Likelihood Estimates for Gaussian Mixtures Are Transcendental. *arXiv* **2015**, arXiv:1508.06958.
38. Hellinger, E. Neue Begründung der Theorie quadratischer Formen von unendlichvielen Veränderlichen. *J. Reine Angew. Math.* **1909**, *136*, 210–271. (In German)
39. Van Erven, T.; Harremos, P. Rényi divergence and Kullback-Leibler divergence. *IEEE Trans. Inf. Theory* **2014**, *60*, 3797–3820.
40. Nielsen, F.; Nock, R. A closed-form expression for the Sharma-Mittal entropy of exponential families. *J. Phys. A Math. Theor.* **2012**, *45*, 032003.
41. Nielsen, F.; Nock, R. On the Chi Square and Higher-Order Chi Distances for Approximating f-Divergences. *IEEE Signal Process. Lett.* **2014**, *21*, 10–13.
42. Nielsen, F.; Boltz, S. The Burbea-Rao and Bhattacharyya centroids. *IEEE Trans. Inf. Theory* **2011**, *57*, 5455–5466.
43. Jarosz, W. Efficient Monte Carlo Methods for Light Transport in Scattering Media. Ph.D. Thesis, University of California, San Diego, CA, USA, 2008.
44. Fujisawa, H.; Eguchi, S. Robust parameter estimation with a small bias against heavy contamination. *J. Multivar. Anal.* **2008**, *99*, 2053–2081.
45. Havrda, J.; Charvát, F. Quantification method of classification processes. Concept of structural α-entropy. *Kybernetika* **1967**, *3*, 30–35.
46. Liang, X. A Note on Divergences. *Neural Comput.* **2016**, *28*, 2045–2062.
47. Lin, J. Divergence measures based on the Shannon entropy. *IEEE Trans. Inf. Theory* **1991**, *37*, 145–151.
48. Endres, D.M.; Schindelin, J.E. A new metric for probability distributions. *IEEE Trans. Inf. Theory* **2003**, *49*, 1858–1860.
49. Nielsen, F.; Boissonnat, J.D.; Nock, R. On Bregman Voronoi diagrams. In Proceedings of the Eighteenth Annual ACM-SIAM Symposium on Discrete Algorithms, New Orleans, LA, USA, 7–9 January 2007; Society for Industrial and Applied Mathematics: Philadelphia, PA, USA, 2007; pp. 746–755.
50. Boissonnat, J.D.; Nielsen, F.; Nock, R. Bregman Voronoi diagrams. *Discret. Comput. Geom.* **2010**, *44*, 281–307.
51. Foster, D.V.; Grassberger, P. Lower bounds on mutual information. *Phys. Rev. E* **2011**, *83*, 010101.
52. Nielsen, F.; Sun, K. PyKLGMM: Python Software for Computing Bounds on the Kullback-Leibler Divergence between Mixture Models. 2016. Available online: https://www.lix.polytechnique.fr/~nielsen/KLGMM/ (accessed on 6 December 2016).
53. Cobb, L.; Koppstein, P.; Chen, N.H. Estimation and moment recursion relations for multimodal distributions of the exponential family. *J. Am. Stat. Assoc.* **1983**, *78*, 124–130.
54. Nielsen, F.; Nock, R. Patch matching with polynomial exponential families and projective divergences. In Proceedings of the 9th International Conference Similarity Search and Applications (SISAP), Tokyo, Japan, 24–26 October 2016.

Riemannian Laplace Distribution on the Space of Symmetric Positive Definite Matrices

Hatem Hajri [1,*,†], **Ioana Ilea** [1,2,†], **Salem Said** [1,†], **Lionel Bombrun** [1,†] and **Yannick Berthoumieu** [1,†]

1 Groupe Signal et Image, CNRS Laboratoire IMS, Institut Polytechnique de Bordeaux, Université de Bordeaux, UMR 5218, Talence 33405, France; ioana.ilea@u-bordeaux.fr (I.I.); salem.said@u-bordeaux.fr (S.S.); lionel.bombrun@u-bordeaux.fr (L.B.); Yannick.Berthoumieu@ims-bordeaux.fr (Y.B.)
2 Communications Department, Technical University of Cluj-Napoca, 71-73 Dorobantilor street, Cluj-Napoca 3400, Romania
* Correspondence: hatem.hajri@ims-bordeaux.fr.
† These authors contributed equally to this work.

Academic Editors: Frédéric Barbaresco and Frank Nielsen

Abstract: The Riemannian geometry of the space $\mathcal{P}_m$, of $m \times m$ symmetric positive definite matrices, has provided effective tools to the fields of medical imaging, computer vision and radar signal processing. Still, an open challenge remains, which consists of extending these tools to correctly handle the presence of outliers (or abnormal data), arising from excessive noise or faulty measurements. The present paper tackles this challenge by introducing new probability distributions, called Riemannian Laplace distributions on the space $\mathcal{P}_m$. First, it shows that these distributions provide a statistical foundation for the concept of the Riemannian median, which offers improved robustness in dealing with outliers (in comparison to the more popular concept of the Riemannian center of mass). Second, it describes an original expectation-maximization algorithm, for estimating mixtures of Riemannian Laplace distributions. This algorithm is applied to the problem of texture classification, in computer vision, which is considered in the presence of outliers. It is shown to give significantly better performance with respect to other recently-proposed approaches.

Keywords: symmetric positive definite matrices; Laplace distribution; expectation-maximization; Bayesian information criterion; texture classification

---

1. Introduction

Data with values in the space $\mathcal{P}_m$, of $m \times m$ symmetric positive definite matrices, play an essential role in many applications, including medical imaging [1,2], computer vision [3–7] and radar signal processing [8,9]. In these applications, the location where a dataset is centered has a special interest. While several definitions of this location are possible, its meaning as a representative of the set should be clear. Perhaps, the most known and well-used quantity to represent a center of a dataset is the Fréchet mean. Given a set of points $Y_1, \cdots, Y_n$ in $\mathcal{P}_m$, their Fréchet mean is defined to be:

$$\text{Mean}(Y_1, \cdots, Y_n) = \text{argmin}_{Y \in \mathcal{P}_m} \sum_{i=1}^{n} d^2(Y, Y_i) \tag{1}$$

where d is Rao's Riemannian distance on $\mathcal{P}_m$ [10,11].

Statistics on general Riemannian manifolds have been powered by the development of different tools for geometric measurements and new probability distributions on manifolds [12,13]. On the manifold $(\mathcal{P}_m, d)$, the major advances in this field have been achieved by the recent papers [14,15], which introduce the Riemannian Gaussian distribution on $(\mathcal{P}_m, d)$. This distribution depends on two parameters $\bar{Y} \in \mathcal{P}_m$ and $\sigma > 0$, and its density with respect to the Riemannian volume form $dv(Y)$ of $\mathcal{P}_m$ (see Formula (13) in Section 2) is:

$$\frac{1}{Z_m(\sigma)} \exp\left[-\frac{d^2(Y,\overline{Y})}{2\sigma^2}\right] \tag{2}$$

where $Z_m(\sigma)$ is a normalizing factor depending only on σ (and not on $\bar{Y}$).

For the Gaussian distribution Equation (2), the maximum likelihood estimate (MLE) for the parameter $\bar{Y}$ based on observations $Y_1, \cdots, Y_n$ corresponds to the mean Equation (1). In [15], a detailed study of statistical inference for this distribution was given and then applied to the classification of data in $\mathcal{P}_m$, showing that it yields better performance, in comparison to recent approaches [2].

When a dataset contains extreme values (or outliers), because of the impact of these values on d^2, the mean becomes less useful. It is usually replaced with the Riemannian median:

$$\text{Median}(Y_1, \cdots, Y_n) = \text{argmin}_{Y \in \mathcal{P}_m} \sum_{i=1}^{n} d(Y, Y_i) \tag{3}$$

Definition Equation (3) corresponds to that of the median in statistics based on ordering of the values of a sequence. However, this interpretation does not continue to hold on $\mathcal{P}_m$. In fact, the Riemannian distance on $\mathcal{P}_m$ is not associated with any norm, and it is therefore only possible to compare distances of a set of matrices to a reference matrix.

In the presence of outliers, the Gaussian distribution on $\mathcal{P}_m$ also loses its robustness properties. The main contribution of the present paper is to remedy this problem by introducing the Riemannian Laplace distribution while maintaining the same one-to-one relation between MLE and the Riemannian median. This will be shown to offer considerable improvement in dealing with outliers.

This paper is organized as follows.

Section 2 reviews the Riemannian geometry of $\mathcal{P}_m$, when this manifold is equipped with the Riemannian metric known as the Rao–Fisher or affine invariant metric [10,11]. In particular, it gives analytic expressions for geodesic curves, Riemannian distance and recalls the invariance of Rao's distance under affine transformations.

Section 3 introduces the Laplace distribution $\mathcal{L}(\bar{Y}, \sigma)$ through its probability density function with respect to the volume form $dv(Y)$:

$$p(Y|\overline{Y}, \sigma) = \frac{1}{\zeta_m(\sigma)} \exp\left[-\frac{d(Y,\overline{Y})}{\sigma}\right]$$

here, σ lies in an interval $]0, \sigma_{\max}[$ with $\sigma_{\max} < \infty$. This is because the normalizing constant $\zeta_m(\sigma)$ becomes infinite for $\sigma \geq \sigma_{\max}$. It will be shown that $\zeta_m(\sigma)$ depend only on σ (and not on $\bar{Y}$) for all $\sigma < \sigma_{\max}$. This important fact leads to simple expressions of MLEs of $\overline{Y}$ and σ. In particular, the MLE of $\bar{Y}$ based on a family of observations $Y_1, \cdots, Y_N$ sampled from $\mathcal{L}(\bar{Y}, \sigma)$ is given by the median of $Y_1, \cdots, Y_N$ defined by Equation (3) where d is Rao's distance.

Section 4 focuses on mixtures of Riemannian Laplace distributions on $\mathcal{P}_m$. A distribution of this kind has a density:

$$p(Y|(\omega_\mu, \overline{Y}_\mu, \sigma_\mu)_{1 \leq \mu \leq M}) = \sum_{\mu=1}^{M} \varpi_\mu p(Y|\overline{Y}_\mu, \sigma_\mu) \tag{4}$$

with respect to the volume form $dv(Y)$. Here, M is the number of mixture components, $\varpi_\mu > 0$, $\overline{Y}_\mu \in \mathcal{P}_m, \sigma_\mu > 0$ for all $1 \leq \mu \leq M$ and $\sum_{\mu=1}^{M} \varpi_\mu = 1$. A new EM (expectation-maximization) algorithm that computes maximum likelihood estimates of the mixture parameters $(\varpi_\mu, \bar{Y}_\mu, \sigma_\mu)_{1 \leq \mu \leq M}$ is provided. The problem of the order selection of the number M in Equation (4) is also discussed and performed using the Bayesian information criterion (BIC) [16].

Section 5 is an application of the previous material to the classification of data with values in $\mathcal{P}_m$, which contain outliers (abnormal data points). Assume to be given a training sequence $Y_1, \cdots, Y_n \in \mathcal{P}_m$. Using the EM algorithm developed in Section 4, it is possible to subdivide this sequence into disjoint classes. To classify new data points, a classification rule is proposed. The robustness of this rule lies in the fact that it is based on the distances between new observations and the respective medians of classes instead of the means [15]. This rule will be illustrated by an application to the problem of texture classification in computer vision. The obtained results show improved performance with respect to recent approaches which use the Riemannian Gaussian distribution [15] and the Wishart distribution [17].

2. Riemannian Geometry of $\mathcal{P}_m$

The geometry of Siegel homogeneous bounded domains, such as Kähler homogeneous manifolds, have been studied by Felix A. Berezin [18] and P. Malliavin [19]. The structure of Kähler homogeneous manifolds has been used in [20,21] to parameterize (Toeplitz–) Block–Toeplitz matrices. This led to a Hessian metric from information geometry theory with a Kähler potential given by entropy and to an algorithm to compute medians of (Toeplitz–)Block–Toeplitz matrices by Karcher flow on Mostow/Berger fibration of a Siegel disk. Optimal numerical schemes of this algorithm in a Siegel disk have been studied, developed and validated in [22–24].

This section introduces the necessary background on the Riemannian geometry of $\mathcal{P}_m$, the space of symmetric positive definite matrices of size $m \times m$. Precisely, $\mathcal{P}_m$ is equipped with the Riemannian metric known as the affine-invariant metric. First, analytic expressions are recalled for geodesic curves and Riemannian distance. Then, two properties are stated, which are fundamental to the following. These are affine-invariance of the Riemannian distance and the existence and uniqueness of Riemannian medians.

The affine-invariant metric, called the Rao–Fisher metric in information geometry, has the following expression:

$$g_Y(A, B) = \mathrm{tr}(Y^{-1}AY^{-1}B) \tag{5}$$

where $Y \in \mathcal{P}_m$ and $A, B \in T_Y\mathcal{P}_m$, the tangent space to $\mathcal{P}_m$ at Y, which is identified with the vector space of $m \times m$ symmetric matrices. The Riemannian metric Equation (5) induces a Riemannian distance on $\mathcal{P}_m$ as follows. The length of a smooth curve $c : [0, 1] \to \mathcal{P}_m$ is given by:

$$L(c) = \int_0^1 \sqrt{g_{c(t)}(\dot{c}(t), \dot{c}(t))}\, dt \tag{6}$$

where $\dot{c}(t) = \frac{dc}{dt}$. For $Y, Z \in \mathcal{P}_m$, the Riemannian distance $d(Y, Z)$, called Rao's distance in information geometry, is defined to be:

$$d(Y, Z) = \inf\{\, L(c), c : [0, 1] \to \mathcal{P}_m \text{ is a smooth curve with } c(0) = Y, c(1) = Z\}\,.$$

This infimum is achieved by a unique curve $c = \gamma$, called the geodesic connecting Y and Z, which has the following equation [10,25]:

$$\gamma(t) = Y^{1/2}\,(Y^{-1/2}ZY^{-1/2})^t\,Y^{1/2} \tag{7}$$

Here, and throughout the following, all matrix functions (for example, square root, logarithm or power) are understood as symmetric matrix functions [26]. By definition, $d(Y, Z)$ coincides with $L(\gamma)$, which turns out to be:

$$d^2(Y, Z) = \mathrm{tr}\,[\log(Y^{-1/2}ZY^{-1/2})]^2 \tag{8}$$

Equipped with the affine-invariant metric Equation (5), the space $\mathcal{P}_m$ enjoys two useful properties, which are the following. The first property is invariance under affine transformations [10,25]. Recall that an affine transformation of $\mathcal{P}_m$ is a mapping $Y \mapsto Y \cdot A$, where A is an invertible real matrix of size $m \times m$,

$$Y \cdot A = A^{\dagger} \, Y \, A \tag{9}$$

and † denotes the transpose. Denote by $\mathrm{GL}(m)$ the group of $m \times m$ invertible real matrices on $\mathcal{P}_m$. Then, the action of $\mathrm{GL}(m)$ on $\mathcal{P}_m$ is transitive. This means that for any $Y, Z \in \mathcal{P}_m$, there exists $A \in \mathrm{GL}(m)$, such that $Y.A = Z$. Moreover, the Riemannian distance Equation (8) is invariant by affine transformations in the sense that for all $Y, Z \in \mathcal{P}_m$:

$$d(Y, Z) = d(Y \cdot A, Z \cdot A) \tag{10}$$

where $Y \cdot A$ and $Z \cdot A$ are defined by Equation (9). The transitivity of the action Equation (9) and the isometry property Equation (10) make $\mathcal{P}_m$ a Riemannian homogeneous space.

The affine-invariant metric Equation (5) turns $\mathcal{P}_m$ into a Riemannian manifold of negative sectional curvature [10,27]. As a result, $\mathcal{P}_m$ enjoys the property of the existence and uniqueness of Riemannian medians. The Riemannian median of N points $Y_1, \cdots, Y_N \in \mathcal{P}_m$ is defined to be:

$$\hat{Y}_N = \mathrm{argmin}_Y \sum_{n=1}^{N} d(Y, Y_n) \tag{11}$$

where $d(Y, Y_n)$ is the Riemannian distance Equation (8). If $Y_1, \cdots, Y_N$ do not belong to the same geodesic, then $\hat{Y}_N$ exists and is unique [28]. More generally, for any probability measure π on $\mathcal{P}_m$, the median of π is defined to be:

$$\hat{Y}_\pi = \mathrm{argmin}_Y \int_{\mathcal{P}_m} d(Y, Z) d\pi(Z) \tag{12}$$

Note that Equation (12) reduces to Equation (11) for $\pi = \frac{1}{N} \sum_{n=1}^{N} \delta_{Y_n}$. If the support of π is not carried by a single geodesic, then again, $\hat{Y}_\pi$ exists and is unique by the main result of [28].

To end this section, consider the Riemannian volume associated with the affine-invariant Riemannian metric [10]:

$$dv(Y) = \det(Y)^{-\frac{m+1}{2}} \prod_{i \leq j} dY_{ij} \tag{13}$$

where the indices denote matrix elements. The Riemannian volume is used to define the integral of a function $f : \mathcal{P}_m \to \mathbb{R}$ as:

$$\int_{\mathcal{P}_m} f(Y) dv(Y) = \int \ldots \int f(Y) \, \det(Y)^{-\frac{m+1}{2}} \prod_{i \leq j} dY_{ij} \tag{14}$$

where the integral on the right-hand side is a multiple integral over the $m(m+1)/2$ variables, Y_{ij} with $i \leq j$. The integral Equation (14) is invariant under affine transformations. Precisely:

$$\int_{\mathcal{P}_m} f(Y \cdot A) dv(Y) = \int_{\mathcal{P}_m} f(Y) dv(Y) \tag{15}$$

where $Y \cdot A$ is the affine transformation given by Equation (9). It takes on a simplified form when $f(Y)$ only depends on the eigenvalues of Y. Precisely, let the spectral decomposition of Y be given by $Y = U^{\dagger} \, \mathrm{diag}(e^{r_1}, \cdots, e^{r_m}) \, U$, where U is an orthogonal matrix and $e^{r_1}, \cdots, e^{r_m}$ are the eigenvalues of Y. Assume that $f(Y) = f(r_1, \ldots, r_m)$, then the invariant integral Equation (14) reduces to:

$$\int_{\mathcal{P}_m} f(Y) dv(Y) = c_m \times \int_{\mathbb{R}^m} f(r_1, \cdots, r_m) \prod_{i<j} \sinh\left(\frac{|r_i - r_j|}{2}\right) dr_1 \cdots dr_m \tag{16}$$

where the constant c_m is given by $c_m = \frac{1}{m!} \times \omega_m \times 8^{\frac{m(m-1)}{4}}$, $\omega_m = \frac{\pi^{m^2/2}}{\Gamma_m(m/2)}$ and Γ_m is the multivariate gamma function given in [29]. See Appendix A for the derivation of Equation (16) from Equation (14).

3. Riemannian Laplace Distribution on $\mathcal{P}_m$

3.1. Definition of $\mathcal{L}(\bar{Y}, \sigma)$

The Riemannian Laplace distribution on $\mathcal{P}_m$ is defined by analogy with the well-known Laplace distribution on $\mathbb{R}$. Recall the density of the Laplace distribution on $\mathbb{R}$,

$$p(x|\bar{x}, \sigma) = \frac{1}{2\sigma} e^{-|x-\bar{x}|/\sigma}$$

where $\bar{x} \in \mathbb{R}$ and $\sigma > 0$. This is a density with respect to the length element dx on $\mathbb{R}$. The density of the Riemannian Laplace distribution on $\mathcal{P}_m$ will be given by:

$$p(Y|\,\bar{Y}, \sigma) = \frac{1}{\zeta_m(\sigma)} \exp\left[-\frac{d(Y, \bar{Y})}{\sigma}\right] \tag{17}$$

here, $\bar{Y} \in \mathcal{P}_m$, $\sigma > 0$, and the density is with respect to the Riemannian volume element Equation (13) on $\mathcal{P}_m$. The normalizing factor $\zeta_m(\sigma)$ appearing in Equation (17) is given by the integral:

$$\int_{\mathcal{P}_m} \exp\left[-\frac{d(Y, \bar{Y})}{\sigma}\right] dv(Y)$$

Assume for now that this integral is finite for some choice of $\bar{Y}$ and σ. It is possible to show that its value does not depend on $\bar{Y}$. To do so, recall that the action of $\mathrm{GL}(m)$ on $\mathcal{P}_m$ is transitive. As a consequence, there exists $A \in \mathcal{P}_m$, such that $\bar{Y} = I.A$, where $I.A$ is defined as in Equation (9). From Equation (10), it follows that $d(Y, \bar{Y}) = d(Y, I.A) = d(Y.A^{-1}, I)$. From the invariance property Equation (15):

$$\int_{\mathcal{P}_m} \exp\left[-\frac{d(Y, \bar{Y})}{\sigma}\right] dv(Y) = \int_{\mathcal{P}_m} \exp\left[-\frac{d(Y, I)}{\sigma}\right] dv(Y) \tag{18}$$

The integral on the right does not depend on $\bar{Y}$, which proves the above claim.
The last integral representation and formula Equation (16) lead to the following explicit expression:

$$\zeta_m(\sigma) = c_m \times \int_{\mathbb{R}^m} e^{-\frac{|r|}{\sigma}} \prod_{i<j} \sinh\left(\frac{|r_i - r_j|}{2}\right) dr_1 \cdots dr_m \tag{19}$$

where $|r| = (r_1^2 + \cdots + r_2^m)^{\frac{1}{2}}$ and c_m is the same constant as in Equation (16) (see Appendix B for more details on the derivation of Equation (19)).

A distinctive feature of the Riemannian Laplace distribution on $\mathcal{P}_m$, in comparison to the Laplace distribution on $\mathbb{R}$ is that there exist certain values of σ for which it cannot be defined. This is because the integral Equation (19) diverges for certain values of this parameter. This leads to the following definition.

Definition 1. *Set $\sigma_m = \sup\{\sigma > 0 : \zeta_m(\sigma) < \infty\}$. Then, for $\bar{Y} \in \mathcal{P}_m$ and $\sigma \in (0, \sigma_m)$, the Riemannian Laplace distribution on $\mathcal{P}_m$, denoted by $\mathcal{L}(\bar{Y}, \sigma)$, is defined as the probability distribution on $\mathcal{P}_m$, whose density with respect to $dv(Y)$ is given by Equation (17), where $\zeta_m(\sigma)$ is defined by Equation (19).*

The constant σ_m in this definition satisfies $0 < \sigma_m < \infty$ for all m and takes the value $\sqrt{2}$ for $m = 2$ (see Appendix C for proofs).

3.2. Sampling from $\mathcal{L}(\bar{Y},\sigma)$

The current section presents a general method for sampling from the Laplace distribution $\mathcal{L}(\bar{Y},\sigma)$. This method relies in part on the following transformation property.

Proposition 1. *Let Y be a random variable in $\mathcal{P}_m$. For all $A \in \mathrm{GL}(m)$,*

$$Y \sim \mathcal{L}(\bar{Y},\sigma) \implies Y \cdot A \sim \mathcal{L}(\bar{Y} \cdot A,\sigma)$$

where $Y \cdot A$ is given by Equation (9).

Proof. Let $\varphi : \mathcal{P}_m \to \mathbb{R}$ be a test function. If $Y \sim \mathcal{L}(\bar{Y},\sigma)$ and $Z = Y \cdot A$, then the expectation of $\varphi(Z)$ is given by:

$$E[\varphi(Z)] = \int_{\mathcal{P}_m} \varphi(X \cdot A)\, p(X|\,\bar{Y},\sigma)\, dv(X) \;=\; \int_{\mathcal{P}_m} \varphi(X)\, p(X \cdot A^{-1}|\,\bar{Y},\sigma)\, dv(X)$$

where the equality is a result of Equation (15). However, $p(X \cdot A^{-1}|\,\bar{Y},\sigma) = p(X|\,\bar{Y} \cdot A,\sigma)$ by Equation (10), which proves the proposition.

□

The following algorithm describes how to sample from $\mathcal{L}(\bar{Y},\sigma)$ where $0 < \sigma < \sigma_m$. For this, it is first required to sample from the density p on $\mathbb{R}^m$ defined by:

$$p(r) = \frac{c_m}{\zeta_m(\sigma)} e^{-\frac{|r|}{\sigma}} \prod_{i<j} \sinh\left(\frac{|r_i - r_j|}{2}\right), \quad r = (r_1, \cdots, r_m).$$

This can be done by a usual Metropolis algorithm [30].

It is also required to sample from the uniform distribution on $\mathrm{O}(m)$, the group of real orthogonal $m \times m$ matrices. This can be done by generating A, an $m \times m$ matrix, whose entries are i.i.d. with normal distribution $\mathcal{N}(0,1)$, then the orthogonal matrix U, in the decomposition $A = UT$ with T upper triangular, is uniformly distributed on $\mathrm{O}(m)$ [29] (p. 70). Sampling from $\mathcal{L}(\bar{Y},\sigma)$ can now be described as follows.

Algorithm 1 Sampling from $\mathcal{L}(\bar{Y},\sigma)$.

1: Generate i.i.d. samples $(r_1, \cdots, r_m) \in \mathbb{R}^m$ with density p
2: Generate U from a uniform distribution on $\mathrm{O}(m)$
3: $X \leftarrow U^{\dagger}\mathrm{diag}(e^{r_1}, \cdots, e^{r_m})U$
4: $Y \leftarrow X.\bar{Y}^{\frac{1}{2}}$

Note that the law of X in Step 3 is $\mathcal{L}(I,\sigma)$; the proof of this fact is given in Appendix D. Finally, the law of Y in Step 4 is $\mathcal{L}(I.\bar{Y}^{\frac{1}{2}} = \bar{Y},\sigma)$ by proposition Equation (1).

3.3. Estimation of $\bar{Y}$ and σ

The current section considers maximum likelihood estimation of the parameters $\bar{Y}$ and σ, based on independent observations $Y_1, \ldots, Y_N$ from the Riemannian Laplace distribution $\mathcal{L}(\bar{Y},\sigma)$. The main results are contained in Propositions 2 and 3 below.

Proposition 2 states the existence and uniqueness of the maximum likelihood estimates $\hat{Y}_N$ and $\hat{\sigma}_N$ of $\bar{Y}$ and σ. In particular, the maximum likelihood estimate $\hat{Y}_N$ of $\bar{Y}$ is the Riemannian median of $Y_1, \ldots, Y_N$, defined by Equation (11). Numerical computation of $\hat{Y}_N$ will be considered and carried out using a Riemannian sub-gradient descent algorithm [8].

Proposition 3 states the convergence of the maximum likelihood estimate $\hat{Y}_N$ to the true value of the parameter $\bar{Y}$. It is based on Lemma 1, which states that the parameter $\bar{Y}$ is the Riemannian median of the distribution $\mathcal{L}(\bar{Y},\sigma)$ in the sense of definition Equation (12).

Proposition 2 (MLE and median). *The maximum likelihood estimate of the parameter $\bar{Y}$ is the Riemannian median $\hat{Y}_N$ of $Y_1,\ldots,Y_N$. Moreover, the maximum likelihood estimate of the parameter σ is the solution $\hat{\sigma}_N$ of:*

$$\sigma^2 \times \frac{d}{d\sigma}\log\zeta_m(\sigma) = \frac{1}{N}\sum_{n=1}^{N} d(\bar{Y},Y_n) \tag{20}$$

Both $\hat{Y}_N$ and $\hat{\sigma}_N$ exist and are unique for any realization of the samples $Y_1,\ldots,Y_N$.

Proof of Proposition 2. The log-likelihood function, of the parameters $\bar{Y}$ and σ, can be written as:

$$\begin{aligned}\sum_{n=1}^{N}\log p(Y_n|\,\bar{Y},\sigma) &= \sum_{n=1}^{N}\log\left(\frac{1}{\zeta_m(\sigma)}e^{-\frac{d(\bar{Y},Y_n)}{\sigma}}\right)\\ &= -N\log\zeta_m(\sigma)-\frac{1}{\sigma}\sum_{n=1}^{N}d(\bar{Y},Y_n)\end{aligned}$$

As the first term in the last expression does not contain $\bar{Y}$,

$$\text{argmax}_{\bar{Y}}\sum_{n=1}^{N}\log p(Y_n|\,\bar{Y},\sigma) = \text{argmin}_{\bar{Y}}\sum_{n=1}^{N}d(\bar{Y},Y_n)$$

The quantity on the right is exactly $\hat{Y}_N$ by Equation (11). This proves the first claim. Now, consider the function:

$$F(\eta) = -N\log(\zeta_m(\frac{-1}{\eta})) + \eta\sum_{n=1}^{N}d(\hat{Y}_N,Y_n),\quad \eta < \frac{-1}{\sigma_m}$$

This function is strictly concave, since it is the logarithm of the moment generating function of a positive measure. Note that $\lim_{\eta\to\frac{-1}{\sigma_m}}F(\eta) = -\infty$, and admit for a moment that $\lim_{\eta\to-\infty}F(\eta) = -\infty$. By the strict concavity of F, there exists a unique $\hat{\eta}_N < \frac{-1}{\sigma_m}$ (which is the maximum of F), such that $F'(\hat{\eta}_N) = 0$. It follows that $\hat{\sigma}_N = \frac{-1}{\hat{\eta}_N}$ lies in $(0,\sigma_m)$ and satisfies Equation (20). The uniqueness of $\hat{\sigma}_N$ is a consequence of the uniqueness of $\hat{\eta}_N$. Thus, the proof is complete. Now, it remains to check that $\lim_{\eta\to-\infty}F(\eta) = -\infty$ or just $\lim_{\sigma\to+\infty}\frac{1}{\sigma}\log(\zeta_m(\frac{1}{\sigma})) = 0$. Clearly:

$$\prod_{i<j}\sinh\left(\frac{|r_i-r_j|}{2}\right) \leq A_m e^{B_m|r|}$$

where A_m and B_m are two constants only depending on m. Using this, it follows that:

$$\frac{1}{\sigma}\log(\zeta_m(\frac{1}{\sigma})) \leq \frac{1}{\sigma}\log(c_m A_m) + \frac{1}{\sigma}\log\left(\int_{\mathbb{R}^m}\exp((-\sigma+B_m)|r|)dr_1\cdots dr_m\right) \tag{21}$$

However, for some constant C_m only depending on m,

$$\int_{\mathbb{R}^m}\exp((-\sigma+B_m)|r|)dr_1\cdots dr_m = C_m\int_0^\infty \exp((-\sigma+B_m)u)u^{m-1}du$$

$$\leq (m-1)!C_m\int_0^\infty \exp((-\sigma+B_m+1)u)du = \frac{(m-1)!C_m}{\sigma-B_m-1}$$

Combining this bound and Equation (21) yields $\lim_{\sigma\to+\infty}\frac{1}{\sigma}\log(\zeta_m(\frac{1}{\sigma})) = 0$. □

Remark 1. *Replacing F in the previous proof with* $F(\eta) = -\log(\zeta_m(\frac{-1}{\eta})) + \eta c$ *where* $c > 0$ *shows that the equation:*

$$\sigma^2 \times \frac{d}{d\sigma} \log \zeta_m(\sigma) = c$$

has a unique solution $\sigma \in (0, \sigma_m)$*. This shows in particular that* $\sigma \mapsto \sigma^2 \times \frac{d}{d\sigma} \log \zeta_m(\sigma)$ *is a bijection from* $(0, \sigma_m)$ *to* $(0, \infty)$*.*

Consider now the numerical computation of the maximum likelihood estimates $\hat{Y}_N$ and $\hat{\sigma}_N$ given by Proposition 2. Computation of $\hat{Y}_N$ consists in finding the Riemannian median of $Y_1, \ldots, Y_N$, defined by Equation (11). This can be done using the Riemannian sub-gradient descent algorithm of [8]. The k-th iteration of this algorithm produces an approximation $\hat{Y}^k_N$ of $\hat{Y}_N$ in the following way.

For $k = 1, 2, \ldots$, let Δ_k be the symmetric matrix:

$$\Delta_k = \frac{1}{N} \sum_{n=1}^{N} \frac{\mathrm{Log}_{\hat{Y}^{k-1}_N}(Y_n)}{||\mathrm{Log}_{\hat{Y}^{k-1}_N}(Y_n)||} \tag{22}$$

Here, Log is the Riemannian logarithm mapping inverse to the the Riemannian exponential mapping:

$$\mathrm{Exp}_Y(\Delta) = Y^{1/2} \exp\left(Y^{-1/2} \Delta Y^{-1/2}\right) Y^{1/2} \tag{23}$$

and $||\mathrm{Log}_a(b)|| = \sqrt{g_a(b,b)}$. Then, $\hat{Y}^k_N$ is defined to be:

$$\hat{Y}^k_N = \mathrm{Exp}_{\hat{Y}^{k-1}_N}(\tau_k \Delta_k) \tag{24}$$

where $\tau_k > 0$ is a step size, which can be determined using a backtracking procedure.

Computation of $\hat{\sigma}_N$ requires solving a non-linear equation in one variable. This is readily done using Newton's method.

It is shown now that the empirical Riemannian median $\hat{Y}_N$ converges almost surely to the true median $\bar{Y}$. This means that $\hat{Y}_N$ is a consistent estimator of $\bar{Y}$. The proof of this fact requires few notations and a preparatory lemma.

For $\bar{Y} \in \mathcal{P}_m$ and $\sigma \in (0, \sigma_m)$, let:

$$\mathcal{E}(Y|\,\bar{Y}, \sigma) = \int_{\mathcal{P}_m} d(Y, Z)\, p(Z|\,\bar{Y}, \sigma) dv(Z)$$

The following lemma shows how to find $\bar{Y}$ and σ from the function $Y \mapsto \mathcal{E}(Y|\,\bar{Y}, \sigma)$.

Lemma 1. *For any* $\bar{Y} \in \mathcal{P}_m$ *and* $\sigma \in (0, \sigma_m)$*, the following properties hold*

(i) $\bar{Y}$ *is given by:*

$$\bar{Y} = \mathrm{argmin}_Y\, \mathcal{E}(Y|\,\bar{Y}, \sigma) \tag{25a}$$

That is, $\bar{Y}$ *is the Riemannian median of* $\mathcal{L}(\bar{Y}, \sigma)$*.*

(ii) σ *is given by:*

$$\sigma = \Phi\left(\mathcal{E}(\bar{Y}|\,\bar{Y}, \sigma)\right) \tag{25b}$$

where the function Φ *is the inverse function of* $\sigma \mapsto \sigma^2 \times d \log \zeta_m(\sigma)/d\sigma$*.*

Proof of Lemma 1. (i) Let $\mathcal{E}(Y) = \mathcal{E}(Y|\,\bar{Y}, \sigma)$. According to Theorem 2.1 in [28], this function has a unique global minimum, which is also a unique stationary point. Thus, to prove that $\bar{Y}$ is the minimum point of $\mathcal{E}$, it will suffice to check that for any geodesic γ starting from $\bar{Y}$, $\frac{d}{dt}|_{t=0}\mathcal{E}(\gamma(t)) = 0$ [31] (p. 76). Note that:

$$\frac{d}{dt}|_{t=0}\mathcal{E}(\gamma(t)) = \int_{\mathcal{P}_m} \frac{d}{dt}|_{t=0} d(\gamma(t), Z)\, p(Z|\,\bar{Y}, \sigma) dv(Z) \tag{26}$$

where for all $Z \neq \bar{Y}$ [32]:

$$\frac{d}{dt}|_{t=0} d(\gamma(t), Z) = -g_{\bar{Y}}(\log_{\bar{Y}}(Z), \gamma'(0)) d(\bar{Y}, Z)^{-1}$$

The integral in Equation (26) is, up to a constant,

$$\frac{d}{dt}|_{t=0} \int_{\mathcal{P}_m} p(Z|\,\gamma(t), \sigma) dv(Z) = 0$$

since $\int_{\mathcal{P}_m} p(Z|\,\gamma(t), \sigma) dv(Z) = 1$.

(ii) Differentiating $\int_{\mathcal{P}_m} \exp(-\frac{d(Z,\bar{Y})}{\sigma}) dv(Z) = \zeta_m(\sigma)$ with respect to σ, it comes that:

$$\sigma^2 \times d \log \zeta_m(\sigma)/d\sigma = \sigma^2 \frac{\zeta_m'(\sigma)}{\zeta_m(\sigma)} = \int_{\mathcal{P}_m} d(Z, \bar{Y}) p(Z|\bar{Y}, \sigma) dv(Z) = \mathcal{E}(\bar{Y}|\,\bar{Y}, \sigma)$$

which proves (ii). □

Proposition 3 (Consistency of $\hat{Y}_N$). *Let $Y_1, Y_2, \cdots$ be independent samples from a Laplace distribution $G(\bar{Y}, \sigma)$. The empirical median $\hat{Y}_N$ of $Y_1, \ldots, Y_N$ converges almost surely to $\bar{Y}$, as $N \to \infty$.*

Proof of Proposition 3. Corollary 3.5 in [33] (p. 49) states that if (Y_n) is a sequence of i.i.d. random variables on $\mathcal{P}_m$ with law π, then the Riemannian median $\hat{Y}_N$ of $Y_1, \cdots, Y_N$ converges almost surely as $N \to \infty$ to $\hat{Y}_\pi$, the Riemannian median of π defined by Equation (12). Applying this result to $\pi = \mathcal{L}(\bar{Y}, \sigma)$ and using $\hat{Y}_\pi = \bar{Y}$, which follows from item (i) of Lemma 1, shows that $\hat{Y}_N$ converges almost surely to $\bar{Y}$. □

4. Mixtures of Laplace Distributions

There are several motivations for considering mixtures of distributions in general. The most natural approach is to envisage a dataset as constituted of several subpopulations. Another approach is the fact that there is a support for the argument that mixtures of distributions provide a good approximation to most distributions in a spirit similar to wavelets.

The present section introduces the class of probability distributions that are finite mixtures of Riemannian Laplace distributions on $\mathcal{P}_m$. These constitute the main theoretical tool, to be used for the target application of the present paper, namely the problem of texture classification in computer vision, which will be treated in Section 5.

A mixture of Riemannian Laplace distributions is a probability distribution on $\mathcal{P}_m$, whose density with respect to the Riemannian volume element Equation (13) has the following expression:

$$p(Y|(\varpi_\mu, \bar{Y}_\mu, \sigma_\mu)_{1 \le \mu \le M}) = \sum_{\mu=1}^{M} \varpi_\mu \times p(Y|\,\bar{Y}_\mu, \sigma_\mu) \tag{27}$$

where ϖ_μ are nonzero weights, whose sum is equal to one, $\bar{Y}_\mu \in \mathcal{P}_m$ and $\sigma_\mu \in (0, \sigma_m)$ for all $1 \le \mu \le M$, and the parameter M is called the number of mixture components.

Section 4.1 describes a new EM algorithm, which computes the maximum likelihood estimates of the mixture parameters $(\varpi_\mu, \bar{Y}_\mu, \sigma_\mu)_{1 \le \mu \le M}$, based on independent observations $Y_1, \ldots, Y_N$ from the mixture distribution Equation (27).

Section 4.2 considers the problem of order selection for mixtures of Riemannian Laplace distributions. Precisely, this consists of finding the number M of mixture components in Equation (27) that realizes the best representation of a given set of data $Y_1, \ldots, Y_N$. This problem is solved by computing the BIC criterion, which is here found in explicit form for the case of mixtures of Riemannian Laplace distributions on $\mathcal{P}_m$.

4.1. Estimation of the Mixture Parameters

In this section, $Y_1, \ldots, Y_N$ are i.i.d. samples from Equation (27). Based on these observations, an EM algorithm is proposed to estimate $(\varpi_\mu, \bar{Y}_\mu, \sigma_\mu)_{1 \leq \mu \leq M}$. The derivation of this algorithm can be carried out similarly to [15].

To explain how this algorithm works, define for all $\vartheta = \{(\varpi_\mu, \bar{Y}_\mu, \sigma_\mu)\}$,

$$\omega_\mu(Y_n, \vartheta) = \frac{\varpi_\mu \times p(Y_n | \bar{Y}_\mu, \sigma_\mu)}{\sum_{s=1}^{M} \varpi_s \times p(Y_n | \bar{Y}_s, \sigma_s)}, \qquad N_\mu(\vartheta) = \sum_{n=1}^{N} \omega_\mu(Y_n) \tag{28}$$

The algorithm iteratively updates $\hat{\vartheta} = \{(\hat{\varpi}_\mu, \hat{Y}_\mu, \hat{\sigma}_\mu)\}$, which is an approximation of the maximum likelihood estimate of the mixture parameters $\vartheta = (\varpi_\mu, \bar{Y}_\mu, \sigma_\mu)$ as follows.

- Update for $\hat{\varpi}_\mu$: Based on the current value of $\hat{\vartheta}$, assign to $\hat{\varpi}_\mu$ the new value $\hat{\varpi}_\mu = N_\mu(\hat{\vartheta})/N$.
- Update for $\hat{Y}_\mu$: Based on the current value of $\hat{\vartheta}$, assign to $\hat{Y}_\mu$ the value:

$$\hat{Y}_\mu = \text{argmin}_Y \sum_{n=1}^{N} \omega_\mu(Y_n, \hat{\vartheta})\, d(Y, Y_n) \tag{29}$$

- Update for $\hat{\sigma}_\mu$: Based on the current value of $\hat{\vartheta}$, assign to $\hat{\sigma}_\mu$ the new value:

$$\hat{\sigma}_\mu = \Phi\left(N_\mu^{-1}(\hat{\vartheta}) \times \textstyle\sum_{n=1}^{N} \omega_\mu(Y_n, \hat{\vartheta})\, d(\hat{Y}_\mu, Y_n)\right) \tag{30}$$

where the function Φ is defined in Proposition 1.

These three update rules should be performed in the above order. Realization of the update rules for $\hat{\varpi}_\mu$ and $\hat{\sigma}_\mu$ is straightforward. The update rule for $\hat{Y}_\mu$ is realized using a slight modification of the sub-gradient descent algorithm described in Section 3.2. More precisely, the factor $1/N$ appearing in Equation (22) is only replaced with $\omega_\mu(Y_n, \hat{\vartheta})$ at each iteration.

In practice, the initial conditions $(\hat{\varpi}_{\mu 0}, \hat{Y}_{\mu 0}, \hat{\sigma}_{\mu 0})$ in this algorithm were chosen in the following way. The weights $(\varpi_{\mu 0})$ are uniform and equal to $1/M$; $(\hat{Y}_{\mu 0})$ are M different observations from the set $\{Y_1, .., Y_N\}$ chosen randomly; and $(\hat{\sigma}_{\mu 0})$ is computed from $(\varpi_{\mu 0})$ and $(\hat{Y}_{\mu 0})$ according to the rule Equation (30). Since the convergence of the algorithm depends on the initial conditions, the EM algorithm is run several times, and the best result is retained, *i.e.*, the one maximizing the log-likelihood function.

4.2. The Bayesian Information Criterion

The BIC was introduced by Schwarz to find the appropriate dimension of a model that will fit a given set of observations [16]. Since then, BIC has been used in many Bayesian modeling problems where priors are hard to set precisely. In large sample settings, the fitted model favored by BIC ideally corresponds to the candidate model that is *a posteriori* most probable; *i.e.*, the model that is rendered most plausible by the data at hand. One of the main features of the BIC is its easy computation, since it is only based on the empirical log-likelihood function.

Given a set of observations $\{Y_1, \cdots, Y_N\}$ arising from Equation (27) where M is unknown, the BIC consists of choosing the parameter:

$$\bar{M} = \text{argmax}_M BIC(M)$$

where:

$$BIC(M) = LL - \frac{1}{2} \times DF \times \log(N) \tag{31}$$

Here, LL is the log-likelihood given by:

$$LL = \sum_{n=1}^{N} \log \left(\sum_{k=1}^{M} \hat{\varpi}_k p(Y_n | \hat{Y}_k, \hat{\sigma}_k) \right) \tag{32}$$

and DF is the number of degrees of freedom of the statistical model:

$$DF = M \times \frac{m(m+1)}{2} + M + M - 1 \tag{33}$$

In Formula (32), $(\hat{\varpi}_k, \hat{Y}_k, \hat{\sigma}_k)_{1 \leq k \leq M}$ are obtained from an EM algorithm as stated in Section 4.1 assuming the exact dimension is M. Finally, note that in Formula (33), $M \times \frac{m(m+1)}{2}$ (respectively M and $M-1$) corresponds to the number of degrees of freedom associated with $(\hat{Y}_k)_{1 \leq k \leq M}$ (respectively $(\hat{\sigma}_k)_{1 \leq k \leq M}$ and $(\hat{\varpi}_k)_{1 \leq k \leq M}$).

5. Application to Classification of Data on $\mathcal{P}_m$

Recently, several approaches have used the Riemannian distance in general as the main innovation in image or signal classification problems [2,15,34]. It turns out that the use of this distance leads to more accurate results (in comparison, for example, with the Euclidean distance). This section proposes an application that follows a similar approach, but in addition to the Riemannian distance, it also relies on a statistical approach. It considers the application of the Riemannian Laplace distribution (RLD) to the classification of data in $\mathcal{P}_m$ and gives an original Laplace classification rule, which can be used to carry out the task of classification, even in the presence of outliers. It also applies this classification rule to the problem of texture classification in computer vision, showing that it leads to improved results in comparison with recent literature.

Section 5.1 considers, from the point of view of statistical learning, the classification of data with values in $\mathcal{P}_m$. Given data points $Y_1, \cdots, Y_N \in \mathcal{P}_m$, this proceeds in two steps, called the learning phase and the classification phase, respectively. The learning phase uncovers the class structure of the data, by estimating a mixture model using the EM algorithm developed in Section 4.1. Once training is accomplished, data points are subdivided into disjoint classes. Classification consists of associating each new data point to the most suitable class. For this, a new classification rule will be established and shown to be optimal.

Section 5.2 is the implementation of the Laplace classification rule together with the BIC criterion to texture classification in computer vision. It highlights the advantage of the Laplace distribution in the presence of outliers and shows its better performance compared to recent approaches.

5.1. Classification Using Mixtures of Laplace Distributions

Assume to be given a set of training data $Y_1, \cdots, Y_N$. These are now modeled as a realization of a mixture of Laplace distributions:

$$p(Y) = \sum_{\mu=1}^{M} \varpi_\mu \times p(Y | \bar{Y}_\mu, \sigma_\mu) \tag{34}$$

In this section, the order M in Equation (34) is considered as known. The training phase of these data consists of learning its structure as a family of M disjoint classes $C_\mu, \mu = 1, \cdots, M$. To be more precise, depending on the family (ϖ_μ), some of these classes may be empty. Training is done by applying the EM algorithm described in Section 4.1. As a result, each class C_μ is represented by a triple $(\hat{\varpi}_\mu, \hat{Y}_\mu, \hat{\sigma}_\mu)$ corresponding to maximum likelihood estimates of $(\varpi_\mu, Y_\mu, \sigma_\mu)$. Each observation Y_n is now associated with the class C_{μ^*} where $\mu^* = \text{argmax}_\mu \omega(Y_n, \hat{\vartheta})$ (recall the definition from Equation (28)). In this way, $\{Y_1, \cdots, Y_N\}$ is subdivided into M disjoint classes.

The classification phase requires a classification rule. Following [15], the optimal rule (in the sense of a Bayesian risk criterion given in [35]) consists of associating any new data Y_t to the class C_{μ^*} where:

$$\mu^* = \text{argmax}_\mu \left\{ \hat{N}_\mu \times p(Y_t | \hat{Y}_\mu, \hat{\sigma}_\mu) \right\} \tag{35}$$

Here, $\hat{N}_\mu$ is the number of elements in C_μ. Replacing $\hat{N}_\mu$ with $N \times \hat{\varpi}_\mu$, Equation (35) becomes $\text{argmax}_\mu \hat{\varpi}_\mu \times p(Y_t | \hat{Y}_\mu, \hat{\sigma}_\mu)$. Note that when the weights ϖ_μ in Equation (34) are assumed to be equal, this rule reduces to a maximum likelihood classification rule $\max_\mu p(Y_t | \hat{Y}_\mu, \hat{\sigma}_\mu)$. A quick look at the expression Equation (17) shows that Equation (35) can also be expressed as:

$$\mu^* = \text{argmin}_\mu \left\{ -\log \hat{\varpi}_\mu + \log \zeta(\hat{\sigma}_\mu) + \frac{d(Y_t, \hat{Y}_\mu)}{\hat{\sigma}_\mu} \right\} \tag{36}$$

The rule Equation (36) will be called the Laplace classification rule. It favors clusters C_μ having a larger number of data points (the minimum contains $-\log \hat{\varpi}_\mu$) or a smaller dispersion away from the median (the minimum contains $\log \zeta(\hat{\sigma}_\mu)$). When choosing between two clusters with the same number of points and the same dispersion, this rule favors the one whose median is closer to Y_t. If the number of data points inside clusters and the respective dispersions are neglected, then Equation (36) reduces to the nearest neighbor rule involving only the Riemannian distance introduced in [2].

The analogous rules of Equation (36) for the Riemannian Gaussian distribution (RGD) [15] and the Wishart distribution (WD) [17] on $\mathcal{P}_m$ can be established by replacing $p(Y_t | \hat{Y}_\mu, \hat{\sigma}_\mu)$ in Equation (35) with the RGD and the WD and then following the same reasoning as before. Recall that a WD depends on an expectation $\Sigma \in \mathcal{P}_m$ and a number of degrees of freedom n [29]. For the WD, Equation (36) becomes:

$$\mu^* = \text{argmin}_\mu \left\{ -2 \log \hat{\varpi}(\mu) - \hat{n}(\mu) \left(\log\det\left(\hat{\Sigma}^{-1}(\mu) Y_t\right) - \text{tr}(\hat{\Sigma}^{-1}(\mu) Y_t) \right) \right\}$$

Here, $\hat{\varpi}(\mu)$, $\hat{\Sigma}(\mu)$ and $\hat{n}(\mu)$ denote maximum likelihood estimates of the true parameters $\varpi(\mu)$, $\Sigma(\mu)$ and $n(\mu)$, which define the mixture model (these estimates can be computed as in [36,37]).

5.2. *Application to Texture Classification*

This section presents an application of the mixture of Laplace distributions to the context of texture classification on the MIT Vision Texture (VisTex) database [38]. The purpose of this experiment is to classify the textures, by taking into consideration the within-class diversity. In addition, the influence of outliers on the classification performances is analyzed. The obtained results for the RLD are compared to those given by the RGD [15] and the WD [17].

The VisTex database contains 40 images, considered as being 40 different texture classes. The database used for the experiment is obtained after several steps. First of all, each texture is decomposed into 169 patches of 128×128 pixels, with an overlap of 32 pixels, giving a total number of 6760 textured patches. Next, some patches are corrupted, in order to introduce abnormal data into the dataset. Therefore, their intensity is modified by applying a gradient of luminosity. For each class, between zero and 60 patches are modified in order to become outliers. An example of a VisTex texture with one of its patches and an outlier patch are shown in Figure 1.

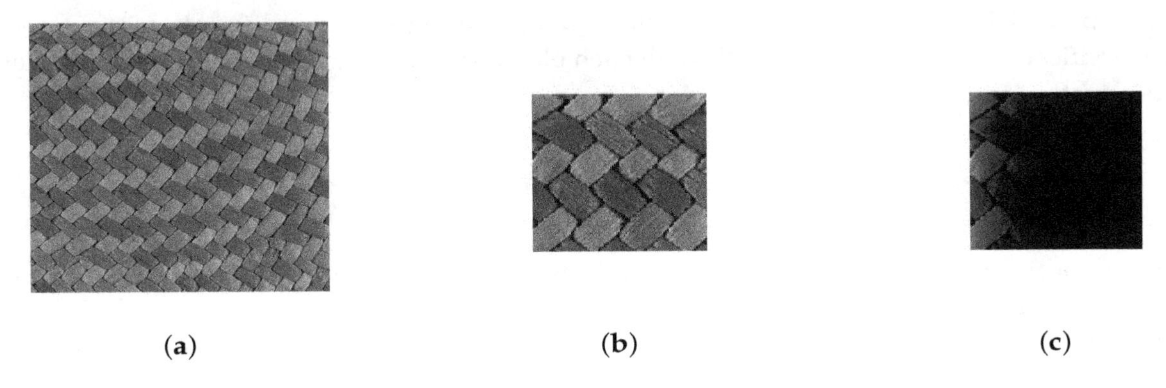

(a) (b) (c)

Figure 1. Example of a texture from the VisTex database (**a**), one of its patches (**b**) and the corresponding outlier (**c**).

Once the database is built, it is 15-times equally and randomly divided in order to obtain the training and the testing sets that are further used in the supervised classification algorithm. Then, for each patch in both databases, a feature vector has to be computed. The luminance channel is first extracted and then normalized in intensity. The grayscale patches are filtered using the stationary wavelet transform Daubechies db4 filter (see [39]), with two scales and three orientations. To model the wavelet sub-bands, various stochastic models have been proposed in the literature. Among them, the univariate generalized Gaussian distribution has been found to accurately model the empirical histogram of wavelet sub-bands [40]. Recently, it has been proposed to model the spatial dependency of wavelet coefficients. To this aim, the wavelet coefficients located in a $p \times q$ spatial neighborhood of the current spatial position are clustered in a random vector. The realizations of these vectors can be further modeled by elliptical distributions [41,42], copula-based models [43,44], *etc.* In this paper, the wavelet coefficients are considered as being realizations of zero-mean multivariate Gaussian distributions. In addition, for this experiment the spatial information is captured by using a vertical (2×1) and a horizontal (1×2) neighborhood. Next, the 2×2 sample covariance matrices are estimated for each wavelet sub-band and each neighborhood. Finally, each patch is represented by a set of $F = 12$ covariance matrices (2 scales $\times$ 3 orientations $\times$ 2 neighborhoods) denoted $Y = [Y_1, \cdots, Y_F]$.

The estimated covariance matrices are elements of $\mathcal{P}_m$, with $m = 2$, and therefore, they can be modeled by Riemannian Laplace distributions. More precisely, in order to take into consideration the within-class diversity, each class in the training set is viewed as a realization of a mixture of Riemannian Laplace distributions (Equation (27)) with M mixture components, characterized by $(\varpi_\mu, \bar{Y}_{\mu,f}, \sigma_{\mu,f})$, having $\bar{Y}_{\mu,f} \in \mathcal{P}_2$, with $\mu = 1, \cdots, M$ and $f = 1, \cdots, F$. Since the sub-bands are assumed to be independent, the probability density function is given by:

$$p(Y|(\varpi_\mu, \bar{Y}_{\mu,f}, \sigma_{\mu,f})_{1 \le \mu \le M, 1 \le f \le F}) = \sum_{\mu=1}^{M} \varpi_\mu \prod_{f=1}^{F} p(Y_f | \bar{Y}_{\mu,f}, \sigma_{\mu,f}) \tag{37}$$

The learning step of the classification is performed using the EM algorithm presented in Section 4, and the number of mixture components is determined using the BIC criterion recalled in Equation (31). Note that for the considered model given in Equation (37), the degree of freedom is expressed as:

$$DF = M - 1 + M \times F \times \left(\frac{m(m+1)}{2} + 1 \right) \tag{38}$$

since one centroid and one dispersion parameter should be estimated per feature and per component of the mixture model. In practice, the number of mixture components M varies between two and five, and the M yielding to the highest BIC criterion is retained. As mentioned earlier, the EM algorithm is sensitive to the initial conditions. In order to minimize this influence, for this experiment, the EM algorithm is repeated 10 times, and the result maximizing the log-likelihood function is retained. Finally, the classification is performed by assigning each element $Y_t \in \mathcal{P}_2$ in the testing set to the class of the closest cluster μ^*, given by:

$$\mu^* = \text{argmin}_{\mu} \left\{ -\log \hat{\varpi}_{\mu} + \sum_{f=1}^{F} \log \zeta(\hat{\sigma}_{\mu,f}) + \sum_{f=1}^{F} \frac{d(Y_t, \hat{Y}_{\mu,f})}{\hat{\sigma}_{\mu,f}} \right\} \tag{39}$$

This expression is obtained starting from Equations (36) and (37), knowing that F features are extracted for each patch.

The classification results of the proposed model (solid red line), expressed in terms of overall accuracy, shown in Figure 2, are compared to those given by a fixed number of mixture components (that is, for $M = 3$, dashed red line) and with those given when the within-class diversity is not considered (that is, for $M = 1$, dotted red line). In addition, the classification performances given by the RGD model (displayed in black) proposed in [15] and the WD model (displayed in blue) proposed in [17] are also considered. For each of these models, the number of mixture components is first computed using the BIC, and next, it is fixed to $M = 3$ and $M = 1$. For all of the considered models, the classification rate is given as a function of the number of outliers, which varies between zero and 60 for each class.

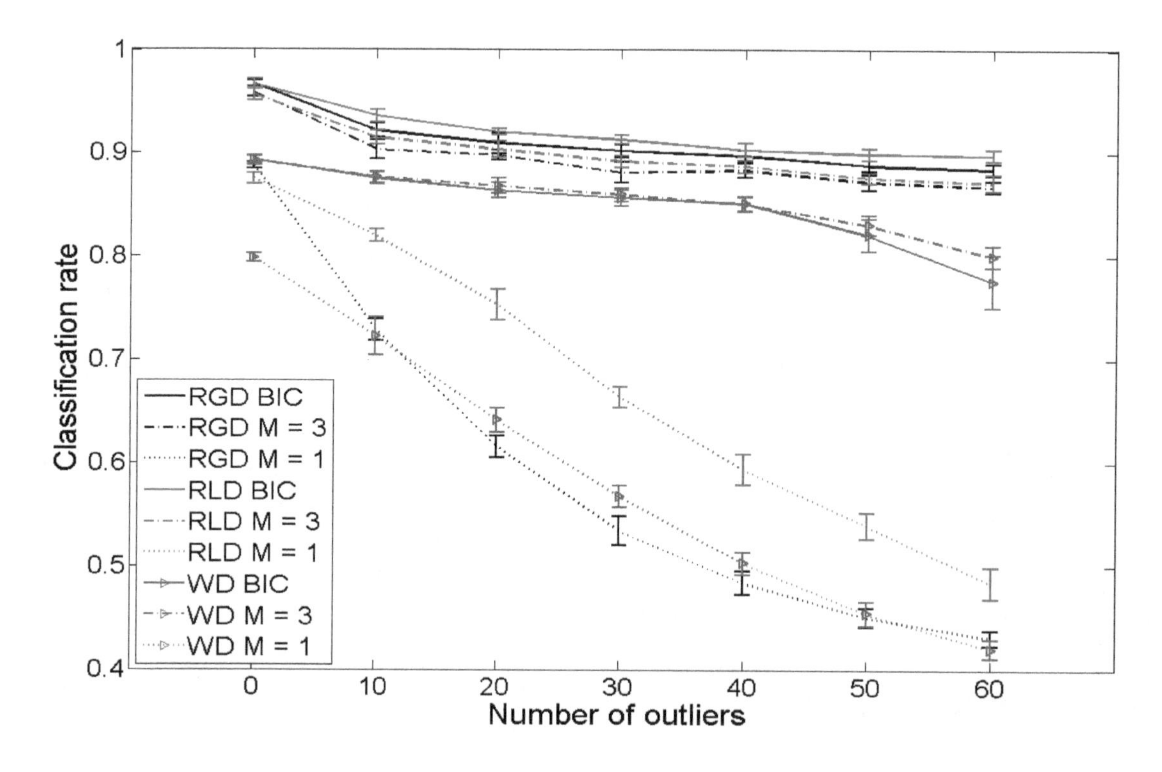

Figure 2. Classification results.

It is shown that, as the number of outliers increases, the RLD gives progressively better results than the RGD and the WD. The results are improved by using the BIC criterion for choosing the suitable number of clusters. In conclusion, the mixture of RLDs combined with the BIC criterion to estimate the best number of mixture components can minimize the influence of abnormal samples present in the dataset, illustrating the relevance of the proposed method.

6. Conclusions

Motivated by the problem of outliers in statistical data, this paper introduces a new distribution on the space $\mathcal{P}_m$ of $m \times m$ symmetric positive definite matrices, called the Riemannian Laplace distribution.

Denoted throughout the paper by $\mathcal{L}(\bar{Y},\sigma)$, where $\bar{Y} \in \mathcal{P}_m$ and $\sigma > 0$ are the indexing parameters, this distribution may be thought of as specifying the law of a family of observations on $\mathcal{P}_m$ concentrated around the location $\bar{Y}$ and having dispersion σ. If d denotes Rao's distance on $\mathcal{P}_m$ and $dv(Y)$ its associated volume form, the density of $\mathcal{L}(\bar{Y},\sigma)$ with respect to $dv(Y)$ is proportional to $\exp(-\frac{d(Y,\bar{Y})}{\sigma})$. Interestingly, the normalizing constant depends only on σ (and not on $\bar{Y}$). This allows us to deduce exact expressions for maximum likelihood estimates of $\bar{Y}$ and σ relying on the Riemannian median on $\mathcal{P}_m$. These estimates are also computed numerically by means of sub-gradient algorithms. The estimation of parameters in mixture models of Laplace distributions are also considered and performed using a new expectation-maximization algorithm. Finally, the main theoretical results are illustrated by an application to texture classification. The proposed experiment consists of introducing abnormal data (outliers) into a set of images from the VisTex database and analyzing their influences on the classification performances. Each image is characterized by a set of 2×2 covariance matrices modeled as mixtures of Riemannian Laplace distributions in the space $\mathcal{P}_2$. The number of mixtures is estimated using the BIC criterion. The obtained results are compared to those given by the Riemannian Gaussian distribution, showing the better performance of the proposed method.

Acknowledgments: This study has been carried out with financial support from the French State, managed by the French National Research Agency (ANR) in the frame of the "Investments for the future" Programme initiative d'excellence (IdEX) Bordeaux-CPU (ANR-10-IDEX-03-02).

Author Contributions: Hatem Hajri and Salem Said carried out the mathematical development and specified the algorithms. Ioana Ilea and Lionel Bombrun conceived and designed the experiments. Yannick Berthoumieu gave the central idea of the paper and managed the main tasks and experiments. Hatem Hajri wrote the paper. All the authors read and approved the final manuscript.

Appendix: Proofs of Some Technical Points

The subsections below provide proofs (using the same notations) of certain points in the paper.

A. Derivation of Equation (16) from Equation (14)

For $U \in \mathrm{O}(m)$ and $r = (r_1, \cdots, r_m) \in \mathbb{R}^m$, let $Y(r,U) = U^{\dagger} \operatorname{diag}(e^{r_1}, \cdots, e^{r_m})\, U$. On $\mathrm{O}(m)$, consider the exterior product $\det(\theta) = \bigwedge_{i<j} \theta_{ij}$, where $\theta_{ij} = \sum_k U_{jk} dU_{ik}$.

Proposition 4. *For all test functions $f : \mathcal{P}_m \to \mathbb{R}$,*

$$\int_{\mathcal{P}_m} f(Y)\, dv(Y) = (m!\, 2^m)^{-1} \times 8^{\frac{m(m-1)}{4}} \int_{\mathrm{O}(m)} \int_{\mathbb{R}^m} f(Y(r,U))\, \det(\theta) \prod_{i<j} \sinh\left(\frac{|r_i - r_j|}{2}\right) \prod_{i=1}^{m} dr_i$$

This proposition allows one to deduce Equation (16) from Equation (14), since $\int_{\mathrm{O}(m)} \det(\theta) = \frac{2^m \pi^{m^2/2}}{\Gamma_m(m/2)}$ (see [29], p. 70).

Sketch of the proof of Proposition 4. In a differential form, the Rao–Fisher metric on $\mathcal{P}_m$ is:

$$ds^2(Y) = \operatorname{tr}[Y^{-1} dY]^2$$

For $U \in \mathrm{O}(m)$ and $(a_1, \cdots, a_m) \in (\mathbb{R}_+^*)^m$, let $Y = U^{\dagger} \operatorname{diag}(a_1, \cdots, a_m)\, U$. Then:

$$ds^2(Y) = \sum_{j=1}^{m} \frac{da_j^2}{a_j^2} + 2 \sum_{1 \le i < j \le m} \frac{(a_i - a_j)^2}{a_i a_j} \theta_{ij}^2$$

(see [10], p. 24). Let $a_i = e^{r_i}$, then simple calculations show that:

$$ds^2(Y) = \sum_{j=1}^{m} dr_j^2 + 8\sum_{i<j} \sinh^2\left(\frac{r_i - r_j}{2}\right)\theta_{ij}^2$$

As a consequence, the volume element $dv(Y)$ is written as:

$$dv(Y) = 8^{\frac{m(m-1)}{4}} \det(\theta) \prod_{i<j} \sinh\left(\frac{|r_i - r_j|}{2}\right) \prod_{i=1}^{m} dr_i$$

This proves the proposition (the factor $m!\,2^m$ comes from the fact that the correspondence between Y and (r, U) is not unique: $m!$ corresponds to all possible reorderings of $r_1, \ldots, r_m$, and 2^m corresponds to the orientation of the columns of U).

B. Derivation of Equation (19)

By Equations (16) and (18), to prove Equation (19), it is sufficient to prove that for all $Y \in \mathcal{P}_m$, $d(Y, I) = (\sum_{i=1}^{m} r_i^2)^{1/2}$ if the spectral decomposition of Y is $Y = U^{\dagger} \operatorname{diag}(e^{r_1}, \cdots, e^{r_m})\, U$, where U is an orthogonal matrix. Note that $d(Y, I) = d(\operatorname{diag}(e^{r_1}, \cdots, e^{r_m}).U, I) = d(\operatorname{diag}(e^{r_1}, \cdots, e^{r_m}).U, I.U)$, where . is the affine transformation given by Equation (9). By Equation (10), it comes that $d(Y, I) = d(\operatorname{diag}(e^{r_1}, \cdots, e^{r_m}), I)$, and so, $d(Y, I) = (\sum_{i=1}^{m} r_i^2)^{1/2}$ holds using the explicit expression Equation (8).

C. The Normalizing Factor $\zeta_m(\sigma)$

The subject of this section is to prove these two claims:

(i) $0 < \sigma_m < \infty$ for all $m \geq 2$;
(ii) $\sigma_2 = \sqrt{2}$.

To check (i), note that $\prod_{i<j} \sinh\left(\frac{|r_i - r_j|}{2}\right) \leq \exp(C|r|)$ for some constant C. Thus, for σ small enough, the integral $I_m(\sigma) = \int_{\mathbb{R}^m} e^{-\frac{|r|}{\sigma}} \prod_{i<j} \sinh\left(\frac{|r_i - r_j|}{2}\right) dr$ given in Equation (19) is finite, and consequently, $\sigma_m > 0$.

Fix $A > 0$, such that $\sinh(\frac{x}{2}) \geq \exp(\frac{x}{4})$ for all $x \geq A$. Then:

$$I_m(\sigma) \geq \int_{\mathcal{C}} \exp\left(\frac{1}{4}\sum_{i<j}(r_j - r_i) - \frac{|r|}{\sigma}\right) dr$$

where $\mathcal{C}$ is the set of infinite Lebesgue measures:

$$\mathcal{C} = \{r = (r_1, \cdots, r_m) \in \mathbb{R}^m : r_i \in [2(i-1)A, (2i-1)A], 1 \leq i \leq m-1, r_m \geq 2(m-1)A\}$$

Now:

$$\frac{1}{4}\sum_{i<j}(r_j - r_i) = \frac{1}{4}r_m + \frac{1}{4}(-r_1 + \sum_{i<j,(i,j)\neq(1,m)} (r_j - r_i))$$

Assume $m \geq 3$ (the case $m = 2$ is easy to deal with separately). Then, on $\mathcal{C}$, $\frac{1}{4}\sum_{i<j}(r_j - r_i) \geq \frac{1}{4}r_m + C'$ and $\frac{|r|}{\sigma} \leq \frac{(C'' + r_m^2)^{\frac{1}{2}}}{\sigma}$, where C' and C'' are two positive constants (not depending on r). However, for σ large enough:

$$\frac{1}{4}\sum_{i<j}(r_j - r_i) - \frac{|r|}{\sigma} \geq \frac{1}{4}r_m + C' - \frac{(C'' + r_m^2)^{\frac{1}{2}}}{\sigma} \geq 0.$$

and so, the integral $I_m(\sigma)$ diverges. This shows that σ_m is finite.

(ii) Note the following easy inequalities $|r_1 - r_2| \leq |r_1| + |r_2| \leq \sqrt{2}|r|$, which yield $\sinh(\frac{|r_1-r_2|}{2}) \leq \frac{1}{2}e^{\frac{|r|}{\sqrt{2}}}$. This last inequality shows that $\zeta_2(\sigma)$ is finite for all $\sigma < \sqrt{2}$. In order to check $\zeta_2(\sqrt{2}) = \infty$, it is necessary to show:

$$\int_{\mathbb{R}^2} \exp(-\frac{|r|}{\sqrt{2}} + \frac{|r_1 - r_2|}{2})dr_1 dr_2 = \infty \tag{40}$$

The last integral is, up to a constant, greater than $\int_{\mathcal{C}} \exp\left(-|r| + \frac{|r_1-r_2|}{\sqrt{2}}\right) dr_1 dr_2$, where:

$$\mathcal{C} = \{(r_1, r_2) \in \mathbb{R}^2 : r_1 \geq -r_2, r_2 \leq 0\} = \{(r_1, r_2) \in \mathbb{R}^2 : r_1 \geq |r_2|, r_2 \leq 0\}.$$

On $\mathcal{C}$,

$$-|r| + \frac{|r_1 - r_2|}{\sqrt{2}} = -|r| + \frac{r_1 - r_2}{\sqrt{2}} \geq -\sqrt{2}r_1 + \frac{r_1 - r_2}{\sqrt{2}} = \frac{-r_1 - r_2}{\sqrt{2}}$$

However, $\int_{\mathcal{C}} \exp\left(\frac{-r_1-r_2}{\sqrt{2}}\right) dr_1 dr_2 = \infty$ by integrating with respect to r_1 and then r_2, which shows Equation (40).

D. The Law of X in Algorithm 1

As stated in Appendix A, the uniform distribution on $\mathrm{O}(m)$ is given by $\frac{1}{\omega'_m}\det(\theta)$, where $\omega'_m = \frac{2^m \pi^{m^2/2}}{\Gamma_m(m/2)}$. Let $Y(s, V) = V^\dagger \operatorname{diag}(e^{s_1}, \cdots, e^{s_m})\, V$, with $s = (s_1, \cdots, s_m)$. Since $X = Y(r, U)$, for any test function $\varphi : \mathcal{P}_m \to \mathbb{R}$,

$$E[\varphi(X)] = \frac{1}{\omega'_m} \int_{\mathrm{O}(m)\times\mathbb{R}^m} \varphi(Y(s,V))p(s)\det(\theta) \prod_{i=1}^{m} ds_i \tag{41}$$

Here, $\det(\theta) = \bigwedge_{i<j} \theta_{ij}$ and $\theta_{ij} = \sum_k V_{jk} dV_{ik}$. On the other hand, by Proposition 4, $\int_{\mathcal{P}_m} \varphi(Y)\, p(Y|\, I, \sigma)\, dv(Y)$ can be expressed as:

$$(m!\,2^m)^{-1} \times 8^{\frac{m(m-1)}{4}} \frac{1}{\zeta_m(\sigma)} \int_{\mathrm{O}(m)} \int_{\mathbb{R}^m} \varphi(Y(s,V))e^{-\frac{|s|}{\sigma}} \det(\theta) \prod_{i<j} \sinh\left(\frac{|s_i - s_j|}{2}\right) \prod_{i=1}^{m} ds_i$$

which coincides with Equation (41).

References

1. Pennec, X.; Fillard, P.; Ayache, N. A Riemannian framework for tensor computing. *Int. J. Comput. Vis.* **2006**, *66*, 41–66.
2. Barachant, A.; Bonnet, S.; Congedo, M.; Jutten, C. Multiclass Brain–Computer Interface Classification by Riemannian Geometry. *IEEE Trans. Biomed. Eng.* **2012**, *59*, 920–928.
3. Jayasumana, S.; Hartley, R.; Salzmann, M.; Li, H.; Harandi, M. Kernel Methods on the Riemannian Manifold of Symmetric Positive Definite Matrices. In Proceedings of the IEEE Conference on Computer Vision and Pattern Recognition (CVPR), Portland, OR, USA, 23–28 June 2013; pp. 73–80.
4. Zheng, L.; Qiu, G.; Huang, J.; Duan, J. Fast and accurate Nearest Neighbor search in the manifolds of symmetric positive definite matrices. In Proceedings of the IEEE International Conference on Acoustics, Speech and Signal Processing (ICASSP), Florence, Italy, 4–9 May 2014; pp. 3804–3808.
5. Dong, G.; Kuang, G. Target recognition in SAR images via classification on Riemannian manifolds. *IEEE Geosci. Remote Sens. Lett.* **2015**, *21*, 199–203.
6. Tuzel, O.; Porikli, F.; Meer, P. Pedestrian detection via classification on Riemannian manifolds. *IEEE Trans. Pattern Anal. Mach. Intell.* **2008**, *30*, 1713–1727.
7. Caseiro, R.; Henriques, J.F.; Martins, P.; Batista, J. A nonparametric Riemannian framework on tensor field with application to foreground segmentation. *Pattern Recognit.* **2012**, *45*, 3997–4017.

8. Arnaudon, M.; Barbaresco, F.; Yang, L. Riemannian Medians and Means With Applications to Radar Signal Processing. *IEEE J. Sel. Top. Signal Process.* **2013**, *7*, 595–604.
9. Arnaudon, M.; Yang, L.; Barbaresco, F. Stochastic algorithms for computing p-means of probability measures, Geometry of Radar Toeplitz covariance matrices and applications to HR Doppler processing. In Proceedings of International International Radar Symposium (IRS), Leipzig, Germany, 7–9 September 2011; pp. 651–656.
10. Terras, A. *Harmonic Analysis on Symmetric Spaces and Applications*; Springer-Verlag: New York, NY, USA, 1988; Volume II.
11. Atkinson, C.; Mitchell, A. Rao's distance measure. *Sankhya Ser. A* **1981**, *43*, 345–365.
12. Pennec, X. Probabilities and statistics on Riemannian manifolds: Basic tools for geometric measurements. In Procedings of the IEEE Workshop on Nonlinear Signal and Image Processing, Antalya, Turkey, 20–23 June 1999; pp. 194–198.
13. Pennec, X. Intrinsic statistics on Riemannian manifolds: Basic tools for geometric measurements. *J. Math. Imaging Vis.* **2006**, *25*, 127–154.
14. Guang, C.; Baba C.V. A Novel Dynamic System in the Space of SPD Matrices with Applications to Appearance Tracking. *SIAM J. Imaging Sci.* **2013**, *6*, 592–615.
15. Said, S.; Bombrun, L.; Berthoumieu, Y.; Manton, J. Riemannian Gaussian distributions on the space of symmetric positive definite matrices. 2015, arXiv:1507.01760.
16. Schwarz, G. Estimating the Dimension of a Model. *Ann. Stat.* **1978**, *6*, 461–464.
17. Lee, J.S.; Grunes, M.R.; Ainsworth, T.L.; Du, L.J.; Schuler, D.L.; Cloude, S.R. Unsupervised classification using polarimetric decomposition and the complex Wishart classifier. *IEEE Trans. Geosci. Remote Sens.* **1999**, *37*, 2249–2258.
18. Berezin, F.A. Quantization in complex symmetric spaces. *Izv. Akad. Nauk SSSR Ser. Mat.* **1975**, *39*, 363–402.
19. Malliavin, P. Invariant or quasi-invariant probability measures for infinite dimensional groups, Part II: Unitarizing measures or Berezinian measures. *Jpn. J. Math.* **2008**, *3*, 19–47.
20. Barbaresco, F. *Information Geometry of Covariance Matrix: Cartan-Siegel Homogeneous Bounded Domains, Mostow/Berger Fibration and Fréchet Median, Matrix Information Geometry*; Bhatia, R., Nielsen, F., Eds.; Springer: New York, NY, USA, 2012; pp. 199–256.
21. Barbaresco, F. Information geometry manifold of Toeplitz Hermitian positive definite covariance matrices: Mostow/Berger fibration and Berezin quantization of Cartan-Siegel domains. *Int. J. Emerg. Trends Signal Process.* **2013**, *1*, 1–11.
22. Jeuris, B.; Vandebril, R. Averaging block-Toeplitz matrices with preservation of Toeplitz block structure. In Proceedings of the SIAM Conference on Applied Linear Algebra (ALA), Atlanta, GA, USA, 20–26 October 2015.
23. Jeuris, B.; Vandebril, R. The Kähler Mean of Block-Toeplitz Matrices with Toeplitz Structured Block. Available online: http://www.cs.kuleuven.be/publicaties/rapporten/tw/TW660.pdf (accessed on 10 March 2016).
24. Jeuris, B. Riemannian Optimization for Averaging Positive Definite Matrices. Ph.D. Thesis, University of Leuven, Leuven, Belgium, 2015.
25. Maass, H. Siegel's modular forms and Dirichlet series. In *Lecture Notes in Mathematics*; Springer-Verlag: New York, NY, USA, 1971; Volume 216.
26. Higham, N.J. *Functions of Matrices, Theory and Computation*; Society for Industrial and Applied Mathematics: Philadelphia, PA, USA, 2008.
27. Helgason, S. *Differential Geometry, Lie Groups, and Symmetric Spaces*; American Mathematical Society: Providence, RI, USA, 2001.
28. Afsari, B. Riemannian L^p center of mass: Existence, uniqueness and convexity. *Proc. Am. Math. Soc.* **2011**, *139*, 655–673.
29. Muirhead, R.J. *Aspects of Multivariate Statistical Theory*; John Wiley & Sons: New York, NY, USA, 1982.
30. Robert, C.P.; Casella, G. *Monte Carlo Statistical Methods*; Springer-Verlag: Berlin, Germany, 2004.
31. Udriste, C. *Convex Functions and Optimization Methods on Riemannian Manifolds*; Mathematics and Its Applications; Kluwer Academic Publishers: Dordrecht, The Netherlands, 1994.
32. Chavel, I. *Riemannian Geometry, a Modern Introduction*; Cambridge University Press: Cambridge, UK, 2006.
33. Yang, L. Médianes de Mesures de Probabilité dans les Variétés Riemanniennes et Applications à la Détection de Cibles Radar. Ph.D. Thesis, L'université de Poitiers, Poitiers, France, 2011. (In French)

34. Li, Y.; Wong, K.M. Riemannian distances for signal classification by power spectral density. *IEEE J. Sel. Top. Sig. Process.* **2013**, *7*, 655–669.
35. Hastie, T.; Tibshirani, R.; Friedman, J. *The Elements of Statistical Learning: Data Mining, Inference, and Prediction*; Springer: Berlin, Germany, 2009.
36. Saint-Jean, C.; Nielsen, F. A new implementation of k-MLE for mixture modeling of Wishart distributions. In *Geometric Science of Information (GSI)*; Springer-Verlag: Berlin/Heidelberg, Germany, 2013; pp. 249–256.
37. Hidot, S.; Saint-Jean, C. An expectation-maximization algorithm for the Wishart mixture model: Application to movement clustering. *Pattern Recognit. Lett.* **2010**, *31*, 2318–2324.
38. VisTex: Vision Texture Database. MIT Media Lab Vision and Modeling Group. Available online: http://vismod.media.mit.edu/pub/ (accessed on 9 March 2016).
39. Daubechies, I. *Ten Lectures on Wavelets*; Society for Industrial and Applied Mathematics: Philadelphia, PA, USA, 1992.
40. Do, M.N.; Vetterli, M. Wavelet-Based Texture Retrieval Using Generalized Gaussian Density and Kullback-Leibler Distance. *IEEE Trans. Image Process.* **2002**, *11*, 146–158.
41. Bombrun, L.; Berthoumieu, Y.; Lasmar, N.-E.; Verdoolaege, G. Mutlivariate Texture Retrieval Using the Geodesic Distance between Elliptically Distributed Random Variables. In Proceedings of 2011 18th IEEE International Conference on Image Processing (ICIP), Brussels, Belgium, 11–14 September 2011.
42. Verdoolaege, G.; Scheunders, P. On the Geometry of Multivariate Generalized Gaussian Models. *J. Math. Imaging Vis.* **2012**, *43*, 180–193.
43. Stitou, Y.; Lasmar, N.-E.; Berthoumieu, Y. Copulas based Multivariate Gamma Modeling for Texture Classification. In Proceedings of the IEEE International Conference on Acoustic Speech and Signal Processing, Taipei, Taiwan, 19–24 April 2009; pp. 1045–1048.
44. Kwitt, R.; Uhl, A. Lightweight Probabilistic Texture Retrieval. *IEEE Trans. Image Process.* **2010**, *19*, 241–253.

10

Non-Asymptotic Confidence Sets for Circular Means

Thomas Hotz *,‡, Florian Kelma ‡ and Johannes Wieditz ‡

Institut für Mathematik, Technische Universität Ilmenau, 98684 Ilmenau, Germany;
florian.kelma@tu-ilmenau.de (F.K.); johannes.wieditz@tu-ilmenau.de (J.W.)

* Correspondence: thomas.hotz@tu-ilmenau.de.

† This paper is an extended version of our paper published in Proceedings of the 2nd International Conference on Geometric Science of Information, Palaiseau, France, 28–30 October 2015; Nielsen, F., Barbaresco, F., Eds.; Lecture Notes in Computer Science, Volume 9389; Springer International Publishing: Cham, Switzerland, 2015; pp. 635–642.

‡ These authors contributed equally to this work.

Academic Editors: Frédéric Barbaresco and Frank Nielsen

Abstract: The mean of data on the unit circle is defined as the minimizer of the average squared Euclidean distance to the data. Based on Hoeffding's mass concentration inequalities, non-asymptotic confidence sets for circular means are constructed which are universal in the sense that they require no distributional assumptions. These are then compared with asymptotic confidence sets in simulations and for a real data set.

Keywords: directional data; circular mean; universal confidence sets; non-asymptotic confidence sets; mass concentration inequalities; Hoeffding's inequality

MSC: 62H11; 62G15

---

1. Introduction

In applications, data assuming values on the circle, i.e., *circular data,* arise frequently, examples being measurements of wind directions, or time of the day that patients are admitted to a hospital unit. We refer to the literature, e.g., [1–5], for an overview of statistical methods for circular data, in particular the ones described in this section.

Here, we will concern ourselves with the arguably simplest statistic, the *mean*. However, given that a circle does not carry a vector space structure, i.e., there is neither a natural addition of points on the circle nor can one divide them by a natural number, what should the meaning of "mean" be?

In order to simplify the exposition, we specifically consider the unit circle in the complex plane, $\mathrm{S}^1 = \{z \in \mathbf{C} : |z| = 1\}$, and we assume the data can be modelled as independent random variables $Z_1, \ldots, Z_n$ which are identically distributed as the random variable Z taking values in S^1. In the literature, however, the circle is often taken to lie in the real plane $\mathbf{R}^2$, i.e., while we denote the point on the circle corresponding to an angle $\theta \in (-\pi, \pi]$ by $\exp(i\theta) = \cos(\theta) + i\sin(\theta) \in \mathbf{C}$ one may take it to be $(\cos\theta, \sin\theta) \in \mathbf{R}^2$.

Of course, $\mathbf{C}$ is a real vector space, so the *Euclidean sample mean* $\bar{Z}_n = \frac{1}{n}\sum_{k=1}^{n} Z_k \in \mathbf{C}$ is well-defined. However, unless all Z_k take identical values, it will (by the strict convexity of the closed unit disc) lie inside the circle, i.e., its *modulus* $|\bar{Z}_n|$ will be less than 1. Though $\bar{Z}_n$ cannot be taken as a mean on the circle, if $\bar{Z}_n \neq 0$, one might say that it specifies a direction; this leads to the idea of calling $\bar{Z}_n / |\bar{Z}_n|$ the circular sample mean of the data.

Observing that the Euclidean sample mean is the minimiser of the sum of squared distances, this can be put in the more general framework of *Fréchet means* [6]: define the *set of circular sample means* to be

$$\hat{\mu}_n = \operatorname*{argmin}_{\zeta \in \mathbf{S}^1} \sum_{k=1}^{n} |Z_k - \zeta|^2, \tag{1}$$

and analoguously define the *set of circular population means* of the random variable Z to be

$$\mu = \operatorname*{argmin}_{\zeta \in \mathbf{S}^1} \mathbf{E}\,|Z - \zeta|^2. \tag{2}$$

Then, as usual, the circular sample means are the circular population means with respect to the empirical distribution of $Z_1, \ldots, Z_n$.

The circular population mean can be related to the Euclidean population mean $\mathbf{E}\,Z$ by noting that $\mathbf{E}\,|Z - \zeta|^2 = \mathbf{E}\,|Z - \mathbf{E}\,Z|^2 + |\,\mathbf{E}\,Z - \zeta|^2$ (in statistics, this is called the *bias-variance decomposition*), so that

$$\mu = \operatorname*{argmin}_{\zeta \in \mathbf{S}^1} |\,\mathbf{E}\,Z - \zeta|^2 \tag{3}$$

is the set of points on the circle closest to $\mathbf{E}\,Z$. It follows that μ is unique if and only if $\mathbf{E}\,Z \neq 0$ in which case it is given by $\mu = \mathbf{E}\,Z/|\,\mathbf{E}\,Z|$, the *orthogonal projection* of $\mathbf{E}\,Z$ onto the circle; otherwise, i.e., if $\mathbf{E}\,Z = 0$, the set of circular population means is all of $\mathbf{S}^1$. We consider the information of whether the circular population mean is not unique, e.g., but not exclusively because Z is uniformly distributed over the circle, to be relevant; it thus should be inferred from the data as well. Analogously, $\hat{\mu}_n$ is either all of $\mathbf{S}^1$ or uniquely given by $\bar{Z}_n/|\bar{Z}_n|$ according to whether $\bar{Z}_n$ is 0 or not. Note that $\bar{Z}_n \neq 0$ a.s. if Z is continuously distributed on the circle, even if $\mathbf{E}\,Z = 0$. $\bar{Z}_n$ is what is known as the *vector resultant*, while $\bar{Z}_n/|\bar{Z}_n|$ is sometimes referred to as the *mean direction*.

The expected squared distances minimised in Equation (2) are given by the metric inherited from the ambient space $\mathbf{C}$; therefore, μ is also called the set of *extrinsic* population means. If we measured distances intrinsically along the circle, i.e., using arc-length instead of chordal distance, we would obtain what is called the set of *intrinsic* population means. We will not consider the latter in the following, see e.g., [7] for a comparison and [8,9] for generalizations of these concepts.

Our aim is to construct *confidence sets* for the circular population mean μ that form a superset of μ with a certain (so-called) *coverage probability* that is required to be not less than some pre-specified significance level $1 - \alpha$ for $\alpha \in (0, 1)$.

The classical approach is to construct an *asymptotic confidence interval* where the coverage probability converges to $1 - \alpha$ when n tends to infinity. This can be done as follows: since Z is a bounded random variable, $\sqrt{n}(\bar{Z}_n - \mathbf{E}\,Z)$ converges to a bivariate normal distribution when identifying $\mathbf{C}$ with $\mathbf{R}^2$. Now, assume $\mathbf{E}\,Z \neq 0$ so μ is unique. Then, the orthogonal projection is differentiable in a neighbourhood of $\mathbf{E}\,Z$, so the δ-method (see e.g., [1] (p. 111) or [4] (Lemma 3.1)) can be applied and one easily obtains

$$\sqrt{n}\,\mathbf{Arg}(\mu^{-1}\hat{\mu}_n) \xrightarrow{\mathcal{D}} \mathcal{N}\left(0, \frac{\mathbf{E}(\mathbf{Im}(\mu^{-1}Z))^2}{|\,\mathbf{E}\,Z|^2}\right), \tag{4}$$

where $\mathbf{Arg} : \mathbf{C} \setminus \{0\} \to (-\pi, \pi] \subset \mathbf{R}$ denotes the *argument* of a complex number (it is defined arbitrarily at $0 \in \mathbf{C}$), while multiplying with μ^{-1} rotates such that $\mathbf{E}\,Z = \mu$ is mapped to $0 \in (-\pi, \pi]$, see e.g., [4] (Proposition 3.1) or [7] (Theorem 5). Estimating the asymptotic variance and applying Slutsky's lemma, one arrives at the asymptotic confidence set $C_A = \{\zeta \in \mathbf{S}^1 : |\,\mathbf{Arg}(\zeta^{-1}\hat{\mu}_n)| < \delta_A\}$ provided $\hat{\mu}_n$ is unique, where the angle determining the interval is given by

$$\delta_A = \frac{q_{1-\frac{\alpha}{2}}}{n|\bar{Z}_n|}\sqrt{\sum_{k=1}^{n}(\mathbf{Im}(\hat{\mu}_n^{-1}Z_k))^2}, \tag{5}$$

with $q_{1-\frac{\alpha}{2}}$ denoting the $(1-\frac{\alpha}{2})$-quantile of the standard normal distribution $\mathcal{N}(0,1)$.

There are two major drawbacks to the use of asymptotic confidence intervals: firstly, by definition, they do not guarantee a coverage probability of at least $1-\alpha$ for finite n, so the coverage probability for a fixed distribution and sample size may be much smaller. Indeed, Simulation 2 in Section 4 demonstrates that, even for $n=100$, the coverage probability may be as low as 64% when constructing the asymptotic confidence set for $1-\alpha=90\%$. Secondly, they *assume* that $\mathbf{E}\,Z \neq 0$, so they are not applicable to all distributions on the circle. Since in practice it is unknown whether this assumption hold, one would have to test the hypothesis $\mathbf{E}\,Z=0$, possibly again by an asymptotic test, and construct the confidence set conditioned on this hypothesis having been rejected, setting $C_A=\mathbb{S}^1$ otherwise. However, this sequential procedure would require some adaptation taking the pre-test into account (cf. e.g., [10])—we come back to this point in Section 5—and it is not commonly implemented in practice.

We therefore aim to construct *non-asymptotic* confidence sets for μ, guaranteeing coverage with at least the desired probability for any sample size n, which in addition are *universal* in the sense that they do not make any distributional assumptions about the circular data besides them being independent and identically distributed. It has been shown in [7] that this is possible; however, the confidence sets that were constructed there were far too large to be of use in practice. Nonetheless, we start by varying that construction in Section 2 but using Hoeffding's inequality instead of Chebyshev's as in [7]. Considerable improvements are possible if one takes the variance $\mathbf{E}(\mathbf{Im}(\mu^{-1}Z))^2$ "perpendicular to $\mathbf{E}\,Z$" into account; this is achieved by a second construction in Section 3. Of course, the latter confidence sets will still be conservative but Proposition 2(iv) shows that they are (for $1-\alpha=95\%$) only a factor $\sim\frac{3}{2}$ longer than the asymptotic ones when the sample size n is large. We further illustrate and compare those confidence sets in simulations and for an application to real data in Section 4, discussing the results obtained in Section 5.

2. Construction Using Hoeffding's Inequality

We will construct a confidence set as the acceptance region of a series of tests. This idea has been used before for the construction of confidence sets for the circular population mean [7] (Section 6); however, we will modify that construction by replacing Chebyshev's inequality—which is too conservative here—by three applications of Hoeffding's inequality [11] (Theorem 1): if $U_1,\ldots,U_n$ are independent random variables taking values in the bounded interval $[a,b]$ with $-\infty<a<b<\infty$. Then, $\bar{U}_n=\frac{1}{n}\sum_{k=1}^{n}U_k$ with $\mathbf{E}\,\bar{U}_n=\nu$ fulfills

$$\mathbf{P}(\bar{U}_n-\nu\geq t)\leq\left[\left(\frac{\nu-a}{\nu-a+t}\right)^{\nu-a+t}\left(\frac{b-\nu}{b-\nu-t}\right)^{b-\nu-t}\right]^{\frac{n}{b-a}} \tag{6}$$

for any $t\in(0,b-\nu)$. The bound on the right-hand side—denoted $\beta(t)$—is continuous and strictly decreasing in t (as expected; see Appendix A) with $\beta(0)=1$ and $\lim_{t\to b-\nu}\beta(t)=\left(\frac{\nu-a}{b-a}\right)^n$ whence a unique solution $t=t(\gamma,\nu,a,b)$ to the equation $\beta(t)=\gamma$ exists for any $\gamma\in\left(\left(\frac{\nu-a}{b-a}\right)^n,1\right)$. Equivalently, $t(\gamma,\nu,a,b)$ is strictly decreasing in γ. Furthermore, $\nu+t(\gamma,\nu,a,b)$ is strictly increasing in ν (see Appendix A again), which is also to be expected. While there is no closed form expression for $t(\gamma,\nu,a,b)$, it can without difficulty be determined numerically.

Note that the estimate

$$\beta(t)\leq\exp(-2nt^2/(b-a)^2) \tag{7}$$

is often used and called Hoeffding's inequality [11]. While this would allow to solve explicitly for t, we prefer to work with β as it is sharper, especially for ν close to b as well as for large t. Nonetheless, it shows that the tail bound $\beta(t)$ tends to zero as fast as if using the central limit theorem which is why it is widely applied for bounded variables, see e.g., [12].

Now, for any $\zeta \in S^1$, we will test the hypothesis that ζ is a circular population mean. This hypothesis is equivalent to saying that there is some $\lambda \in [0,1]$ such that $\mathbf{E}\,Z = \lambda\zeta$. Multiplication by ζ^{-1} then rotates $\mathbf{E}\,Z$ onto the non-negative real axis: $\mathbf{E}\,\zeta^{-1}Z = \lambda \geq 0$.

Now, fix ζ and consider $X_k = \mathbf{Re}(\zeta^{-1}Z_k)$, $Y_k = \mathbf{Im}(\zeta^{-1}Z_k)$ for $k = 1, \ldots, n$ which may be viewed as the projection of $Z_1, \ldots, Z_k$ onto the line in the direction of ζ and onto the line perpendicular to it. Both are sequences of independent random variables taking values in $[-1,1]$ with $\mathbf{E}\,X_k = \lambda$ and $\mathbf{E}\,Y_k = 0$ under the hypothesis. They thus fulfill the conditions for Hoeffding's inequality with $a = -1$, $b = 1$ and $\nu = \lambda$ or 0, respectively.

We will first consider the case of non-uniqueness of the circular mean, i.e., $\mu = S^1$, or equivalently $\lambda = 0$. Then, the critical value $s_0 = t(\frac{\alpha}{4}, 0, -1, 1)$ is well-defined for any $\frac{\alpha}{4} > 2^{-n}$, and we get $\mathbf{P}(\bar{X}_n \geq s_0) \leq \frac{\alpha}{4}$, and also, by considering $-X_1, \ldots, -X_n$, that $\mathbf{P}(-\bar{X}_n \geq s_0) \leq \frac{\alpha}{4}$. Analogously, $\mathbf{P}(|\bar{Y}_n| \geq s_0) \leq 2\frac{\alpha}{4} = \frac{\alpha}{2}$. We conclude that

$$\mathbf{P}(|\bar{Z}_n| \geq \sqrt{2}s_0) = \mathbf{P}(|\bar{X}_n|^2 + |\bar{Y}_n|^2 \geq 2s_0^2) \leq \mathbf{P}(|\bar{X}_n|^2 \geq s_0^2) + \mathbf{P}(|\bar{Y}_n|^2 \geq s_0^2) \leq \alpha.$$

Rejecting the hypothesis $\mu = S^1$, i.e., $\mathbf{E}\,Z = 0$, if $|\bar{Z}_n| \geq \sqrt{2}s_0$ thus leads to a test whose probability of false rejection is at most α (see Figure 1). Of course, one may work with $|\bar{X}_n|^2 \geq s_0^2$ and $|\bar{Y}_n|^2 \geq s_0^2$ as criterions for rejection; however, we prefer working with $|\bar{Z}_n| \geq \sqrt{2}s_0$ since it is independent of the chosen ζ.

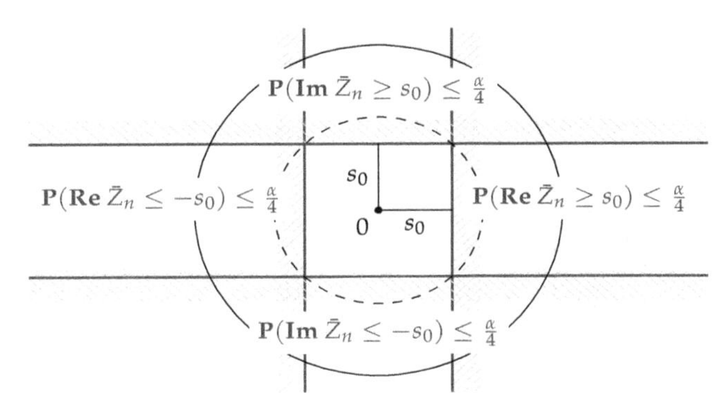

Figure 1. The construction for the test of the hypothesis $\mu = S^1$, or equivalently $\mathbf{E}\,Z = 0$.

In the case of uniqueness of the circular mean, i.e., for the hypothesis $\lambda > 0$, we use the monotonicity of $\nu + t(\gamma, \nu, a, b)$ in ν and obtain

$$\mathbf{P}(\bar{X}_n \leq -s_0) = \mathbf{P}(-\bar{X}_n \geq t(\tfrac{\alpha}{4}, 0, -1, 1)) \leq \mathbf{P}(-\bar{X}_n \geq -\lambda + t(\tfrac{\alpha}{4}, -\lambda, -1, 1)) \leq \frac{\alpha}{4}$$

as well. For the direction perpendicular to the direction of ζ (see Figure 2), however, we may now work with $\frac{3}{8}\alpha$, so for $s_p = t(\frac{3}{8}\alpha, 0, -1, 1)$—which is well-defined whenever s_0 is since $\frac{3}{8}\alpha > \frac{\alpha}{4} > 2^{-n}$—we obtain

$$\mathbf{P}(\bar{Y}_n \geq s_p) + \mathbf{P}(\bar{Y}_n \leq -s_p) \leq 2 \cdot \tfrac{3}{8}\alpha.$$

Rejecting if $\bar{X}_n \leq -s_0$ or $|\bar{Y}_n| \geq s_p$, then, will happen with probability at most $\frac{\alpha}{4} + 2 \cdot \frac{3}{8}\alpha = \alpha$ under the hypothesis $\mu = \zeta$. In case that we already rejected the hypothesis $\mu = S^1$, i.e., if $|\bar{Z}_n| \geq \sqrt{2}s_0$, ζ will not be rejected if and only if $\bar{X}_n > s_0 > 0$ and $|\bar{Y}_n| < s_p < s_0$ which is then equivalent to $|\mathbf{Arg}(\zeta^{-1}\bar{Z}_n)| = \arcsin(|\bar{Y}_n|/|\bar{Z}_n|) < \arcsin(s_p/|\bar{Z}_n|) = \delta_H$ (see Figure 3).

Define C_H as all ζ which we could not reject, i.e.,

$$C_H = \begin{cases} S^1, & \text{if } \alpha \leq 2^{-n+2} \text{ or } |\bar{Z}_n| \leq \sqrt{2}s_0, \\ \{\zeta \in S^1 : |\mathbf{Arg}(\zeta^{-1}\hat{\mu}_n)| < \delta_H\} & \text{otherwise.} \end{cases} \tag{8}$$

Then, we obtain the following result:

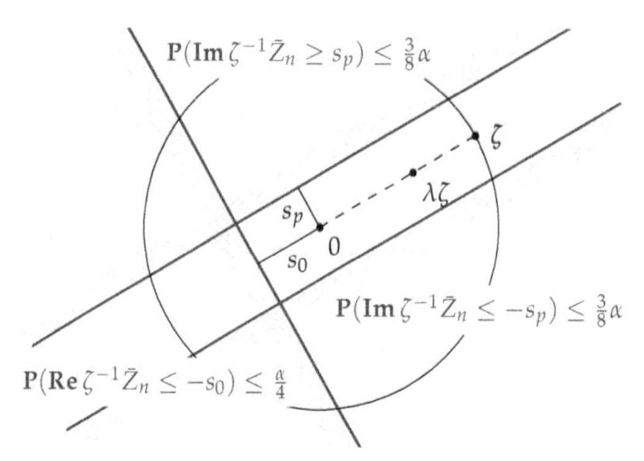

Figure 2. The construction for the test of the hypothesis $\mathbf{E}\,Z = \lambda\zeta$ with $\lambda > 0$.

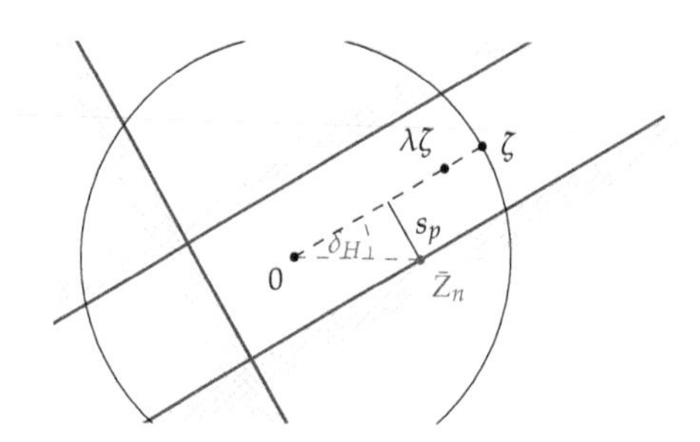

Figure 3. The critical $\bar{Z}_n$ regarding the rejection of ζ. δ_H bounds the angle between $\hat{\mu}_n$ and any accepted ζ.

Proposition 1. *Let $Z_1, \ldots, Z_n$ be random variables taking values on the unit circle S^1, $\alpha \in (0,1)$, and let C_H be defined as in Equation* (8).

(i) *C_H is a $(1-\alpha)$-confidence set for the circular population mean set. In particular, if $\mathbf{E}\,Z = 0$, i.e., the circular population mean set equals S^1, then $|\bar{Z}_n| > \sqrt{2}s_0$ with probability at most α, so indeed $C_H = \mathrm{S}^1$ with probability at least $1-\alpha$.*

(ii) *s_0 and s_p are of order $n^{-\frac{1}{2}}$.*

(iii) *If $\mathbf{E}\,Z \neq 0$, then $\sqrt{n}\delta_H \to 0$ in probability and the probability of obtaining the trivial confidence set, i.e., $\mathbf{P}(C_H = \mathrm{S}^1) = \mathbf{P}(|\bar{Z}_n| \leq \sqrt{2}s_0)$, goes to 0 exponentially fast.*

Proof. (i) holds by construction.

For (ii), recall Equation (7), from which we obtain the estimates $\frac{\alpha}{4} \leq \exp(-ns_0^2/2)$ resp. $\frac{3}{8}\alpha \leq \exp(-ns_p^2/2)$, implying that s_0 and s_p are of order $n^{-\frac{1}{2}}$; the same holds stochastically for δ_H since $\bar{Z}_n \to \mathbf{E}\,Z$ a.s. Regarding the second statement of (iii), if μ is unique, consider $\zeta = -\mu$; then, $\tau = \mathbf{E}\,\bar{X}_n < 0$ and $-\sqrt{2}s_0$ is eventually less than $\frac{\tau}{2}$ and also $\alpha > 2^{-n+2}$ eventually. Hence, the probability of obtaining the trivial confidence set $C_H = \mathrm{S}^1$ is eventually bounded by $\mathbf{P}(\zeta \in C_H) \leq \mathbf{P}(\bar{X}_n > -s_0) \leq \mathbf{P}(\bar{X}_n > \frac{\tau}{2}) = \mathbf{P}(\bar{X}_n - \mathbf{E}\,\bar{X}_n > -\frac{\tau}{2}) \leq \exp(-n\tau^2/8)$, and thus will go to zero exponentially fast as n tends to infinity. □

3. Estimating the Variance

From the central limit theorem for $\hat{\mu}_n$ in case of unique μ, cf. Equation (4), we see that the aymptotic variance of $\hat{\mu}_n$ gets small if $|\,\mathbf{E}\,Z|$ is close to 1 (then $\mathbf{E}\,Z$ is close to the boundary S^1 of the unit disc, which is possible only if the distribution is very concentrated) or if the variance $\mathbf{E}(\mathbf{Im}(\mu^{-1}Z))^2$ in the direction perpendicular to μ is small (if the distribution were concentrated on $\pm\mu$, this variance would be zero and $\hat{\mu}_n$ would equal μ with large probability). While δ_H ($|\bar{Z}_n|$ being the denominator of its sine) takes the former into account, the latter has not been exploited yet. To do so, we need to estimate $\mathbf{E}(\mathbf{Im}(\mu^{-1}Z))^2$.

Consider $V_n = \frac{1}{n}\sum_{k=1}^{n} Y_k^2$ that is under the hypothesis that the corresponding ζ is the unique circular population mean has expectation $\sigma^2 = \mathbf{Var}(Y_k) = \mathbf{E}(\mathbf{Im}(\zeta^{-1}Z))^2$. Now, $1 - V_n = \frac{1}{n}\sum_{k=1}^{n}(1 - Y_k^2)$ is the mean of n independent random variables taking values in $[0,1]$ and having expectation $1-\sigma^2$. By another application of Equation (6), we obtain $\mathbf{P}(\sigma^2 \geq V_n + t) = \mathbf{P}(1 - V_n \geq 1 - \sigma^2 + t) \leq \frac{\alpha}{4}$ for $t = t(\frac{\alpha}{4}, 1-\sigma^2, 0, 1)$, the latter existing if $\frac{\alpha}{4} > (1-\sigma^2)^n$. Since $1 - \sigma^2 + t(\frac{\alpha}{4}, 1-\sigma^2, 0, 1)$ increases with $1-\sigma^2$, there is a minimal σ^2 for which $1 - V_n \geq 1 - \sigma^2 + t(\frac{\alpha}{4}, 1-\sigma^2, 0, 1)$ holds and becomes an equality; we denote it by $\widehat{\sigma^2} = V_n + t(\frac{\alpha}{4}, 1-\widehat{\sigma^2}, 0, 1)$. Inserting into Equation (6), it by construction fulfills

$$\frac{\alpha}{4} = \left[\left(\frac{1-\widehat{\sigma^2}}{1-V_n}\right)^{1-V_n}\left(\frac{\widehat{\sigma^2}}{V_n}\right)^{V_n}\right]^n. \tag{9}$$

It is easy to see that the right-hand side depends continuously on and is strictly decreasing in $\widehat{\sigma^2} \in [V_n, 1]$ (see Appendix A), thereby traversing the interval $[0,1]$ so that one can again solve the equation numerically. We then may, with an error probability of at most $\frac{\alpha}{4}$, use $\widehat{\sigma^2}$ as an upper bound for σ^2. Note that $\widehat{\sigma^2} > V_n$ exists if $\frac{\alpha}{4} > (1-\widehat{\sigma^2})^n$. The latter is fulfilled for any $V_n < 1$ since Equation (9) is equivalent to

$$\frac{\alpha}{4} = (1-\widehat{\sigma^2})^n\Bigg[\underbrace{\left(\frac{1}{1-V_n}\right)}_{>1}\underbrace{\left(\frac{1-\widehat{\sigma^2}}{1-V_n}\right)^{-V_n}}_{>1}\underbrace{\left(\frac{\widehat{\sigma^2}}{V_n}\right)^{V_n}}_{>1}\Bigg]^n.$$

For $V_n = 1$, let $\widehat{\sigma^2} = 1$ be the trivial bound.

With such an upper bound on its variance, we now can get a better estimate for $\mathbf{P}(\bar{Y}_n > t)$. Indeed, one may use another inequality by Hoeffding [11] (Theorem 3): the mean $\bar{W}_n = \frac{1}{n}\sum_{k=1}^{n} W_k$ of a sequence $W_1, \ldots, W_n$ of independent random variables taking values in $(-\infty, 1]$, each having zero expectation as well as variance ρ^2 fulfills

$$\mathbf{P}(\bar{W}_n \geq w) \leq \left[\left(1+\frac{w}{\rho^2}\right)^{-\rho^2-w}\left(1-w\right)^{w-1}\right]^{\frac{n}{1+\rho^2}}, \tag{10}$$

$$\leq \exp(-nt[(1+\tfrac{\rho^2}{t})\ln(1+\tfrac{t}{\rho^2})-1]). \tag{11}$$

for any $w \in (0,1)$. Again, an elementary calculation (analogous to Lemma A1) shows that the right-hand side of Equation (10) is strictly decreasing in w, continuously ranging between 1 and $(\frac{\rho^2}{1+\rho^2})^n$ as w varies in $(0,1)$, so that there exists a unique $w = w(\gamma, \rho^2)$ for which the right-hand side equals γ, provided $\gamma \in \left((\frac{\rho^2}{1+\rho^2})^n, 1\right)$. Moreover, the right-hand side increases with ρ^2 (as expected), so that $w(\gamma, \rho^2)$ is increasing in ρ^2, too (cf. Appendix A).

Therefore, under the hypothesis that the corresponding ζ is the unique circular population mean, $\mathbf{P}(|\bar{Y}_n| \geq w(\frac{\alpha}{4}, \sigma^2)) \leq 2\frac{\alpha}{4} = \frac{\alpha}{2}$. Now, since $\mathbf{P}(w(\frac{\alpha}{4}, \sigma^2) \geq w(\frac{\alpha}{4}, \widehat{\sigma^2})) = \mathbf{P}(\sigma^2 \geq \widehat{\sigma^2}) \leq \frac{\alpha}{4}$, setting $s_V = w(\frac{\alpha}{4}, \widehat{\sigma^2})$ we get $\mathbf{P}(|\bar{Y}_n| \geq s_V) \leq \frac{3}{4}\alpha$. Note that $\frac{\rho^2}{1+\rho^2}$ increases with ρ^2, so in case s_0 exists, $\widehat{\sigma^2} \leq 1$ implies $\frac{\alpha}{4} > 2^{-n} \geq \left(\frac{\widehat{\sigma^2}}{1+\widehat{\sigma^2}}\right)^n$, i.e., the existence of s_V.

Following the construction for C_H from Section 2, we can again obtain a confidence set for μ with coverage probability at least $1-\alpha$ as shown in our previous article [13]. In practice however, this confidence set is hard to calculate since $\widehat{\sigma^2} = \widehat{\sigma^2}(\zeta)$ has to be calculated for every $\zeta \in \mathrm{S}^1$. Though these confidence sets can be approximated by using a grid as in [13], we suggest using a simultaneous upper bound for the variance of $\mathbf{Im}\,\zeta^{-1}Z_k$.

We obtain a (conservative) connected, symmetric confidence set $C_V \subseteq C_H$ by testing $\zeta \in C_H$ with $\widehat{\sigma^2_{\max}} = \sup_{\zeta \in C_H} \widehat{\sigma^2}$ as a common upper bound for the variance perpendicular to any $\zeta \in C_H$. Note that $\widehat{\sigma^2_{\max}}$ can be obtained as the solution of Equation (9) with

$$\tilde{V}_n = \sup_{\zeta \in C_H} \tfrac{1}{n} \sum_{k=1}^{n} \left(\mathbf{Im}\, \zeta^{-1} Z_k\right)^2.$$

Furthermore, we can shorten C_V by iteratively redefining $\tilde{V}_n = \sup_{\zeta \in C_V} \frac{1}{n} \sum_{k=1}^{n} \left(\mathbf{Im}\, \zeta^{-1} Z_k\right)^2$ and recalculating C_V (see Algorithm 1). The resulting opening angle will be denoted by $\delta_V = \arcsin \frac{s_V}{|\bar{Z}_n|}$.

Algorithm 1: Algorithm for computation of C_V.

Data: observations $Z_1, \ldots, Z_n \in \mathrm{S}^1$; significance level α; stop criterion ε
Result: a non-asymptotic confidence set C_V for the circular population mean

```
1  compute the confidence set C_H;
2  if C_H = S^1 then
3  |  C_V ← S^1
4  else
5  |  C_V ← C_H; σ²_max ← 1;
6  |  while sup_{ζ∈C_V} σ̂² < σ²_max − ε do
7  |  |  σ²_max ← sup_{ζ∈C_V} σ̂²;
8  |  |  s_V ← w(α/4, σ̂²);
9  |  |  C_V ← {ζ ∈ S^1 : |Arg(ζ^{-1} μ̂_n)| < arcsin (s_V / |Z̄_n|)}
10 |  end
11 end
```

Proposition 2. *Let $Z_1, \ldots, Z_n$ be random variables taking values on the unit circle S^1, and let $\alpha \in (0,1)$.*

(i) The set C_V resulting from Algorithm 1 is a $(1-\alpha)$-confidence set for the circular population mean set. In particular, if $\mathbf{E}\, Z = 0$, i.e., the circular population mean set equals S^1, then $|\bar{Z}_n| > \sqrt{2} s_0$ with probability at most α, so indeed $C_V = \mathrm{S}^1$ with probability of at least $1-\alpha$.

(ii) s_V is of order $n^{-\frac{1}{2}}$.

(iii) If $\mathbf{E}\, Z \neq 0$, i.e., if the circular population mean is unique, then $\sqrt{n}\delta_V \to 0$ in probability, and the probability of obtaining a trivial confidence set, i.e., $\mathbf{P}(C_H = \mathrm{S}^1) = \mathbf{P}(|\bar{Z}_n| \leq \sqrt{2} s_0)$, goes to 0 exponentially fast.

(iv) If $\mathbf{E}\, Z \neq 0$, then

$$\limsup_{n \to \infty} \frac{\delta_V}{\delta_A} \leq \frac{\sqrt{-2 \ln \frac{\alpha}{4}}}{q_{1-\frac{\alpha}{2}}} \quad a.s.$$

with $q_{1-\frac{\alpha}{2}}$ denoting the $(1-\frac{\alpha}{2})$-quantile of the standard normal distribution $\mathcal{N}(0,1)$.

Proof. Again, (i) follows by construction, while (iii) is shown as in Proposition 1.

For (ii), note that $s_V \leq s_0$ since the bound in Equation (10) for $\rho^2 = 1$ agrees with the bound in Equation (6) for $a = -1$, $b = 1$ and $v = 0$, thus s_V and δ_V are at least of the order $n^{-\frac{1}{2}}$.

For (iv), we will use the estimate in Equation (11). Recall that $\ln(1+x) = x - \frac{x^2}{2} + o(x^2)$; therefore, for large n and hence small s_V a.s.

$$\begin{aligned}\frac{\alpha}{4} &\leq \exp\Big(-ns_V\Big[\Big(1+\frac{\widehat{\sigma^2_{\max}}}{s_V}\Big)\Big(\frac{s_V}{\widehat{\sigma^2_{\max}}} - \frac{s_V^2}{2(\widehat{\sigma^2_{\max}})^2} + o(s_V^2)\Big) - 1\Big]\Big) \\ &= \exp(-ns_V^2/2\widehat{\sigma^2_{\max}} + o(s_V^2)),\end{aligned}$$

thus $s_V \leq \sqrt{-2\widehat{\sigma^2_{\max}}\ln(\frac{\alpha}{4})/n} + o(n^{-\frac{1}{2}})$. Additionally, $\arcsin x = x + o(x)$ for x close to 0 which gives $\delta_V = s_V/|\bar{Z}_n| + o(s_V) \leq \sqrt{-2\widehat{\sigma^2_{\max}}\ln\frac{\alpha}{4}}/(\sqrt{n}|\bar{Z}_n|) + o(n^{-\frac{1}{2}})$ a.s.

Furthermore, $\widehat{\sigma^2_{\max}} \to \sigma^2$ a.s. for $n \to \infty$, and we obtain

$$\limsup_{n\to\infty} \frac{\delta_V}{\delta_A} \leq \frac{\sqrt{-2\ln\frac{\alpha}{4}}}{q_{1-\frac{\alpha}{2}}} \quad \text{a.s.}$$

since

$$\delta_A = \frac{q_{1-\frac{\alpha}{2}}}{\sqrt{n}|\bar{Z}_n|}\underbrace{\sqrt{\frac{1}{n}\sum_{k=1}^{n}\left(\mathbf{Im}(\hat{\mu}_n^{-1}Z_k)\right)^2}}_{\to\sqrt{\sigma^2}}$$

(see Equation (5)). □

4. Simulation and Application to Real Data

We will compare the asymptotic confidence set C_A, the confidence set C_H constructed directly using Hoeffding's inequality in Section 2, and the confidence set C_V resulting from Algorithm 1 by reporting their corresponding opening angles δ_A, δ_H, and δ_V in degrees (°) as well as their coverage frequencies in simulations.

All computations have been performed using our own code based on the software package R (version 2.15.3) [14] .

4.1. Simulation 1: Two Points of Equal Mass at ±10°

First, we consider a rather favourable situation: $n = 400$ independent draws from the distribution with $\mathbf{P}(Z = \exp(10\pi i/180)) = \mathbf{P}(Z = \exp(-10\pi i/180)) = \frac{1}{2}$. Then, we have $|\mathbf{E}Z| = \mathbf{E}Z = \cos(10\pi i/180) \approx 0.985$, implying that the data are highly concentrated, $\mu = 1$ is unique, and the variance of Z in the direction of μ is 0; there is only variation perpendicular to μ, i.e., in the direction of the imaginary axis (see Figure 4).

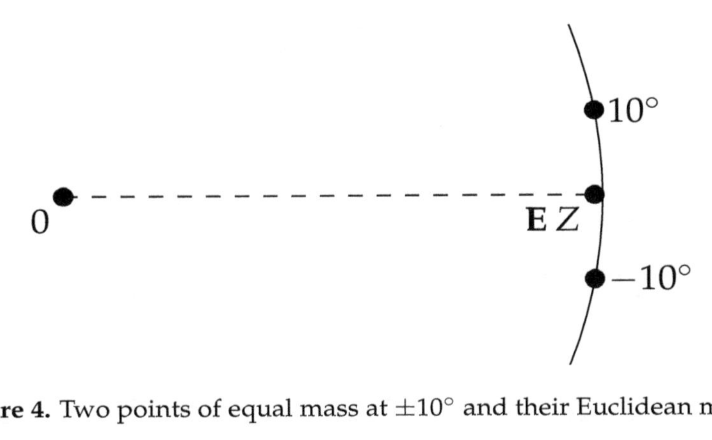

Figure 4. Two points of equal mass at ±10° and their Euclidean mean.

Table 1. Results for simulation 1 (two points of equal mass at $\pm 10°$) based on 10,000 repetitions with $n = 400$ observations each: average observed δ_H, δ_V, and δ_A (with corresponding standard deviation), as well as frequency (with corresponding standard error) with which $\mu = 1$ was covered by C_H, C_V, and C_A, respectively; the nominal coverage probability was $1 - \alpha = 95\%$.

Confidence Set	Mean δ ($\pm$s.d.)	Coverage Frequency ($\pm$s.e.)
C_H	8.2° ($\pm$0.0005°)	100.0% ($\pm$0.0%)
C_V	2.4° ($\pm$0.0025°)	100.0% ($\pm$0.0%)
C_A	1.0° ($\pm$0.0019°)	94.8% ($\pm$0.2%)

Table 1 shows the results based on 10,000 repetitions for a nominal coverage probability of $1 - \alpha = 95\%$: the average δ_H is about 3.5 times larger than δ_V, which is about twice as large as δ_A. As expected, the asymptotics are rather precise in this situation: C_A did cover the true mean in about 95% of the cases, which implies that the other confidence sets are quite conservative; indeed C_H and C_V covered the true mean in all repetitions. One may also note that the angles varied only a little between repetitions.

4.2. Simulation 2: Three Points Placed Asymmetrically

Secondly, we consider a situation which has been designed to show that even a considerably large sample size ($n = 100$) does not guarantee approximate coverage for the asymptotic confidence set C_A: the distribution of Z is concentrated on three points, $\xi_j = \exp(\theta_j \pi i/180)$, $j = 1, 2, 3$ with weights $\omega_j = \mathbf{P}(Z = \xi_j)$ chosen such that $\mathbf{E}\,Z = |\,\mathbf{E}\,Z| = 0.9$ (implying a small variance and $\mu = 1$), $\omega_1 = 1\%$ and $\mathbf{Arg}\,\xi_1 > 0$, while $\mathbf{Arg}\,\xi_2, \mathbf{Arg}\,\xi_3 < 0$. In numbers, $\theta_1 \approx 25.8$, $\theta_2 \approx -0.3$, and $\theta_3 \approx -179.7$ (in °) while $\omega_2 \approx 94\%$, and $\omega_3 \approx 5\%$ (see Figure 5).

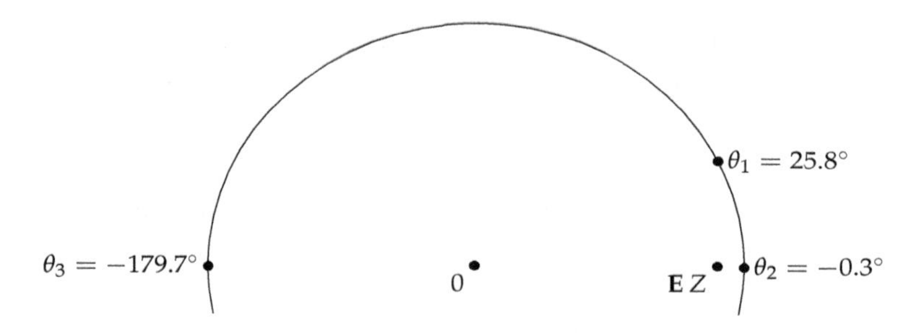

Figure 5. Three points placed asymmetrically with different masses and their Euclidean mean.

The results based on 10,000 repetitions are shown in Table 2 where a nominal coverage probability of $1 - \alpha = 90\%$ was prescribed. Clearly, C_A with its coverage probability of less than 64% performs quite poorly while the others are conservative; $\delta_V \approx 5°$ still appears small enough to be useful in practice, though.

Table 2. Results for simulation 2 (three points placed asymmetrically) based on 10,000 repetitions with $n = 100$ observations each: average observed δ_H, δ_V, and δ_A (with corresponding standard deviation), as well as frequency (with corresponding standard error) with which $\mu = 1$ was covered by C_H, C_V, and C_A, respectively; the nominal coverage probability was $1 - \alpha = 90\%$.

Confidence Set	Mean δ ($\pm$s.d.)	Coverage Frequency ($\pm$s.e.)
C_H	16.5° ($\pm$0.85°)	100.0% ($\pm$0.0%)
C_V	5.0° ($\pm$0.38°)	100.0% ($\pm$0.0%)
C_A	0.4° ($\pm$0.28°)	62.8% ($\pm$0.5%)

4.3. Real Data: Movements of Ants

Fisher [3] (Example 4.4) describes a data set of the directions 100 ants took in response to an illuminated target placed at 180° for which it may be of interest to know whether the ants indeed (on average) move towards that target (see [15] for the original publication). The data set is available as `Ants_radians` within the R package `CircNNTSR` [16].

The circular sample mean for this data set is about $-176.9°$; for a nominal coverage probability of $1-\alpha = 95\%$, one gets $\delta_H \approx 27.3°$, $\delta_V \approx 20.5°$, and $\delta_A \approx 9.6°$ so that all confidence sets contain $\pm 180°$ (see Figure 6). The data set's concentration is not very high, however, so the circular population mean could—according to C_V—also be $-156.4°$ or $162.6°$.

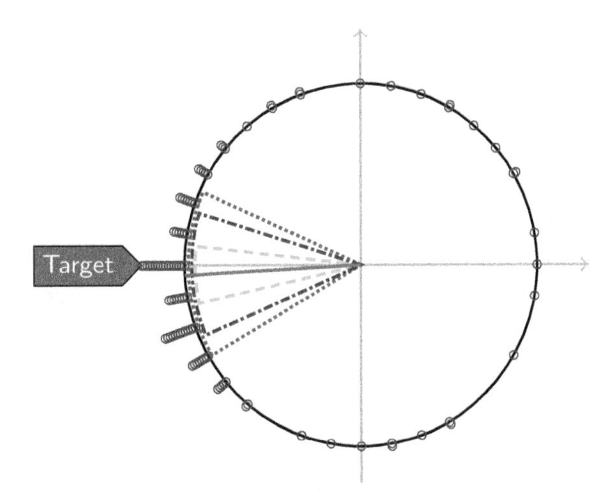

Figure 6. Ant data (○) placed at increasing radii to visually resolve ties; in addition, the circular mean direction (—) as well as confidence sets C_H (·····), C_V (-·-), and C_A (- -) are shown.

5. Discussion

We have derived two confidence sets, C_H and C_V, for the set of circular sample means. Both guarantee coverage for any finite sample size without making any assumptions on the distribution of the data (besides that they are independent and identically distributed) at the cost of potentially being quite conservative; they are non-asymptotic and universal in this sense. Judging from the simulations and the real data set, C_V—which estimates the variance perpendicular to the mean direction—appears to be preferable over C_H (as expected) and small enough to be useful in practice.

While the asymptotic confidence set's opening angle is less than half (asymptotically about 2/3 for $\alpha = 5\%$) of the one for C_V in our simulations and application, it has the drawback that even for a sample size of $n = 100$, it may fail to give a coverage probability close to the nominal one; in addition, one has to assume that the circular population mean is unique. Of course, one could also devise an asymptotically justified test for the latter but this would entail a correction for multiple testing (for example working with $\frac{\alpha}{2}$ each time), which would also render the asymptotic confidence set conservative.

Further improvements would require sharper "universal" mass concentration inequalities taking the first or the first two moments into account; however, this is beyond the scope of this article.

Acknowledgments: T. Hotz wishes to thank Stephan Huckemann from the Georgia Augusta University of Göttingen for fruitful discussions concerning the first construction of confidence regions described in Section 2. We acknowledge support for the Article Processing Charge by the German Research Foundation and the Open Access Publication Fund of the Technische Universität Ilmenau. F. Kelma acknowledges support by the Klaus Tschira Stiftung, gemeinnützige Gesellschaft, Projekt 03.126.2016.

Author Contributions: All authors contributed to the theoretical and numerical results as well as to the writing of the article. All authors have read and approved the final manuscript.

Appendix A. Proofs of Monotonicity

Lemma A1. $\beta(t) = \left[\left(\frac{v-a}{v-a+t}\right)^{v-a+t}\left(\frac{b-v}{b-v-t}\right)^{b-v-t}\right]^{\frac{n}{b-a}}$ *is strictly decreasing in* t.

Proof. We show the equivalent statement that $\tilde{\beta}(t) = \ln\left[\left(\frac{v-a}{v-a+t}\right)^{v-a+t}\left(\frac{b-v}{b-v-t}\right)^{b-v-t}\right]$ is strictly decreasing in t:

$$\begin{aligned}\frac{d}{dt}\tilde{\beta}(t) &= \frac{d}{dt}\Big((\ln(v-a)-\ln(v-a+t))(v-a+t)+(\ln(b-v)-\ln(b-v-t))(b-v-t)\Big)\\ &= \ln(v-a)-\ln(v-a+t)-\tfrac{1}{v-a+t}(v-a+t)-\ln(b-v)+\ln(b-v-t)+\tfrac{1}{b-v-t}(b-v-t)\\ &= \ln\Big(\underbrace{\frac{b-v-t}{b-v}}_{<1}\cdot\underbrace{\frac{v-a}{v-a+t}}_{<1}\Big)<0.\end{aligned}$$

Hence, $\tilde{\beta}(t)$ and thus $\beta(t)$ are strictly decreasing in t. □

Lemma A2. *Let* $t = t(\gamma, v, a, b)$ *be the solution to the equation* $\beta(t) = \gamma$. *Then,* $v + t$ *is strictly increasing in* v.

Proof. t is the solution of the equation

$$(v-a+t)\ln\left(\frac{v-a}{v-a+t}\right)+(b-v-t)\ln\left(\frac{b-v}{b-v-t}\right)=\frac{b-a}{n}\ln\gamma. \tag{A1}$$

The derivatives of the left-hand side of Equation (A1) w.r.t. v and t exist and are continuous. Furthermore, the derivative w.r.t. t does not vanish for any $t \in (0, b-v)$, cf. the proof of Lemma A1, whence the derivative $t' = \frac{dt}{dv}$ exists by the implicit function theorem. When differentiating Equation (A1) with respect to v, one obtains

$$\begin{aligned}(1+t')\ln\left(\frac{v-a}{v-a+t}\right)+(v-a+t)\left(\frac{1}{v-a}-\frac{1+t'}{v-a+t}\right)&\\ -(1+t')\ln\left(\frac{b-v}{b-v-t}\right)+(b-v-t)\left(-\frac{1}{b-v}+\frac{1+t'}{b-v-t}\right)&=0,\end{aligned}$$

or equivalently

$$(1+t')\Big[\underbrace{\ln\left(\frac{v-a}{v-a+t}\right)}_{<0}-\underbrace{\ln\left(\frac{b-v}{b-v-t}\right)}_{>0}\Big]=\frac{t(a-b)}{(v-a)(b-v)}<0,$$

whence $1+t' = \frac{d}{dv}(v+t) > 0$ finishes the proof. □

Lemma A3. *The function*

$$\xi(\widehat{\sigma^2}) = \left[\left(\frac{1-\widehat{\sigma^2}}{1-V_n}\right)^{1-V_n}\left(\frac{\widehat{\sigma^2}}{V_n}\right)^{V_n}\right]^n$$

is strictly decreasing in $\widehat{\sigma^2} \in [V_n, 1]$.

Proof. We show the equivalent statement that $n^{-1}\ln\xi(\widehat{\sigma^2})$ is strictly decreasing in $\widehat{\sigma^2}$:

$$\begin{aligned}\frac{d}{d\widehat{\sigma^2}}\Big[n^{-1}\ln\xi(\widehat{\sigma^2})\Big] &= \frac{d}{d\widehat{\sigma^2}}\Big[(1-V_n)(\ln(1-\widehat{\sigma^2})-\ln(1-V_n))+V_n(\ln(\widehat{\sigma^2})-\ln(V_n))\Big]\\ &= -\underbrace{\frac{1-V_n}{1-\widehat{\sigma^2}}}_{>1}+\underbrace{\frac{V_n}{\widehat{\sigma^2}}}_{<1}<0.\end{aligned}$$

□

Lemma A4. *Let $w = w(\gamma,\rho^2)$ be the solution of the equation*

$$\left[\left(1+\frac{w}{\rho^2}\right)^{-\rho^2-w}\left(1-w\right)^{w-1}\right]^{\frac{n}{1+\rho^2}}=\gamma.$$

Then, w is increasing in ρ^2.

Proof. w is the solution of the equation

$$\frac{\rho^2+w}{1+\rho^2}\ln\left(1+\frac{w}{\rho^2}\right)+\frac{1-w}{1+\rho^2}\ln(1-w)=-\frac{\ln\gamma}{n}. \tag{A2}$$

The derivatives of the left-hand side of Equation (A2) w.r.t. ρ^2 and w exist and are continuous. Furthermore, the derivative w.r.t. w does not vanish for any $w\in(0,1)$: this derivative is

$$\frac{1}{1+\rho^2}\left[\ln\left(1+\frac{w}{\rho^2}\right)+\frac{\rho^2+w}{\rho^2(1+\frac{w}{\rho^2})}-\ln(1-w)-1\right]=\frac{1}{1+\rho^2}\left[\ln\left(1+\frac{w}{\rho^2}\right)-\ln(1-w)\right],$$

vanishing if and only if $1+\frac{w}{\rho^2}=1-w$, i.e., if and only if $w(1+\frac{1}{\rho^2})=0$, which does not happen for $w,\rho^2>0$. Now, the derivative $w'=\frac{dw}{d\rho^2}$ exists by the implicit function theorem. When differentiating Equation (A2) with respect to ρ^2, one obtains

$$\frac{(1+w')(1+\rho^2)-(\rho^2+w)}{(1+\rho^2)^2}\ln\left(1+\frac{w}{\rho^2}\right)+\underbrace{\frac{\rho^2+w}{1+\rho^2}\cdot\frac{\frac{w'}{\rho^2}-\frac{w}{\rho^4}}{1+\frac{w}{\rho^2}}}_{\frac{w'\rho^2-w}{\rho^2(1+\rho^2)}}-\frac{w'(1+\rho^2)+(1-w)}{(1+\rho^2)^2}\ln(1-w)-\frac{w'}{1+\rho^2}=0,$$

or equivalently

$$w'\underbrace{\left[\ln\left(1+\frac{w}{\rho^2}\right)-\ln(1-w)\right]}_{>0}=\frac{w}{\rho^2}-\frac{1-w}{1+\rho^2}\ln\left(\frac{\rho^2+w}{\rho^2(1-w)}\right).$$

Hence, $w'\geq 0$ if and only if $\frac{w}{\rho^2}\geq\frac{1-w}{1+\rho^2}\ln(\frac{\rho^2+w}{\rho^2(1-w)})$, which holds since $\ln(\frac{\rho^2+w}{\rho^2(1-w)})=\ln(1+\frac{w(1+\rho^2)}{\rho^2(1-w)})\leq\frac{w}{\rho^2}\frac{1+\rho^2}{1-w}$, finishing the proof. □

References

1. Mardia, K.V. *Directional Statistics*; Academic Press: London, UK, 1972.

2. Watson, G.S. *Statistics on Spheres*; Wiley: New York, NY, USA, 1983.
3. Fisher, N.I. *Statistical Analysis of Circular Data*; Cambridge University Press: Cambridge, UK, 1993.
4. Jammalamadaka, S.R.; SenGupta, A. *Topics in Circular Statistics*; Series on Multivariate Analysis; World Scientific: Singapore, 2001.
5. Mardia, K.V.; Jupp, P.E. *Directional Statistics*; Wiley: New York, NY, USA, 2000.
6. Fréchet, M. Les éléments aléatoires de nature quelconque dans un espace distancié. *Annales de l'Institut Henri Poincaré* **1948**, *10*, 215–310. (In French)
7. Hotz, T. Extrinsic vs. Intrinsic Means on the Circle. In Proceedings of the 1st Conference on Geometric Science of Information, Paris, France, 28–30 October 2013; Lecture Notes in Computer Science, Volume 8085; Springer-Verlag: Heidelberg, Germany, 2013; pp. 433–440.
8. Afsari, B. Riemannian L^p center of mass: Existence, uniqueness, and convexity. *Proc. Am. Math. Soc.* **2011**, *139*, 655–673.
9. Arnaudon, M.; Miclo, L. A stochastic algorithm finding p-means on the circle. *Bernoulli* **2016**, *22*, 2237–2300.
10. Leeb, H.; Pötscher, B.M. Model selection and inference: Facts and fiction. *Econ. Theory* **2005**, *21*, 21–59.
11. Hoeffding, W. Probability Inequalities for Sums of Bounded Random Variables. *J. Am. Stat. Assoc.* **1963**, *58*, 13–30.
12. Boucheron, S.; Lugosi, G.; Massart, P. *Concentration Inequalities : A Nonasymptotic Theory of Independence*; Oxford University Press: Oxford, UK, 2013.
13. Hotz, T.; Kelma, F.; Wieditz, J. Universal, Non-asymptotic Confidence Sets for Circular Means. In Proceedings of the 2nd International Conference on Geometric Science of Information, Palaiseau, France, 28–30 October 2015; Nielsen, F., Barbaresco, F., Eds.; Lecture Notes in Computer Science, Volume 9389; Springer International Publishing: Cham, Switzerland, 2015; pp. 635–642.
14. R Core Team. *R: A Language and Environment for Statistical Computing*; version 2.15.3; R Foundation for Statistical Computing: Vienna, Austria, 2013.
15. Jander, R. Die optische Richtungsorientierung der Roten Waldameise (Formica Rufa L.). *Zeitschrift für Vergleichende Physiologie* **1957**, *40*, 162–238. (In German)
16. Fernandez-Duran, J.J.; Gregorio-Dominguez, M.M. *CircNNTSR: An R Package for the Statistical Analysis of Circular Data Using Nonnegative Trigonometric Sums (NNTS) Models*, version 2.1; 2013.

Permissions

All chapters in this book were first published in MDPI; hereby published with permission under the Creative Commons Attribution License or equivalent. Every chapter published in this book has been scrutinized by our experts. Their significance has been extensively debated. The topics covered herein carry significant findings which will fuel the growth of the discipline. They may even be implemented as practical applications or may be referred to as a beginning point for another development.

The contributors of this book come from diverse backgrounds, making this book a truly international effort. This book will bring forth new frontiers with its revolutionizing research information and detailed analysis of the nascent developments around the world.

We would like to thank all the contributing authors for lending their expertise to make the book truly unique. They have played a crucial role in the development of this book. Without their invaluable contributions this book wouldn't have been possible. They have made vital efforts to compile up to date information on the varied aspects of this subject to make this book a valuable addition to the collection of many professionals and students.

This book was conceptualized with the vision of imparting up-to-date information and advanced data in this field. To ensure the same, a matchless editorial board was set up. Every individual on the board went through rigorous rounds of assessment to prove their worth. After which they invested a large part of their time researching and compiling the most relevant data for our readers.

The editorial board has been involved in producing this book since its inception. They have spent rigorous hours researching and exploring the diverse topics which have resulted in the successful publishing of this book. They have passed on their knowledge of decades through this book. To expedite this challenging task, the publisher supported the team at every step. A small team of assistant editors was also appointed to further simplify the editing procedure and attain best results for the readers.

Apart from the editorial board, the designing team has also invested a significant amount of their time in understanding the subject and creating the most relevant covers. They scrutinized every image to scout for the most suitable representation of the subject and create an appropriate cover for the book.

The publishing team has been an ardent support to the editorial, designing and production team. Their endless efforts to recruit the best for this project, has resulted in the accomplishment of this book. They are a veteran in the field of academics and their pool of knowledge is as vast as their experience in printing. Their expertise and guidance has proved useful at every step. Their uncompromising quality standards have made this book an exceptional effort. Their encouragement from time to time has been an inspiration for everyone.

The publisher and the editorial board hope that this book will prove to be a valuable piece of knowledge for researchers, students, practitioners and scholars across the globe.

List of Contributors

Rainer Hollerbach
Department of Applied Mathematics, University of Leeds, Leeds LS2 9JT, UK

Eun-jin Kim
School of Mathematics and Statistics, University of Sheffield, Sheffield S3 7RH, UK

Donovan Dimanche
School of Mathematics and Statistics, University of Sheffield, Sheffield S3 7RH, UK
Institut National des Sciences Appliquées de Rouen, 76801 Saint-Étienne-du-Rouvray CEDEX, France

Hideyuki Ishi
Graduate School of Mathematics, Nagoya University, Nagoya 464-8602, Japan

Paul Marriott
Department of Statistics and Actuarial Science, University of Waterloo, 200 University Avenue West, Waterloo, ON N2L 3G1, Canada

Radka Sabolová and Frank Critchley
School of Mathematics and Statistics, The Open University, Walton Hall, Milton Keynes, Buckinghamshire MK7 6AA, UK

Germain Van Bever
Department of Mathematics & ECARES, Université libre de Bruxelles, Avenue F.D. Roosevelt 42, 1050 Brussels, Belgium

Stefan Sommer
Department of Computer Science, University of Copenhagen, DK-2100 Copenhagen E, Denmark

Frédéric Barbaresco
Advanced Radar Concepts Business Unit, Thales Air Systems, Limours 91470, France

Matilde Marcolli
Department of Mathematics, California Institute of Technology, Pasadena, CA 91125, USA

Frank Nielsen
Computer Science Department LIX, École Polytechnique, 91128 Palaiseau Cedex, France
Sony Computer Science Laboratories Inc, Tokyo 141-0022, Japan

Ke Sun
King Abdullah University of Science and Technology, Thuwal 23955, Saudi Arabia

Hatem Hajri, Salem Said, Lionel Bombrun and Yannick Berthoumieu
Groupe Signal et Image, CNRS Laboratoire IMS, Institut Polytechnique de Bordeaux, Université de Bordeaux, UMR 5218, Talence 33405, France

Ioana Ilea
Groupe Signal et Image, CNRS Laboratoire IMS, Institut Polytechnique de Bordeaux, Université de Bordeaux, UMR 5218, Talence 33405, France
Communications Department, Technical University of Cluj-Napoca, 71-73 Dorobantilor street, Cluj-Napoca 3400, Romania

Thomas Hotz, Florian Kelma and Johannes Wieditz
Institut für Mathematik, Technische Universität Ilmenau, 98684 Ilmenau, Germany

Index